Primer
of Quantum
Mechanics

THE MACHINERY OF QUANTUM MECHANICS

Wave-particle duality, the de Broglie relations . . . . $\mathbf{p} = \hbar\mathbf{k}$, $E = \hbar\omega$

The spectrum of values cataloging the measurement results . . . . the q

The operator corresponding to the observable being measured . . . . $\hat{q}$

Its eigenvalues = the measurement spectrum . . . . $\hat{q}|q\rangle = q|q\rangle$

In its home space the operator wears no hat . . . . $\langle q|\hat{q} = q\langle q|$

Home space defines operator functions . . . . $\Omega(\hat{q})|q\rangle = \Omega(q)|q\rangle$

Remove the hats in home space . . . . $\langle q|\Omega(\hat{q}) = \Omega(q)\langle q|$

Index-space labels discrete spectra . . . . $|q\rangle = |n\rangle$

Discrete spectrum q-sum . . . . $\displaystyle\sum_q = \sum_n \quad d^v q = dn = 1$

Continuous spectrum q-sum . . . . $\displaystyle\sum_q = \int d^v q = \int d(\text{volume in } q\text{-space})$

The self-space bracket, $\langle Q|q\rangle$; the amplitude to find $q = Q$ in q-space.

$$\sum_q \langle Q|q\rangle f(q) = f(Q) \text{ and } \langle Q|q\rangle \, d^v q = \begin{cases} 0 \text{ if } Q \neq q \text{ in } dq \\ 1 \text{ if } Q = q \text{ in } dq \end{cases}$$

Superposition = the ket-bra sum theorem . . . . $1 = \displaystyle\sum_q |q\rangle\langle q|$

For physical measurements
    where

$$\hat{q}|q\rangle = q|q\rangle \text{ and } \hat{r}|r\rangle = r|r\rangle, \text{ all } q \text{ and } r \text{ are real}$$

$$\hat{r} = \hat{r}\dagger, \qquad \hat{q} = \hat{q}\dagger \quad \text{and} \quad \langle q|r\rangle = \langle r|q\rangle^*$$

the probability, given r, to find q in $\int d^v q$, is $\int d^v q |\langle q|r\rangle|^2 / \langle r|r\rangle$,
the simultaneous eigenvalues of a compound state, $|a,b\rangle$, come
from compatible observables: $[\hat{a}, \hat{b}] = 0$ and in time t the state $|\alpha\rangle$
evolves into the state $\exp(-it\hat{H}/\hbar)|\alpha\rangle$.

Primer
of Quantum
Mechanics

Marvin Chester

Professor Emeritus
University of California, Los Angeles

DOVER PUBLICATIONS, INC.
Mineola, New York

Bibliographical Note

This Dover edition, first published in 2003, is an unabridged reprint of the 1992 corrected Krieger edition of the work first published by John Wiley & Sons, Inc., in 1987.

Library of Congress Cataloging-in-Publication Data

Chester, Marvin, 1930–
    Primer of quantum mechanics / Marvin Chester.
        p. cm.
    "This...is an unabridged reprint of the 1992 corrected Krieger edition of the work first published by John Wiley & Sons, Inc., in 1987"—T.p. verso.
    Includes index.
    ISBN 0-486-42878-8 (pbk.)
        1. Quantum theory. I. Title.

QC174.12.C46 2003
530.12—dc21

2003040901

Manufactured in the United States of America
Dover Publications, Inc., 31 East 2nd Street, Mineola, N.Y. 11501

*Blessed fate, abetted by
my precious wife, Elfi,
created this book*

Preface

Quantum mechanics is the fundamental theoretical infrastructure upon which all understanding of the nature of the physical world is built. This book is an exposition of that theory presented at the level of a junior-year undergraduate physics student. The reader must have attained a level of technical training that includes these essentials:

1. A freshman-year survey course in physics. The survey should have included an overview of electricity and magnetism, classical mechanics, and a historical introduction to quantum mechanics.

2. A good grasp of mathematics that includes integral calculus, vector analysis, differential equations, complex numbers, and fourier analysis. The reader should be able to evaluate a determinant and to perform elementary matrix addition and multiplication.

3. A first course in classical Newtonian mechanics. The concept of a Hamiltonian and its significance should have been explored.

In writing this book, I set as my primary task this one: to fuse the mathematical machinery of quantum mechanics with the philosophical world view embedded in it. That the mathematics appear to grow organically out of the philosophy was my aim in the exposition. My object was to present a consistent physical imagery that tightly parallels the mathematics and thus, with verbal threads, to weave the philosophical tapestry into the mathematical formalism.

It is incontrovertible that quantum mechanics faithfully describes nature's behavior. To accommodate one's world view to what is nature's behavior is the reason for a philosophical tapestry. The tapestry allows us to have a way of perceiving nature as natural.

In his classic work, *Principles of Quantum Mechanics,* P. A. M. Dirac initiated the fusion process. To spotlight the physics within the mathematics, he invented a special notation. Dirac notation embodies the quantum mechanical way of thinking about nature.

This book is entirely in the language of Dirac notation. My finding has been that even junior-year undergraduates can acquire an impressive facility with Dirac notation when the physical imagery is embedded in the mathematics.

That imagery must necessarily treat the question "What does quantum mechanics tell us about the nature of nature?" This question serves as the underlying theme in the book.

Primer is meant to be a step in the process that Dirac initiated: to build a quantum intuition; to perceive the quantum universe as self-evident. That the endeavor appears in the format of a primer on the subject is natural: the place to expose the philosophical tapestry is at the first encounter with the formalism.

The study is pursued by examining what quantum mechanics says about the key model physical systems of nature. In the examination are encountered all the practical mathematical techniques useful to a student of the subject.

Marvin Chester

Contents

CHAPTER 9
INDISTINGUISHABLE PARTICLES: IDENTICAL BOSONS, AND IDENTICAL FERMIONS   249

CHAPTER 10
STATIONARY-STATE PERTURBATION THEORY   285

EPILOGUE
SO WHAT?   303

INDEX   311

Primer
of Quantum
Mechanics

Chapter 1

Readers' Orientation:
Premise, and Design of the Study

1.1 Philosophical Orientation

For what purpose, dear reader, do you study physics?

To use it technologically? Physics can be put to use: so can art and music. But that's not why you study them.

It isn't their social relevance that attracts you. The most precious things in life are the irrelevant ones. It is a meager life, indeed, that is consumed only by the relevant, by the problems of mere survival.

You study physics because you find it fascinating. You find poetry in conceptual structures. You find it romantic to understand the workings of nature. You study physics to acquire an intimacy with nature's way.

Our entire understanding of nature's way is founded on the subject called *quantum mechanics*. No fact of nature has ever been discovered that contradicts quantum mechanics. In its existence of over 60 years, quantum theory has experienced only success in describing the physical world. It has survived a stunning multitude of tests on its validity. We must accept it as the soul of nature. Quantum mechanics is nature's way.

What's the attraction of studying quantum mechanics? It lies in perceiving the conceptual structure of the subject. I don't mean by this the *mathematical structure*. The *conceptual structure* is more. It confronts the question, When we use the mathematics, what are we saying about the nature of nature? It is in this pursuit—the pursuit of meaning—that both the difficulty and the pleasure lie. The idea is best captured in an aphorism attributed to Niels Bohr, one of the founding fathers of the subject: "Those who are not shocked when they first come across quantum theory cannot possibly have understood it."

Prepare for a shock, advises Bohr. If you are not shocked, you ought to question whether you have understood. The idea is this: the theory completely confounds our classical common sense about nature.

1

To understand nature, we must readjust our common sense. That is the central object of this study: to readjust common sense so as to accommodate quantum theory. We make whatever sense we can of quantum mechanics so that after a while it becomes our common sense. That is how we become intimate with nature's way.

This revision of thinking is not a mere mathematical matter. It is not the mathematical apparatus that shocks us. The mathematics is a beautiful machinery. It is, however, all formality. It can be mastered in shockless serenity. The shock to which Bohr refers is in the interpretation of the mathematics: the physical meaning is shocking.

One of the great savants of quantum mechanics, P. A. M. Dirac, set himself this special task: to mate the conceptual structure with the mathematical. He invented a mathematical notation that directly embeds the philosophy of the subject into the calculational methods. It exhibits what is being said about the nature of nature. Dirac notation exposes the internal logic of quantum mechanics. It displays the sense—the meaning—of the theory in every equation one writes.

The characterization Dirac *notation* is an infelicitous misnomer for one of the great contributions to modern thought. It is not merely a way of writing; a way of writing expresses a way of thinking. Dirac notation is a *way of thinking*.

Consider the notation change from roman to arabic numerals. It was one of the most significant advances in the history of mathematical thought. The mechanics of the change in writing was not the issue; the importance lay in the new way of thinking that accompanied the notational change.

So it is with Dirac notation. It expresses the quantum mechanical way of thinking. With it, one can proceed from the philosophy of the subject to its mathematical expression rather than the other way around. That shall be the direction this primer takes in the study of quantum mechanics. The subject will be developed through the philosophy underlying it. The object is to proceed from meaning. Because Dirac notation is exactly suited to this goal, it is used exclusively in the text.

Meaning does not reside in the mathematical symbols. It resides in the cloud of thought enveloping these symbols. It is conveyed in the words; these assign meaning to the symbols.

Here is an example. The most important dictum of quantum mechanics is this one:

WHAT YOU CAN MEASURE IS WHAT YOU CAN KNOW

This precept can be recast as a formal mathematical statement. In Dirac notation it's particularly simple.

Most of quantum mechanics proceeds from this precept in combination with others. They all can be expressed in both words and in symbols. The weight of thought is in the words; the mathematics and its symbols enables us to probe the consequences of the thought.

One deduces from such dicta the most intriguing (*shocking*) notions. Here are two of them:

1. There exist particles that are fundamentally indistinguishable from one another: no matter how refined experiments become, none will ever perceive a difference

between, say, two electrons. Among a swarm of them, you may be able to recognize an individual fruit fly, but never an individual electron among electrons.

2. Only discrete spectra of atomic energies can be found in nature. There are values of energy that for an orbiting system, no measurement can ever detect! The detectable ones can be calculated.

In Dirac notation each of these notions is a simple mathematical statement. Dirac allows us, while doing the practical calculations, to focus on the meaning of the theory. This brings out its conceptual poetry. It responds to the *why* of this study. The romance and fascination are preserved while we learn the formal mathematics of the subject.

1.2 Mathematical Orientation

Our study does require a certain level of mathematical sophistication. You need to be proficient in calculus, and you must have been exposed to complex numbers. You must be familiar with the wonderful deduction of Gauss that the unit of imaginary numbers, $i = \sqrt{-1}$, may be used to define the two-dimensional complex plane in which every point corresponds to a complex number.

The complex number may be called z, its real part x, and its imaginary part y, as shown in Figure 1-1. The complex conjugate of z is called z^*; it results from changing i to $-i$ everywhere in z. And the magnitude of z, written symbolically as $|z|$, is never complex, nor even negative: it is the length of the line connecting the point z to the origin—always nonnegative real.

What is important to our study is this most stunning of notions: that the exponential function of imaginary argument is related to the common trigonometric functions (see Problem 1.3).

$$e^{i\theta} = \cos\theta + i\sin\theta \qquad (1\text{-}1)$$

The exponential function of a real argument, $\exp x$, is a monotonic function; it changes steadily in one direction. The exponential function of an imaginary argument is

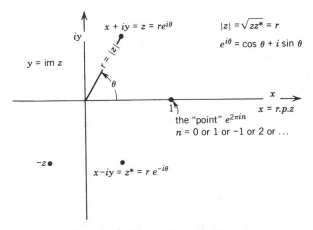

Figure 1-1 The complex plane.

periodic. It has the same period as the sin and the cos trigonometric function: 2π. Thus both the complex numbers $r \exp i\theta$ and $r \exp i(2\pi + \theta)$ correspond to precisely the same point in the complex plane. The point $z = 1$ is represented by an infinity of complex numbers. For all real integers n, the points $\exp 2\pi in$ coincide with $z = 1$ (see Problem 1.4).

This has a profound consequence in fourier analysis: fourier expansions may be cast in terms of exponential functions.

Any continuous single-valued function of a real variable, $f(x)$, can be represented in a fourier expansion. If the function repeats itself along the x-axis, like the one shown in Figure 1-2, then Fourier teaches us that it may be written as a sum of *pure waves*. The sum is over all the possible wavelengths that fit into the repetition length.

$$f(x) = \sum_{\lambda} [a_\lambda \cos \frac{2\pi x}{\lambda} + b_\lambda \sin \frac{2\pi x}{\lambda}] \qquad (1\text{-}2)$$

If the repetition length is L, then the permissible wavelengths, λ, are indexed by the nonnegative integer n in the formula

$$n\lambda = L \qquad (1\text{-}3)$$

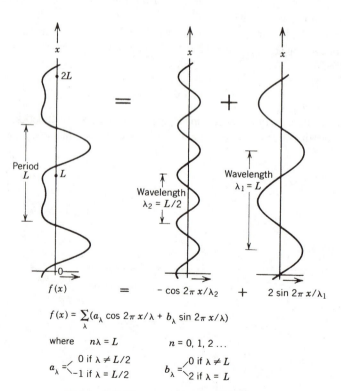

$$f(x) = \sum_{\lambda}(a_\lambda \cos 2\pi x/\lambda + b_\lambda \sin 2\pi x/\lambda)$$

where $n\lambda = L$ $n = 0, 1, 2 \dots$

$a_\lambda = \begin{cases} 0 \text{ if } \lambda \neq L/2 \\ -1 \text{ if } \lambda = L/2 \end{cases}$ $b_\lambda = \begin{cases} 0 \text{ if } \lambda \neq L \\ 2 \text{ if } \lambda = L \end{cases}$

***Figure 1-2* The Fourier Expansion of f(x).**

How do we know that the expansion in Equation 1-2 can be made? "We know," explained Fourier, "because if you give me any $f(x)$ that you want expanded, I'll produce for you every a_λ and b_λ that you need." Here are the formulas for them:

$$a_\lambda = \frac{2}{L} \int_0^L dx\, f(x) \cos \frac{2\pi x}{\lambda} \qquad \frac{L}{\lambda} = n > 0$$

$$a_{n=0} = \frac{1}{L} \int_0^L dx\, f(x) \qquad\qquad \frac{L}{\lambda} = n = 0 \qquad (1\text{-}4)$$

$$b_\lambda = \frac{2}{L} \int_0^L dx\, f(x) \sin \frac{2\pi x}{\lambda} \qquad \frac{L}{\lambda} = n \geqslant 0$$

Each a_λ represents how much of $f(x)$ comes from the pure wave $\cos 2\pi x/\lambda$. Each b_λ gives the strength of the pure wave, $\sin 2\pi x/\lambda$, in the function $f(x)$. In this view, the function $f(x)$ is a superposition—a sum—of all possible pure waves. By *pure waves* I mean the special trigonometric functions of x called sin and cos.

WAVELENGTH: THE PERIODICITY OF A PURE WAVE. ONLY PURE WAVES HAVE WAVELENGTH

The important thing is not to confuse *periodicity* with *wavelength*. The wavelength, λ, is something that characterizes *only* the sin and cos trigonometric functions of x. The periodicity of $f(x)$ along the x-axis is its repetition length, L. The function $f(x)$ does not in general, have a wavelength; it has only a periodicity. For the special functions $\sin 2\pi x/\lambda$ and $\cos 2\pi x/\lambda$, there is a wavelength. It is the periodicity of these special functions that is denoted by the term *wavelength*.

But according to Equation 1-1 these special trigonometric functions can just as well be written as exponential ones. Instead of the pair $\sin 2\pi x/\lambda$ and $\cos 2\pi x/\lambda$, we can just as well use the pair $\exp 2\pi i x/\lambda$ and $\exp -2\pi i x/\lambda$ in the fourier expansion (see Problem 1.6.

The elegance of the result is spotlighted if we speak in terms of the wave vector k instead of the wavelength λ. They are related.

$$|k| = 2\pi/\lambda \qquad (1\text{-}5)$$

The advantage of using k is that it can be negative as well as positive: wavelength is always positive. The negative k's take care of the negative exponentials: $\exp -2\pi i x/\lambda = \exp ikx$ where k is the negative number $-2\pi/\lambda$.

With this notion it is an easy step, using (1-1), to rewrite the fourier expansion of $f(x)$ given in (1-2).

$$f(x) = \sum_k c_k e^{ikx} \qquad (1\text{-}6)$$

Equation 1-6 says as much as (1-2) does, but with fewer symbols; it is more elegant.

The sum is over permissible wave vectors, k. These are indexed by the integer n in the formula

$$kL = 2\pi n \qquad (1\text{-}7)$$

Unlike those of (1-3), these integers take on all values—negative as well as nonnegative—from minus infinity to plus infinity.

Like Fourier, we know the expansion in (1-6) is possible because we can give the formula for every coefficient c_k:

$$c_k = \frac{1}{L} \int_0^L dx \, e^{-ikx} f(x) \qquad (1\text{-}8)$$

The c_k in the expansion (1-6) have this meaning: each one is the strength of the function $f(x)$ in the pure mode of wave vector k. Each c_k measures the extent to which $f(x)$ behaves like exp ikx. The expansion portrays $f(x)$ as a superposition of pure waves, just as before. But here each pure wave is the periodic function exp ikx. This pure wave function has as its period the wavelength $\lambda = 2\pi/|k|$, just as do the sin and cos pure waves. This kind of pure wave is one of fixed wave vector. The pure wave exp $-i|k|x$ has the same wavelength as does exp $i|k|x$, but it is characterized by a different wave vector: one of opposite sign.

Any analytic $f(x)$ is a superposition of pure waves. The pair sin $2\pi x/\lambda$ and cos $2\pi x/\lambda$ both are pure wave functions of wavelength λ, and so the superposition expansion may be made in terms of these. But the pair exp ikx, $k = +2\pi/\lambda$ and $k = -2\pi/\lambda$, are also available as independent pure waves of wavelength λ; so the superposition expansion may equally well be made in terms of this pair. Equations 1-6 through 1-8 effect this expansion.

1.3 Physics Orientation: The Classical Bead on a Track

To understand the behavior of the physical world around us, it is helpful to have in mind particular examples—model systems—from that world. We will examine the theory and behavior of the model system. Every model has its limits; every one is a special case of a more general thing. Each time we reach the limits of a particular model, we will replace it with a new, more sophisticated model, That's how we'll proceed from the particular to the general. We begin with something simple.

Figure 1-3 is the simplest mechanical system I can imagine. A very small bead is constrained to move on a very large circular track.

In the classical realm the bead might have a mass m of about one-half gram. It

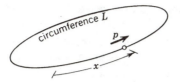

Figure 1-3 The classical bead-on-a-wire-track.

might be five millimeters in diameter. It can move freely without friction anywhere along the circular wire track. The circle has a large circumference, L, of about 1 meter.

The position of the bead is measured from some arbitrarily chosen fixed point on the wire. The distance around the track between the origin and the bead is x. It takes only one number—x—to locate the bead: that fact makes this system one-dimensional. This bead-on-a-wire is an archetypical one-dimensional system. It has one degree of freedom; one variable is enough to define where the bead is.

A full classical description of the motion of this bead is founded not on one variable but on two. You must be given not only the initial position of the bead but also its initial velocity if you are to know its ensuing motion. Newton's laws fix how the position and momentum at the time t—$x(t)$ and $p(t)$—evolve from their initial values—$x(t=0)$ and $p(t=0)$.

Two dynamical properties of the bead are its position x and its momentum, p: a third such property is its energy, E. In our model there is no potential energy; the entire energy is kinetic. In terms of momentum, the energy is given by the famous hamiltonian function, H, of classical physics.

$$H = \frac{p^2}{2\,m} = E \qquad (1\text{-}9)$$

The dynamical properties of the bead are things that you can measure: they are measurables or observables. Every measurable has a domain of possible values; a spectrum of possible measurement results.

THE SPECTRUM OF MEASUREMENT RESULTS IS THE DOMAIN OF THE OBSERVABLE

Consider the bead's position, x. You could find the bead at any value of x between zero and L. The spectrum of possible position measurement results comprises all of the points on a line extending from zero to L, all possible physical places along the ring. The *domain* of x runs from zero to L.

A value of $x = 5$ cm $+ L$ is outside the domain; it corresponds to the physical position $x = 5$ cm inside the domain. Values of x outside 0 to L offer no information not already present within that domain.

The points $O \leqslant x < L$ are sufficient to describe any possible measurement of the bead's position; other values are not needed. If you insist upon using values outside the domain, then an interpretation must be made to bring them into the domain. For example: the position -5 cm, though given by an x outside the domain, is a valid measurement result. We interpret it as the position $x = L - 5$ cm.

The set of points, $0 \leqslant x < L$ are, of course, infinite in number. It is a nondenumerable set. This set of points is the spectrum of possible position measurement results.

In actual experiments one doesn't locate the bead *at x* but, rather, *at x within dx*. One can never say that the bead is precisely located at 15.0 cm; rather, one finds that the bead is at 15.0 cm $\pm$ 0.3 cm. Here the *dx* is 0.6 cm. When the spectrum of the observable is continuous, there is always some differential range connected to any measurement result.

Experiments measuring momentum also exhibit a spectrum of possible measurement results. The momentum of the bead might be $+10.6$ gm$-$cm/sec $\pm$ 0.05 gm$-$cm/sec. It might be -27.3 gm$-$cm/sec $\pm$ 0.05 gm$-$cm/sec, that is, moving in the negative x direction. It might be anything at all between $-\infty$ and $+\infty$. The momentum, p, is a continuous variable like x; it has a different domain.

The spectrum of possible results for a measurement of the energy of the bead ranges continuously from zero to infinity and spans yet a third domain.

Suppose you kicked the bead to the right ($+x$ direction). It moves along the wire. By the laws of classical mechanics it continues indefinitely with momentum $p =$ (perhaps) 10.6 gm$-$cm/sec. Now you make a series of measurements. You make them nondestructively. You take a series of stroboscopic snapshots and analyze the film results. Classical mechanics allows the idealization that measurements can be made without disturbing the system.

You measure the bead's position. You find $x = 7.6$ cm $\pm$ 0.3 cm.

Then you measure its momentum. You find $p = 10.6(1 \pm 0.005)$ gm$-$cm/sec.

You wait and measure position again. You find $x = 18.2$ cm $\pm$ 0.3 cm.

You measure momentum again. You find $10.6(1 \pm 0.005)$ gm$-$cm/sec.

You measure position again. You find 28.8 cm $\pm$ 0.3 cm.

Your measurement results span a whole gamut of positions. You find the same momentum, 10.6 gm$-$cm/sec, over and over again, the one (arbitrary) with which the bead started. These measurement results exemplify what you might find for a half-gram bead on a slippery track; they illustrate the classical picture of the bead-on-a-wire.

MICROSCOPIC IS NOT MINIATURIZED MACROSCOPIC

Now suppose that the wire track is of atomic dimension and the bead an atomic mass. This system is not directly discernible on the human scale of experience. The bead might be one of the electrons (10^{-27} gm) free to move within a long chain, polymerlike conducting ring ($L = 10$Å). Chemical species called polynuclear aromatic molecules, like benzine, napthaline, anthracene, tetracene, and the like form such conducting rings for their free electrons. These electrons are inaccessible to direct human experience. They can't be watched with human eyes. They can't be held in the hand. To examine them, the intermediary of instruments is necessary, and the readings from these instruments must be interpreted.

We are inclined to interpret what we measure in terms of what is familiar to us. We imagine the electron to be a smaller version of the half-gram bead. It is quite a bit smaller, smaller than a cherry would be in relation to a giant's thumb the size of our planet. We midgets on the planet can see the cherry clearly enough; it's accessible on our size scale. But the cherry is just too small for the giant to see.

We might imagine that we are such giants for the atoms. We might expect that the same classical newtonian mechanics that describes beads can be used to describe atoms.

We might imagine it, but it isn't true! Remarkably enough, the difference between macroscopic and microscopic is not merely a matter of size. The microscopic world is *not* just a miniaturized version of the macroscopic.

Our only access to the atomic bead is through experiments on many atoms like it. The results of a multitude of such experiments show that the classical picture does not apply on the scale of the atomic particles in nature. The experimental results flatly prohibit any description of an atomic particle in which its position and momentum are followed in time. The physical world on the microscopic level cannot be described by classical physics. *The newtonian description contradicts experiment.* Newtonian mechanics must, therefore, be an approximation, valid only for macroscopic things, of a broader mechanics. That broader theory is quantum mechanics. The quantum mechanical picture of the bead-on-a-track has replaced the classical picture.

PROBLEMS

1.1 Show the following points in the complex plane

a. $r = 1$ and $\theta = 0$; that is, $z = 1$

b. $z = \exp i\pi/6$

c. $z = \exp 7i\pi/4$

d. $z = \exp 9i\pi/4$

e. $z = 2 \exp i\pi/6$

f. $z = \exp i\pi/4$

g. $z = \exp -i\pi/4$

h. $z = \exp 2\pi i$

1.2 Exhibit in three columns the real part of z, the imaginary part of z, and the magnitude of z for each of the eight points of Problem 1-1.

1.3 Convince yourself of the truth of Equation 1-1, by consulting a mathematics textbook or by constructing a proof yourself. You expand the three functions in the equation as infinite Taylor series in θ and compare the results term by term with Equation 1-1.

1.4 Prove, using Equation 1-1, that all the complex numbers, $r \exp i(\theta + 2\pi n)$, represent precisely the same point in the complex plane for any integer n— positive or negative.

1.5 To prove that the amplitudes a_λ and b_λ given by Equation 1-4, do make (1-2) true takes some sophistication. Prove, instead, that (1-2) makes (1-4) true.

1.6 Prove, founded on the result of Problem 1.3 that (1-1) is true, that

$$e^{-i\theta} = \cos\theta - i\sin\theta$$

$$\cos\theta = \frac{1}{2}(e^{i\theta} + e^{-i\theta})$$

$$\sin\theta = \frac{1}{2i}(e^{i\theta} - e^{-i\theta})$$

1.7 Prove that the c's of Equation 1-6 are related to the a's and b's of (1-2), as follows:

$$c_{n>0} = \frac{1}{2}(a_n - ib_n)$$

$$c_{n=0} = a_{n=0}$$

$$c_{n<0} = \frac{1}{2}(a_{|n|} + ib_{|n|})$$

Note: The a's and b's are defined only for positive n's. Thus for $n < 0$ the c_n can be functions only of the a and b indexed by $-n = |n| > 0$.

1.8 The integral

$$g(\alpha) = \int_0^L e^{i\alpha x}\, dx$$

is a function of α. Evaluate it for

a. real nonzero α

b. the set of α's given by $\alpha L = 2\pi n$ where n is any nonzero integer

c. $\alpha = 0$

Answers:

a. $(e^{i\alpha L} - 1)/i\alpha$

b. zero

c. L

1.9 Using the results of the previous problem in conjunction with the allowed values of k in Equation 1-7, show that the relation between c_k and $f(x)$ given in Equation 1-8 is consistent with that given in Equation 1-6; that is, substitute $f(x)$ from the latter into the former and check the validity of the equality. Critical features: Don't confuse the dummy variable of the summation with the k of the integration. The order of summation and integration are interchangeable.

1.10 Consider the function:

$$f(x) = \sqrt{8/3L} \cos^2 \frac{\pi x}{L}$$

a. What is its period?
Answer: L.

b. Does it have a wavelength?
Answer: No.

c. Find all the coefficients $c_n = c_k$ for this $f(x)$.
  Answer: $c_0 = \sqrt{2/3L}$, $c_1 = c_{-1} = 1/\sqrt{6L}$, $c_{\text{other } n} = 0$

Chapter 2

The Quantum Bead on a Track:
Its States and Representations

2.1 WHAT YOU MEASURE IS WHAT YOU KNOW

Newtonian mechanics is an inadequate description of nature. We know this through measurement. We build our picture of nature from measurement results. *Measurement* is the fundamental consideration in quantum mechanics.

WHAT YOU MEASURE IS AN OBSERVABLE

Something about the bead is measured. An observable is measured, like the position of the bead or its momentum or its energy. An observable is some physically accessible attribute, something that might be measured.

In the notational scheme invented by P. A. M. Dirac there is a special way to record the measurement results. The last measurement made is put in a *ket;* you put what you know in a ket.

PUT WHAT YOU KNOW IN THE KET

Here's an example. Suppose that a measurement just showed that the bead's momentum is $p = 0.2$ eV-sec/Å. The measurement result momentum $p = 0.2$ eV-sec/Å is written $|0.2$ eV-sec/Å$\rangle$. A ket is the enclosure of measurement data by a vertical line and an angle bracket as shown.

BEAD HAS MOMENTUM *p:* IT'S DENOTED $|p\rangle$
BEAD IS AT POSITION *x:* IT'S DENOTED $|x\rangle$

A ket is a catologue entry. It is a descriptive document. In a one-dimensional system the ket contains the result of what has just been measured. You register the last measurement result—the one just made—in the ket. That is what you *know* about the system. The ket tells the condition of the system as an immediate measurement would reveal it to be.

2.2 Degenerate and Nondegenerate Measurements

There is a vast variety of measurements one might make, even on our simple one-dimensional bead-on-a-track. Take position as an example. I have considered only particular place measurements; ones yielding particular values of x. This is an example: bead is at $x = L/4$ in $dx = 0.003L$. It is written in a ket as $|x=L/4\rangle$.

But suppose you don't measure the particular place where the bead is located. Suppose, instead, you make a measurement to answer this question: In which half of the track is the bead located? Say you find that it is in the right half $(0 \leqslant x < L/2)$. You know that it is not in the left half.

This is a perfectly legitimate measurement result. It is expressed by the ket $|0\leqslant x<L/2\rangle$.

These two findings, $|x=L/4\rangle$ and $|0\leqslant x<L/2\rangle$, exemplify a crucial distinction among measurement results: their degeneracy. The half-track result is a degenerate one; the single position result is not. (see Figure 2-1).

A STATE IS A NONDEGENERATE MEASUREMENT RESULT

In a one-dimensional system a measurement result that is nondegenerate defines what is called a *state*. A degenerate measurement result doesn't represent a particular state of the system; rather, it is a superposition of many states.

Degeneracy is the technical word meaning that a particular measurement result characterizes many states rather than just one. Some measurements yield degenerate

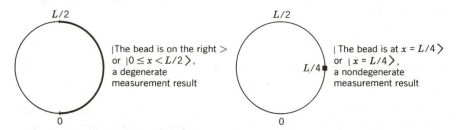

Figure 2.1 A degenerate measurement result is a superposition of states. A nondegenerate measurement result defines a state.

results, ones that fit many states. Some measurements yield nondegenerate results, ones that fit only one state. *The state of a one-dimensional system is the last nondegenerate measurement result obtained.*

What the half-track measurement says is that the bead may be in the state $|x=L/3\rangle$ or the state $|x=L/4\rangle$, and so forth. It cannot be in the state $|x=3L/4\rangle$. It may be in any one of the states $|x=\xi\rangle$ where $0 \leqslant \xi < L/2$. The degeneracy of the measurement result lies in this: there are many distinguishable conditions of the system, all of which fit the same measurement result. Many positions are in the right half-track!

A DEGENERATE MEASUREMENT RESULT
FITS MANY STATES

Here's another example of the idea. A precise measurement of x (within dx = small) defines a state, $|x\rangle$: a precise measurement of p (within dp) defines a state, $|p\rangle$; but a precise measurement of energy E (within dE), $|E\rangle$, does *not* define a state.

The reason is this: an energy measurement of the bead-on-a-track is degenerate. There are two states that can have the same energy. They are $|p=\sqrt{2\,m\,E}\rangle$ and $|p=-\sqrt{2\,m\,E}\rangle$. Thus a measurement of energy with the result E, although a legitimate measurement finding for an observable, is insufficient to define the state of the bead on the track completely.

A state is always a measurement result, but a measurement result is not always a state. In a system with one degree of freedom you need only one measurement to define its state, but you must choose the right one.

How can you be sure which measurements define a state and which don't? Rest easy, in practice it will never be a problem. You can know this way: by probing for degeneracy.

REFER ALL MATTERS TO MEASUREMENT
FURTHER EXPERIMENT UNCOVERS DEGENERACY

The basis on which degeneracy is decided is *further experiment*. When you have a measurement result that is degenerate, further experiment must eventually demonstrate it; further experiment will reveal that the same result applies to two or more different states.

In our examples this is quite obvious; it will be equally obvious in every practical case you encounter. The half-ring experimental result is degenerate because by further measurement you could establish that the finding "bead at $x = L/4$" (state $|x=L/4\rangle$) and the finding "bead at $x = L/3$" (different state $|x=L/3\rangle$) both describe a bead in the right-hand half-ring ($0 \leqslant x < L/2$). You have thus proved that the measurement result—in the right half of the ring—encompasses more than one state of position. It is a degenerate finding with respect to the spectrum of positions.

On the other hand, the measurement result $|x=L/4\rangle$ does define a state. Further measurement confirms it. You can't find two distinguishable states of position both of which have the property $x = L/4$. A one-dimensional state is an indivisible measurement result.

A STATE IS A KET
NOT ALL KETS ARE STATES

The bead's energy is a degenerate measurement result. Further measurement will show that there are two states with the same value of energy. The forward-going bead (state $|p = +\sqrt{2\,m\,E}\rangle$) has the same energy as does the reverse-going bead (state $|p = -\sqrt{2\,m\,E}\rangle$). Thus an energy measurement alone does not describe a bead-on-a-ring state.

A measurement of momentum does define a state. The bead's momentum is a nondegenerate observable, further measurement does not uncover two different states with the same momentum.

AN EVENT: TWO STATE MEASUREMENTS
INSTANTANEOUSLY SUBSEQUENT

An *event* is a measurement result on a prepared system. Here is an example of an event. The bead is known to have momentum *p*: this state has been prepared or selected. A measurement of position is made and found to be *x*. That's the event: the bead with momentum *p* is found at position *x*.

In an event two states are involved: the *prepared*, or initial, one and the *measured*, or final, one. To know the prepared state is to have measured it. To know the final state is to have measured it. To qualify as an event the two measurements must be instantaneously subsequent. The idea is that the prepared state upon which the final state measurement is made must be a *known* one.

Consider this event: the bead is known to be at position $x = 6.0$ Å when you measure its momentum. You find its momentum to be $p = 0.2$ eV-sec/Å. In this event the prepared state is one of position.

Suppose you had waited some time after the position measurement before you made your momentum measurement. Then the sequential pair of measurement results ($x = 6.0$ Å first, $p = 0.2$eV-sec/Å second) would *not* refer to an event. The reason is simple. You cannot be sure that the bead was at position 6.0 Å at the time of the momentum measurement. By the time you got around to the momentum measurement, the bead might have been found at $x = 16.3$ Å! The initial or prepared state might change into something else if you allow time to elapse between the two measurements.

An event is a measurement finding for a system when it is in a known state. The system is prepared in such a way that it is known to be in a certain condition when the new measurement is made on it. In principle, you can know the prepared state only if you measure it. That's why we need the concept of *instantaneously subseq·ent* measurements. It is a useful conceptual fiction.

The central issue governing an event is this: that a measurement result is obtained on a system that is in a known condition; a system, known to be in a prepared state, is found in some newly measured state. That's an event.

AN EVENT MAY HAPPEN
A FINDING HAS A PROBABILITY

For any event there is some probability that it will happen. Many findings are possible in a given measurement; there is a probability for any particular one of them. The bead with momentum p (say, 0.20 eV-sec/Å within $dp = 0.01$ eV-sec/Å) might be anywhere on the track. When its position is measured, it may or may not be found at the point x (say, 6.0 Å) within the interval dx (say, 0.05 Å). There is some probability to find the bead at x within dx. We'll call it $dx\,P(x|p)\,dp$.

The probability itself is $dx\,P(x|p)\,dp$: the probability *density* is $P(x|p)$ (see Figure 2-2). The relative density of particles in a gas $(N(x)/N)$ is a probability density: the number of particles within dx at x $(N(x)dx)$ divided by the total number in the gas (N) is just the probability for finding one of the particles in dx. Probabilities and probability densities are classical notions (see Problem 2.1). But classically there is no such thing

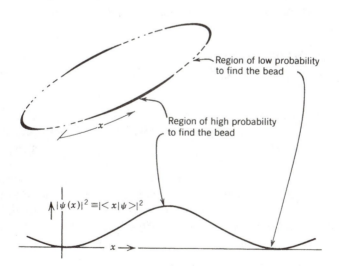

Region of low probability to find the bead

Region of high probability to find the bead

$$|\psi(x)|^2 \equiv |< x|\psi >|^2$$

$x \rightarrow$

Figure 2.2 A probability density of $|\psi(x)|^2$ for the bead-on-a-track. It is a continuous, single-valued function of x.

as a *probability amplitude*. The probability amplitude is a function whose absolute value squared is the probability density!

Quantum Principle 1

2.3 AN EVENT MAY HAPPEN: THERE'S AN AMPLITUDE FOR IT

The physical universe is governed by *amplitudes for events*. With this notion we enter into quantum mechanics. Up to this point all the ideas we have discussed apply to classical systems as well as quantum ones. But in classical physics there exists no concept called the amplitude for an event. The amplitude for an event is what nature directs: the probability is derived from the amplitude.

In Dirac notation there is a special way to indicate amplitudes. Consider this event: bead with momentum p is found at x. The amplitude for it is written $\langle x|p \rangle$. Dirac called $\langle \alpha|$ a *bra*, and $|\beta\rangle$ a *ket*. Together $\langle \alpha|\beta \rangle$, they are a bra-ket or bracket. The amplitude for the event being considered is the bracket $\langle x|p \rangle$.

The essential property of the amplitude for an event is that its absolute square yields the probability density for that event. The probability density, $P(x|p)$, is a function of x. Thus $\langle x|p \rangle$ is another way of writing something that is a function of x. It is that function of x whose absolute value squared is the probability density $P(x|p)$.

$$|\langle x|p \rangle|^2 = P(x|p) \tag{2-1}$$

This equation illustrates the most basic property of the amplitude for an event. The amplitude governs the probability density for the event.

THE WAVE FUNCTION: THE AMPLITUDE FOR POSITION

When the event ends in a measurement of position, the Dirac bracket has an alternative name: the *wave function*. The event we have been considering is a case in point: the position is measured of a bead prepared in a state of momentum. Thus $\langle x|p \rangle$ is the wave function for a bead in the state of momentum p.

A state of momentum is a particular example of a state in general. When the bead is in the state $|\psi\rangle$ there is an amplitude to find it at position x: it's called $\langle x|\psi \rangle$. But this is that function of x whose absolute square is the position probability density. When exhibited in traditional mathematical notation as a function, $\psi(x)$, this is called the *wave function of the bead*.

$$\langle x|\psi \rangle = \psi(x) \qquad |\psi(x)|^2 = P(x|\psi) \tag{2-2}$$

The wave function of a bead in a state of momentum p is $\langle x|p \rangle = \psi_p(x)$.

THE AMPLITUDE FOR POSITION:
A CONTINUOUS, SINGLE-VALUED FUNCTION OF x

The Dirac bracket for an event ending in a position state is another way of writing a function of x. This function of x refers to physical findings. For such functions there are certain behavioral constraints.

The probability density $|\langle x|\psi\rangle|^2$ cannot have more than one value at each point, x, on the ring. More than one value wouldn't make sense physically. For each value of x we can have only one value of the probability density, not two or three. Put mathematically, $|\langle x|\psi\rangle|^2$ must be a *single-valued* function of x. That the wave function itself, $\langle x|\psi\rangle$, is a single-valued function of x is an axiom of quantum mechanics.

On the same grounds—physical interpretation—the wave function, $\langle x|\psi\rangle$, must be a *continuous* function of x. It is continuous because we expect the probability density to vary smoothly from place to place and not to jump wildly around. Discontinuities would imply a strange physical world. It is axiomatic in quantum mechanics that the amplitude is continuous for events that end in continuous variables. A wave function is a particular example of an amplitude for an event ending in a continuous coordinate. The event ends in a position measurement; x is a continuous coordinate.

2.4 THE PILLAR OF QUANTUM MECHANICS IS WAVE-PARTICLE DUALITY

In his doctoral thesis in 1925 Louis de Broglie framed the key idea of quantum mechanics: *wave-particle duality*. This contribution places him among the greatest of physicists. Wave-particle duality is the critical concept on which all of quantum mechanics is built. De Broglie said that to the particle properties of momentum and energy correspond the wave properties of wavelength and frequency.

It is wave-particle duality that destroys our picture of atoms as merely very small beads. Atoms exhibit wave properties. Wave properties of macroscopic beads are never seen! Only for sufficiently small particles does their wave-particle duality become observable.

It is as if nature has played a trick on us. We cannot see the wave-particle duality; we can only deduce it from instrumental measurements. Our five senses are prohibited from directly experiencing nature's fundamental laws. Only our abstract thought processes are allowed entry, and these permit us to represent nature's laws but not to perceive them. These laws are not directly evident in the objects of our ordinary experience.

THE SIGNATURE OF QUANTUM MECHANICS:
PLANCK'S CONSTANT, h

The power of de Broglie's hypothesis does not reside in the mere statement of wave-particle duality. Its power is in his *quantitative formulation* of it. He said that a particle that has a momentum p must have a wave associated with it of wavelength λ where the two are connected quantitatively. The magnitude of the momentum and the wavelength relate inversely.

$$|p| = \frac{h}{\lambda} \tag{2-3}$$

The critical quantity, h, relating the two is Planck's constant. It is one of the fundamental constants of nature. A helpful way to commit its value to memory is in relation to the value of the velocity of light, c. The product of h and c is an energy times a wavelength; the typical chemical reaction energy 2 eV multiplied by the typical wavelength of radiation reaching us from the sun, 6000 Å. More exactly,

$$hc = 1.24 \text{ micron-electron volts} = 1.24 \text{ } \mu\text{-eV} \tag{2-4}$$
$$c = 3 \times 10^8 \text{ meters/sec} = 3 \times 10^{14} \text{ microns/sec} \tag{2-5}$$

TO TEST FOR THEIR WAVES, DIFFRACT THE PARTICLES

De Broglie's equation (2-3) suggests a simple and straightforward test of his wave-particle duality hypothesis: Look for particle diffraction.

Let a beam of particles pass through a grating (see Figure 2-3). Arrange the grating spacing relative to the expected wavelength so that diffraction effects would be evident if they occurred. Diffraction effects are evident when the wavelength is of the order of the grating spacing.

By virtue of the interference effects that characterize waves, the diffracted beam propagates only in certain directions and not in others. The waves add in some directions

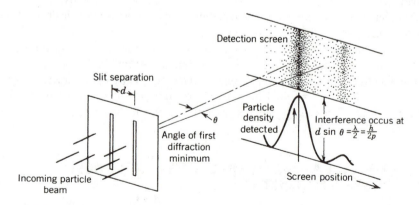

Figure 2.3 Particles of momentum p diffract as if they were waves of wavelength $\lambda = h/p$.

and cancel in others. From the pattern of the diffracted beam intensity, the wavelength can be deduced, as shown in Figure 2-3.

An incident beam of particles, all moving with about the same momentum, is easily prepared. Thus, the momentum of an incident particle is a known quantity. If de Broglie is correct, the incident beam momentum and its wavelength should correlate, as in Equation 2-3. They do.

Electrons can be diffracted just as light waves or sound waves or water waves can be diffracted. Electrons are waves! That electrons are particles is established by this fact: they can be individually counted; they come in units.

The diffraction of electrons was first seen in an experiment by Davisson and Germer in 1927. The effect is, nowadays, a common laboratory one (see Problem 2.2).

Quantum Principle 2
WAVE VECTOR EXPRESSES PARTICLE MOMENTUM

Equation 2-3 connects wavelength to the magnitude of the momentum. But momentum has direction as well as magnitude. Its directional properties are just the same as those of the wave vector, k, introduced in Equation 1-5. If positive values mean *forward*, then negative values will mean *backward*. When you exchange $-x$ for x, both momentum and wave vector acquire the opposite sign. Together with the evident correspondence between (1-5) and (2-3), this fact makes it clear that de Broglie's idea is better and more completely expressed in this way:

$$p = \hbar k \tag{2-6}$$

The symbol $\hbar$ used here follows the standard convention in physics. It goes by the name of *h-bar* and means

$$\hbar = \frac{h}{2\pi} \tag{2-7}$$

In Equation 2-6, the particle property is the momentum, p, and the wave property is k. One is proportional to the other, and they both have the same direction.

2.5 The Pure Momentum State Wave Function

We are now in a position to deduce the pure momentum state wave function, $\psi_p(x) = \langle x|p \rangle$, for our bead-on-a-wire. It must be this function of x:

$$\langle x|p \rangle = \frac{1}{\sqrt{L}} \exp\left(\frac{ixp}{\hbar}\right) \tag{2-8}$$

Here's why.

1. The bracket $\langle x|p \rangle$ is some function of x, the one that characterizes a bead with a particular momentum p. This function meets exactly the same mathematical conditions as did the $f(x)$ discussed in Chapter 1. It, too, can be resolved into pure waves, those characterized by particular wavelengths or particular wave vectors.

2. But according to de Broglie, a fixed momentum corresponds to a particular *single* pure wave, one of fixed wave vector $k = p/\hbar$.

3. The only fuction of x characterized by a fixed wave vector $k = p/\hbar$ has precisely the form of Equation 2-8. Its x dependence must be the exponential one shown.

4. The function $\langle x|p \rangle$ is single valued. The position x and the position $L + x$ represent the same physical place. The bead can't have two different amplitudes at this same place. Thus $\langle x|p \rangle$ must be a periodic function of x: it has the period of the ring circumference L. It has this property in common with the $f(x)$ discussed in Chapter 1.

Because not every possible value of k is permitted for such functions (see Equation 1-7) neither is every value of p permitted. The only way that the amplitude $\langle x|p \rangle$ of (2-8) can be single valued is if p is constrained to have only the discrete set of values indexed by the integer n:

$$p = \frac{nh}{L}, \qquad n = \ldots -2, -1, 0, +1, \ldots \qquad (2\text{-}9)$$

Functions using values of p other than these cannot qualify as wave functions. They are not the same at $L + x$ as at x; they are not single valued (see Problem 2.3).

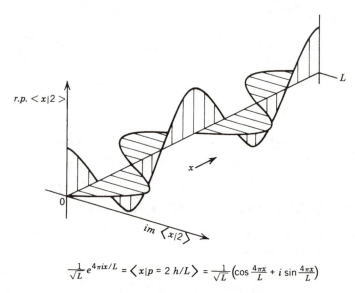

$$\frac{1}{\sqrt{L}} e^{4\pi i x/L} = \left\langle x | p = 2\,h/L \right\rangle = \frac{1}{\sqrt{L}}\left(\cos \frac{4\pi x}{L} + i \sin \frac{4\pi x}{L}\right)$$

Figure 2.4 The pure momentum state wave function $\langle x|p \rangle$ for the case $p = 2h/L$.

5. When a momentum state is prepared, it is one of those characterized by an index. You prepare the state of momentum $|p = 3h/L\rangle$ and ask for the probability to find the bead at x within dx. That probability is $dx|\langle x|p = 3h/L\rangle|^2$.

The probability of finding the bead somewhere on the ring is surely 1: the bead is certainly somewhere on the ring. Thus the wave function $\langle x|p = nh/L\rangle$ must satisfy this condition:

$$\int_0^L dx\ |\langle x|p = nh/L\rangle|^2 = 1 \qquad (2\text{-}10)$$

It is this condition—the certainty condition—by which the factor $1/\sqrt{L}$ is chosen in Equation 2-8. The factor, $1/\sqrt{L}$, guarantees that it is a certainty to find the bead somewhere on the ring (see Problems 2.4 through 2.6).

This completes the reasoning by which the pure momentum state wave function, $\langle x|p\rangle$, is deduced. An example of such a state is shown in Figure 2.4 for the case $n = Lp/h = 2$.

MEASUREMENTS YIELD ONLY POINTS IN THE SPECTRUM

The physical significance of this result is stunning. It means that measurements of the momentum of the bead-on-a-ring can yield only the values shown in (2-9). No measurement can ever yield $3.5h/L$! There exists no state for such a value of p: $|p = 3.5h/L\rangle = 0$. A measurement of momentum can produce only one out of the spectrum of allowed p's: you might find $Lp/h = -3$ or $+1$ or $+17$. That is the physical significance of Equation 2-9. It asserts that the probability density $|\langle x|p \neq nh/L\rangle|^2$ is zero. There is no state $|p \neq nh/L\rangle$ that you could ever prepare.

Classically the momentum of the bead might be anything at all: $p = 3.5h/L$ is a conceivable measurement result according to classical physics. Not so quantum mechanically. This dramatic effect makes observable the distinction between classical physics and quantum physics.

Imagine a researcher making momentum measurements on the bead. No matter how he jiggles the system, he can never find $p = 3.5h/L$. Certain momentum findings are forbidden by quantum mechanics; any p not obeying (2-9) can never be measured.

The quantum mechanical result is the one found in nature; the classical result is not.

2.6 Discrete and Continuous Spectra

Suppose you decide to measure the position of the bead. You could find it anywhere on the ring. No position along the ring is forbidden. The bead we are given can be anywhere on the track. A measurement of position must yield some state, $|x\rangle$, belonging to the spectrum of possible position measurement results.

The spectrum of position measurement results is continuous. The spectrum of momentum measurement results is discrete. There is no other possibility for measurement results: you get either a continuous spectrum or a discrete spectrum.

One of the profound attributes of Dirac notation is that it can embrace both types of spectra in a single formalism. It applies equally well to discrete and to continuous spectra. But it does so only if we have a clear understanding regarding the symbols. It is to this understanding that we turn our attention.

A DISCRETE SPECTRUM OF MEASUREMENT RESULTS CAN BE INDEXED

The essential attribute of a discrete spectrum is that you can enumerate the states. You can designate a first state and then a second one, and then a third one, and so forth. You might also include a minus-ninth state and a minus-eighth one, and so on. A discrete variable is one that you can put into a one-to-one correspondence with integers. The variable x, which spans the spectrum of position measurements, cannot be put into such a correspondence. The momentum measurement variable, p, can.

Energy measurement results also comprise a discrete spectrum. Having the momentum spectrum (2-9), you can deduce the energy spectrum. The energy increases as the square of its index: $E = n^2 h^2 / 2 m L^2$. A discrete variable need not be proportional to integers for a one-to-one correspondence to be made; even nonproportional ones can be indexed. The energy is indexed by n, just as is the momentum (see Figure 2-5).

This is the key mathematical idea: discrete spectrum measurement results always come indexed. Each can be assigned an index. The seventh one is number 7, and so on.

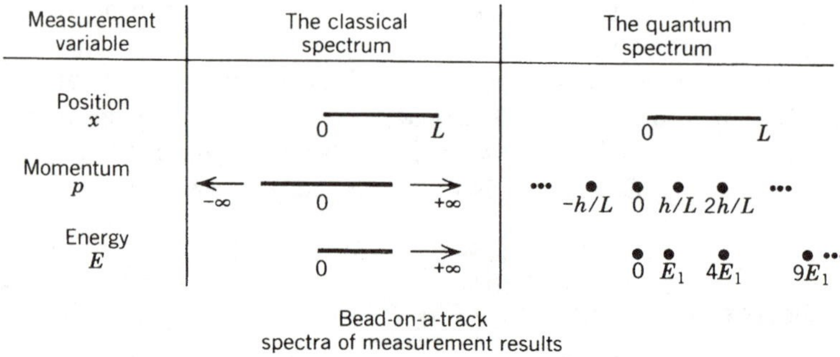

Measurement variable	The classical spectrum	The quantum spectrum
Position x	0 ——— L	0 ——— L
Momentum p	$-\infty$ ←——— 0 ———→ $+\infty$	$\cdots$ • • • • $\cdots$ $-h/L$ 0 h/L $2h/L$
Energy E	0 ——————→ $+\infty$	• • • • $\cdots$ 0 E_1 $4E_1$ $9E_1$

Bead-on-a-track
spectra of measurement results

Figure 2.5 Bead-on-a-track momentum and energy quantum measurement results are enumerable: they come indexed.

Symbol Rule 1

NO MATTER WHAT YOU CALL THE STATE, ITS INDEX IS ITS NAME

Consider the state of pure momentum $p = 3h/L$. Whether we call it the $n = 3$ state, the $p = 3h/L$ state, or the state of $k = 6\pi/L$, we are always speaking of the third state. The symbols $|p\rangle$, $|n\rangle$, and $|k\rangle$ refer to the same thing. Thus, for our bead-on-a-ring

$$|p\rangle = |n\rangle = |k\rangle \qquad \text{(If } p = \hbar k \text{ is a discrete variable)} \qquad (2\text{-}11)$$

The labeling doesn't matter. They all mean the nth state; the p and k are the nth ones. The wave function for momentum state number 3 is

$$\langle x|p = 3h/L\rangle = \frac{1}{\sqrt{L}}\exp\left(\frac{6\pi ix}{L}\right) = \langle x|n = 3\rangle \qquad (2\text{-}12)$$

This equivalent labeling is valid *only* for a discrete set of states. Only states for which there is some enumeration integer, n, have this equivalent labeling property. States characterized by a continuous variable do not.

The significance of the index is that it governs the state. It is the key label among all possible labels with which to label a state. No matter what you choose to call the state, its real name is its index.

You must always treat the state label as if it were the index: treat $|p\rangle$ and $|k\rangle$ as if they were $|n\rangle$. For indexable measurement spectra, you must think in index space.

This rule is simply an understanding on the use of symbols. It is the first of three such understandings which allow a single mathematical formalism to embrace all types of measurement spectra, both the discrete and the continuous kinds.

Symbol Rule 2

IN INDEX SPACE, THE DIFFERENTIAL INCREMENT IS UNITY

Consider a measurement of momentum. There is no dp. Only for a classical bead is there a probability to find its momentum at p within dp. For the quantum bead you find $3h/L$ or $4h/L$, and so forth. There is a probability to find one *or* the other, nothing in between.

For a continous variable like x, the phrase at x within dx makes immediate sense. For a discrete variable, like p, the phrase needs interpretation. The interpretation is evident if you calculate in index space. By the phrase *to find momentum* $p = 3h/L$, you mean *to find the index n* = 3 *within the interval* 2.5 to 3.5. The idea is this: you will get the right answer when you compute with the continuous variable formula. But you must use it in index space, and there you must interpret *dn* to mean one step.

$$dn = 1 \tag{2-13}$$

The differential increment of index space is unity. This is the second of the three rules; the second one of our understandings on how to interpret symbols.

Here's how it works. Because p is a discrete variable, we reinterpret $dx\,|\langle x|p\rangle|^2\,dp = dx\,P(x|p)dp$. This expression is a probability, the probability—if its momentum is p within dp—that we will find the bead at x within dx. The expression would be meaningful and appropriate as it stands were p a continuous variable. But because p is discrete, we must *think* in index space. Instead of $dx\,|\langle x|p\rangle|^2\,dp$, we must use $dx\,|\langle x|n\rangle|^2\,dn = dx\,P(x|n)\,dn$. Because $dn = 1$, this is the same as $dx\,|\langle x|n\rangle|^2$: the probability—if a momentum state $p = nh/L$ is prepared—of finding the bead at x within dx. We arrive at exactly the correct result, that used in Equation 2-10.

AMONG ENUMERABLE STATES, PROBABILITY EQUALS PROBABILITY DENSITY

In index space the unit length is 1, dimensionless. Thus among discrete variables the probability for an event is exactly the same as the probability density for the event. Probabilities for any discrete variable are always in index space where the unit length is dimensionless unity.

When the measurable is a discrete variable, you must be sure to use the index to label a state rather than the variable itself. When it is a continuous variable, it has no integer index, and so you use the variable itself to label the state. In x-space, $P(x|n)$ remains a probability density. But in reference to momentum, the probability density is in n-space, and so it collapses to the probability itself, the probability for a particular index value, n (within $dn = 1$).

A SUM IS AN AREA IN INDEX SPACE

The third synthesizing rule is the implementation in symbols of this basic idea. An integral in continuous variable space and a sum in discrete variable space represent the same thing, the area under a curve.

That an integral is an area you well know. How it works out for a sum is best

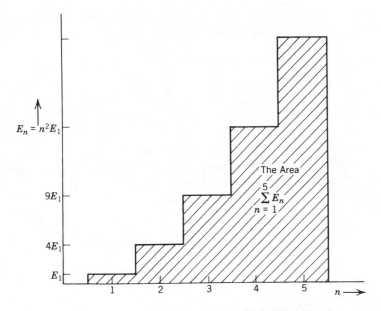

***Figure 2.6* A summation is an area in index space.**

appreciated by examining an example. Consider the discrete variable $E = E_n$. The energy of our bead is such a variable; it depends on n quadratically.

$$E_n = E = n^2 E_1 \tag{2-14}$$

Now, merely as an instructional mathematical exercise, consider the summation over the first five values of E. Not equal increments of E but, rather, equal increments of n separate the terms in the sum. There are five terms in the sum, each issuing from one of the indices.

You can see in Figure 2-6 that the sum is an area. It is the sum of areal rectangles $E_n \, dn$ where $dn = 1$. There is one rectangle for each term in the sum. The sum of terms being the sum of areas is the whole area.

Symbol Rule 3

A CONTINUOUS VARIABLE SUM MEANS AN INTEGRAL

Because they both represent areas, we erase the distinction between sums and integrals. We assign a meaning to *the summation over a continuous variable*. In traditional mathematical notation this has no meaning, for us it shall mean the integral over the domain of the continous variable. Thus

$$\sum_x = \int dx \tag{2-15}$$

This equation exemplifies the rule. If the variable is continuous, then we are to interpret the summation sign as an integral over the variable. Of course, if the variable is discrete, then it will come with an index, and we can perform the summation over its index in the traditional way. We use summation signs to mean either sums or integrals, whichever is appropriate: that's Rule 3 of the three understandings regarding symbols.

In the same spirit a meaning is attached to the *integral over a discrete variable:* it means the sum over that variable.

$$\int dn = \sum_n \tag{2-16}$$

A SPECTRUM OF STATES
IS A MEASUREMENT SPACE

The spectrum of all possible results that you can get from a measurement of position constitutes a space: the measurement space of position. Every point of this space represents a position state. The variable x labels the state because it is the position measurement result. The role that x plays in the ket $|x\rangle$ is to designate a state of position. It has another role.

The other role is that of a language variable. In the *language of position* the state $|p\rangle$ is represented by $\langle x|p\rangle$. Any state, $|\psi\rangle$, may be represented in the language of position: $\langle x|\psi\rangle = \psi(x)$. The wave function, $\langle x|\psi\rangle$, is the state $|\psi\rangle$ represented in the space of position states. The symbol through which x plays its role as a language variable is the bra $\langle x|$. The space of x in the bra and that in the ket is the same: its domain comprises the points on the line between 0 and L.

STATE SPACE EQUALS MEASUREMENT SPACE
STATE SPACE EQUALS REPRESENTATION SPACE

The spectrum of all possible results that you can get from measurements of momentum also constitutes a space, the measurement space of momentum. In this space each point represents a momentum state.

We have seen the variable p in its role designating a state written as the ket $|p\rangle$.

But p can also be a language variable. The amplitude "when in state ψ to find momentum p" is written $\langle p|\psi\rangle$. This is the state $|\psi\rangle$ represented as some function of p. It is $|\psi\rangle$ represented in the *language of momentum*. The bra $\langle p|$ captures the language role of p.

That $\langle p|\psi\rangle$ is, indeed, some function of p is easily appreciated. The fundamental attribute of an amplitude is that its absolute square yields the probability density. Thus

$$|\langle p|\psi\rangle|^2 = P(p|\psi)$$

Because the probability in this equation is a function of p, so must be the amplitude. Thus $\langle p|\psi\rangle$ is the *representative* of the state $|\psi\rangle$ in the space of momentum. It is that function of p which describes the state, the description in p language. To be a *function of p* and to be *represented in the language of momentum* are the same thing, both are notated by the bra $\langle p|$.

The possible values of p in the bra $\langle p|$ are the same as those for the ket $|p\rangle$: the discrete set $p = nh/L$. The domain of the space is the same, whether in a bra or in a ket.

The proof comes from this observation: the amplitude to find $p = 3.5h/L$ is zero no matter what state $|\psi\rangle$ you prepare. Put this statement mathematically. It is $\langle p = 3.5h/L|\psi\rangle = 0$ for any ψ. Because it is true for any state $|\psi\rangle$, we may write $\langle p = 3.5h/L| = 0$. Obviously the generalization is true: $\langle p \neq nh/L| = 0$. Hence the same discrete set of values that make up the state space of momentum forms the representation space of momentum.

Whether they be a continuous set, a discrete set, infinite in number, or finite in number, the universe of measurement results obtained for an observable constitutes a space. Each point in the spectrum of measurement results is a point in that measurement space.

If all of the measurement results are nondegenerate ones, the measurement space is a *state space*. Each point of the space corresponds to a state of the system. It is also a *language space,* one in which any state can be represented. The formal mathematical term for such a space is a *basis*.

The spaces of measurement are not spaces in the conventional sense of physical position; rather, they are abstract spaces. The state spaces of measurement are called Hilbert spaces after the great mathematician David O. Hilbert (1862–1943) who devoted himself to the theory of these generalized spaces.

Quantum Principle 3

THE AMPLITUDES FOR PHYSICAL EVENTS FORWARD* = REVERSE

Suppose the state to be represented in p-space is that of position. The state $|x\rangle$ represented in p-space is a function of p. It is the bracket $\langle p|x\rangle$. This bracket is also an event amplitude. It is the amplitude that the bead at position x have the momentum p. The event represented by the bracket $\langle p|x \rangle$ is just the reverse of that represented by the bracket $\langle x|p \rangle$.

A key axiom of quantum mechanics is that the amplitudes for reverse events are related. This relationship is easily put mathematically: *interchanging the bra and ket is equivalent to complex conjugation.*

$$\langle p|x \rangle = \langle x|p \rangle^* \tag{2-17}$$

From this equation it follows that the probability density to-find-x-given-p is the same as that to-find-p-given-x.

$$P(p|x) = P(x|p) \tag{2-18}$$

It must be accepted as a rule of nature that the reverse amplitude for an event is the complex conjugate of the forward amplitude for that event. This is the event reversal theorem. It is one of the elemental axioms underpinning quantum mechanics. It is true for physical amplitude brackets quite generally.

By this rule we can deduce the function of p that represents the state of position labeled by a given x: the amplitude $\langle p|x \rangle$. It is

$$\langle p|x \rangle = \frac{1}{\sqrt{L}} \exp\left(\frac{-ixp}{\hbar}\right) \tag{2-19}$$

STATE PREPARATION IS ALWAYS DIFFERENTIAL

Having just measured it, you know the bead's momentum, $p = nh/L$. The prepared state is $|n\rangle$. Now measure the bead's position. The probability of finding it at $x = X$ in $dx = \varepsilon$, where ε is a small length of ring, is $dx\, P(X|n) = \varepsilon\, |\langle X|n\rangle|^2$.

From its meaning it is evident that this probability distribution may be used to answer a broader question: you can assess a range of possible results. The probability of finding the bead between, say, $x = L/6$ and $L/2$, two positions on the ring distinctly removed from each other, is

$$\int_{L/6}^{L/2} dx\, P(x|n) = \int_{L/6}^{L/2} dx\, |\langle x|n\rangle|^2 \tag{2-20}$$

The probability of one *or* the other of many possible results is the sum of the probabilities for each. The sum becomes an integral since the variable x is continuous.

The sum over a group of possible measurement results has physical meaning: it is the probability of *detecting any of the group* of states.

A sum over a group of prepared states has no meaning: one does not *prepare any of a group* of states. The phrase makes no sense. One procedes from a known prepared state not an unknown assembly of them. The prepared state on which measurements are made is a specific single state: $|n\rangle$. The state preparation interval is always differential: $dn = 1$.

Now consider the inverse event: to find n having prepared x. In the laboratory one cannot prepare a specific state $|x = X\rangle$. You can prepare a single one of a discrete set of states but to prepare a single one of a continuous set the preparation interval must shrink to zero. To calculate the probability of a result issuing from a position state we must take the preparation interval dx equal to zero.

In practice the bead is placed on the ring within an interval $dx = \varepsilon$ covering the position $x = X$. But the probability, $P(n|X)\, dx$, of finding this bead with momentum index $n = pL/h$ is simply zero. It is not $P(n|X)\, \varepsilon$. (See Problem 2.18) That we do not know within ε in which of the states around X the bead was placed is irrelevant. We calculate for each particular one of these states.

This result illustrates the rule regarding state preparation.

When the prepared state is one among a continuous distribution the preparation interval dx is zero. The prepared state is $|X\rangle$ in $dx = 0$. For a discrete variable the preparation interval is $dn = 1$. As will be made clear later at Equation 2-33 the formal rendering for the state preparation intervals are

$$dn = \frac{1}{\langle n|n\rangle} = 1 \qquad \begin{array}{c}\textit{State}\\ \textit{Preparation Intervals}\end{array} \qquad dx = \frac{1}{\langle x|x\rangle} = 0 \qquad (2\text{-}21)$$

Thus the probability, if the bead is placed on the ring at X, that its momentum index will be found to be 1, 2 or 3 is

$$[P(1|X) + P(2|X) + P(3|X)]\,\frac{1}{\langle X|X\rangle} = 0 \qquad (2\text{-}22)$$

(See Problem 2.18)

2.7 The Superposition of States

The set of momenta $p_n = nh/L$ where $n = 0, \pm 1, \pm 2 \ldots$ comprises the entire spectrum of measurable momenta possible for the bead. No matter what state you prepare when you measure momentum, you must find one of these values. Let's call the state that you prepare—whatever it is—by the symbol $|\psi\rangle$.

If you now measure the momentum of the bead, there is an amplitude to find $p = 0$. That amplitude is $\langle p = 0|\psi\rangle$. But then you might, instead, find $p = h/L$. Because there is a probability for this event to happen, there is an amplitude for it: $\langle p = h/L|\psi\rangle$. It is also possible to find $-h/L$ or $2h/L$ or $-2h/L$, and so on. And for each of these possibilities there is an amplitude (a probability) for the event to happen. The state $|\psi\rangle$ can be experimentally resolved into its momentum state components.

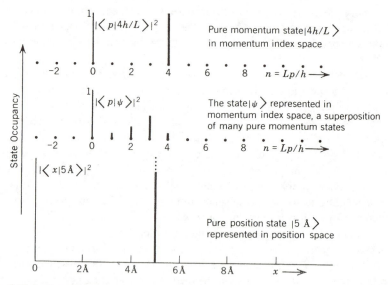

Figure 2.7 A state is represented in the space of measurement results. A state in its own space is one point occupied.

In quantum mechanics we think of the state $|\psi\rangle$ as a little bit of $|p=0\rangle$ plus a little bit of $|p=h/L\rangle$ plus a little bit of $|p=-2h/L\rangle$, and so forth (see Figure 2-7). The amount of $|\psi\rangle$ in the state $|p=0\rangle$ is $\langle p=0|\psi\rangle$: the bead has this amplitude to be in $|p=0\rangle$. It has amplitude $\langle p=h/L|\psi\rangle$ to be (found) in $|p=h/L\rangle$. The state $|\psi\rangle$ is a linear superposition of all momentum states. The amplitude for being in a particular one of the states $|p\rangle$ is $\langle p|\psi\rangle$. The state $|\psi\rangle$ is viewed as a superposition of the states $|p=nh/L\rangle$, with amplitude $\langle p=nh/L|\psi\rangle$ to be in each state $|p=nh/L\rangle$. Here's the mathematical expression of this idea:

$$|\psi\rangle = \sum_{p} |p\rangle \langle p|\psi\rangle \qquad (2\text{-}23)$$

This sum is understood to be over the entire domain of p: it is a sum over all the integers, n, from $-\infty$ to $+\infty$.

The same state $|\psi\rangle$ can equally well be perceived as a superposition of position states. The physical reason is evident. If having prepared the state $|\psi\rangle$, you measure the position of the bead, you must find some value of x between 0 and L. You might find any of the states $|x\rangle$. The amplitude to find $|x\rangle$ is $\langle x|\psi\rangle$. The state $|\psi\rangle$ has some amplitude at each $|x\rangle$. Thus

$$|\psi\rangle = \sum_{x} |x\rangle \langle x|\psi\rangle \qquad (2\text{-}24)$$

Of course, x being a continuous variable, the summation is to be carried out, in practice, by integration. The integral is over the whole domain of x, from zero to L.

A STATE MEASUREMENT SPACE
IS A BASIS AND A COMPLETE SET

Because any possible wave function is a superposition of pure position states $|x\rangle$, these states form a *complete set*. The totality of momentum states is also a complete set. A complete set of states is enough of them so that any state at all—an arbitrary state—can be represented as a superposition of them. A complete set of states is all that you need. Such a set is formally called a *basis set*.

Quantum Principle 4

THE KET-BRA SUM IS ALWAYS ONE

Both Equations 2-23 and 2-24 are special cases of a general rule. This rule is at the foundation of quantum mechanics. It's often called the *completeness theorem*. It embodies the idea of superposition: the idea that any state at all is always a linear combination of all the states in a nondegenerate measurement space. Here is the symbolic statement of the superposition theorem:

$$\sum_q |q\rangle \langle q| = 1 \tag{2-25}$$

The variable q could be x or p (really $n = Lp/h$). It catalogues a spectrum of non-degenerate measurement results. It is a state variable. It has a domain, the totality of all the q's possible. In Equation 2-25 the sum must be performed over the entire domain of q for the theorem to be valid.

The theorem says that the ket-bra sum on the left-hand side of the equation may always be treated as if it were unity, as if it were the number 1. Wherever you can apply the number one you can just as well apply the ket-bra sum. When using (2-25) in algebraic manipulation, you view the vertical line in a ket, bra, or bracket as if it were unity: so $| = 1 = \|$, a pair of verticals is the same as one of them.

Look at Equation 2-24. The ket–bra sum $\sum_x |x\rangle\langle x|$ has the effect of multiplying $|\psi\rangle$ by 1: the equation says $|\psi\rangle = 1|\psi\rangle$. In (2-23) the same thing is expressed, but there it is $\sum_p |p\rangle\langle p|$ that replaces 1.

FOURIER EXPANSIONS ARE KET-BRA SUMS

The entire theory of fourier analysis is a special case of the ket-bra sum rule. Here's how to see that. Apply the ket-bra sum in momentum space to split the bracket $\langle x|\psi\rangle$. Here is the result:

$$\langle x|\psi\rangle = \sum_p \langle x|p\rangle \langle p|\psi\rangle \tag{2-26}$$

You get it by casting (2-23) into x-space. You *cast* something into x-space by applying the bra $\langle x|$ from the left: any state $|S\rangle$ represented in x-space is $\langle x|S\rangle$. Applying $\langle x|$ to the left of every term in (2-23) produces (2-26).

To make apparent the association between (2-26) and fourier analysis, we must rewrite (2-26) in traditional notation: replace $\langle x|\psi\rangle$ by $\psi(x)$ and use the functional form of $\langle x|p\rangle$ given by (2-8). Here is (2-26) rewritten:

$$\psi(x) = \sum_{p} (1/\sqrt{L}) \exp\left(\frac{ixp}{\hbar}\right) \langle p|\psi\rangle \qquad (2\text{-}27)$$

To compare this result with fourier analysis, we apply the fourier theorem of Equation 1-6 to the expansion for $\psi(x)$: take $\psi(x) = f(x)$. Keeping in mind that $p = \hbar k = nh/L$, compare these two equations. They are the same! Equations 2-26 and 1-6 are alternative notations for the same conceptual statement. By comparing the two, it immediately becomes evident that the fourier coefficients c_k of (1-6) are given by

$$c_k\sqrt{L} = \langle p|\psi\rangle \qquad (2\text{-}28)$$

The coefficients, c_k, may equally well be written c_p or c_n: no matter how you write it, what indexes c is always n. From Equation 2-28 we see that these coefficients have a clear physical interpretation: they are event amplitudes. Given the state $|\psi\rangle$ followed by an instantaneously subsequent measurement of momentum, the quantity $\sqrt{L}\ c_k$, is the amplitude to find the bead in its nth momentum state—the one with $p = \hbar k = nh/L$. Equation 2-28 shows us that fourier coefficients are event amplitudes.

Now look at the inverse transformation. In the traditional notation of Chapter 1 it's given by the formula for c_k in terms of $f(x) = \psi(x)$: Equation 1-8. In Dirac notation this same statement is merely the position space ket-bra sum $\sum_x |x\rangle \langle x|$ applied to $|\psi\rangle$ in the bracket $\langle p|\psi\rangle$: you cast (2-24) into momentum space.

$$\langle p|\psi\rangle = \sum_{x} \langle p|x\rangle \langle x|\psi\rangle \qquad (2\text{-}29)$$

The correspondence between this equation and (1-8) is straightforward: just rewrite it in traditional notation. Write $\psi(x)$ instead of $\langle x|\psi\rangle$, and use for $\langle p|x\rangle$ its functional form: the complex conjugate of $\langle x|p\rangle$, given in (2-19). You will find immediately that (2-29) says exactly the same thing as (1-8) does, that we must accept the association given in (2-28) between the fourier coefficients c_k and the amplitudes $\langle p|\psi\rangle$.

Fourier expansions are examples of the ket-bra sum theorem.

THE IDENTITY OF MEANING IS A KET-BRA SUM

The state $|\psi\rangle$ can be represented in x-space: its *representative* is $\langle x|\psi\rangle$. The very same state can also be represented in momentum space by $\langle p|\psi\rangle$. The two representations are related. They are descriptions of the same state in two different languages.

The key idea here is that the state $|\psi\rangle$ has an existence no matter how you represent it. In fourier analysis the position representation is written $\psi(x)$, and the momentum representation is in c_k. These symbols don't suggest that they, in fact, refer to the *same thing*. Traditional notation doesn't capture this idea. Dirac notation is constructed so as to emphasize this essential notion, that $\psi(x)$ and c_k are merely two representations of the same thing. This *sameness* is the crucial idea. In the notation $\langle x|\psi\rangle$ and $\langle p|\psi\rangle$, instead of $\psi(x)$ and c_k, the sameness is made evident. That the ket-bra sum is always unity is the way that sameness, or the identity of meaning, is implemented.

2.8 A STATE IS A POINT IN SOME BASIS SPACE

The space generated by the measurement results for a nondegenerate observable is a basis space. Basis points are points in the spectrum of measurement results. Thus any state is a point in some basis space.

The state $|\psi\rangle$ that we have been talking about must belong to some set, because it could have come only from a measurement of some observable. The other possible measurement results for this observable label the basis points in the space to which $|\psi\rangle$ belongs.

Thus a direct physical meaning attaches to $\langle\psi|p\rangle$, the complex conjugate of $\langle p|\psi\rangle$. It is the state of momentum p expressed in the basis to which ψ belongs. The bracket, $\langle p|\psi\rangle$, is the state characterized by ψ expressed in the basis to which p belongs.

A BASIS SPACE IS A LANGUAGE

Together all of the points in the basis called x-space constitute a language in which to express the state $|\psi\rangle$. Each point in the basis is like a single word in the language. You may need all of the words to express the condition of the system in the state $|\psi\rangle$. But the language of x is not the only way to express the condition of the bead. You may equally well use the language of momentum.

A basis is the medium through which you describe something. Position space, x, characterizes physical location; p-space describes momentum measurement results. In a language an incomplete vocabulary is insufficient to describe everything. A basis space is a complete vocabulary; with it you can describe everything.

The central fact of language is that it expresses your perception and experience. To speak another language is to alter your *Weltanschauung*, your view of reality. This was the thesis of Benjamin Lee Whorf in his famous tract, "An American Indian Model of the Universe" (see the book entitled *Language, Thought and Reality*, Wiley, 1956). He found that "the Hopi language contains no reference to *time* either explicit or implicit . . . [yet it] is capable of accounting for . . . all observable phenomena of the universe" (p. 58). What Whorf showed was that time is not one of the measurement observables that the Hopi Indians employ. They use other means to speak of the universe. Their language expresses their perception, and it doesn't include time!

A language encompasses all of the perceptions of the people who speak it. Our measurement spectra encompass our perception of the physical world. Measurements are the only windows we have through which to look at the physical world. Thus the measurements we choose to make define our perception of reality. They fix what we choose to look at and therefore what we see. The measurement results constitute the language in which we portray our perception. To record your perceptions is to specify the state in which you find the world.

The superposition principle relates to translations between languages. Consider the German word *gemütlich*. As anyone who has lived in Germany knows, there is no single English word to describe it. Many English words are necessary. The single word *cozy* is sometimes offered. But in fact, to be *gemütlich* and to be *cozy* are two quite different things. *Gemütlich* needs many words in English. *Cozy* needs many words in German. The words of the language are the analogue of the points in the basis space. *To express a concept in a language is to represent a state in a basis.* They both are a superposition of words. On the other hand, every state is, in some language, only a single word.

THE TRANSFORMATION MATRIX ELEMENTS ARE LANGUAGE TRANSLATION DICTIONARY ENTRIES

Look at the sums in Equations 2-26 and 2-29. The first equation tells how you get $\langle x|\psi\rangle$, the x-basis representative of the state, if you know it in p-space, if you know $\langle p|\psi\rangle$. It effects the translation from p-language into x-language. The other equation effects the reverse translation, from x-language into p-language.

To transform between bases you must know the transformation coefficients $\langle p|x\rangle$ or $\langle x|p\rangle$. Because they have this function, Dirac brackets are also called *transformation matrix elements* (see Figure 2-8). They are matrix elements because they can be displayed as a matrix array. That they effect the transformation between bases explains the term *transformation*. The matrix of all $\langle p|x\rangle$ amounts to a dictionary: with it you

---

$$\langle q|r\rangle$$

1. The *amplitude* for the *event* to find q when you know r. Its absolute square is the probability density for the event.

2. The *representation* of the state $|r\rangle$ in the language of q, the *function* of q that portrays the state $|r\rangle$.

3. The *transformation matrix elements* between the bases r and q; the language-dictionary entries for translation from r-language into q-language.

Figure 2-8. The three simultaneous meanings of the Dirac bracket $\langle q|r\rangle$.

can speak in *p*-language if you have the state in *x*-language. Each $\langle p|x \rangle$ is a dictionary entry.

THE AMPLITUDE FOR A STATE IN ITS OWN SPACE
IS A SELF-SPACE BRACKET

We have discussed the representation of states in foreign languages; now we shall examine the representation of a state in its own language, the self-space representation. A self-space bracket is the amplitude for a special kind of event: the same measurement is performed twice in rapid succession.

Suppose we prepare a particular bead momentum state $|P\rangle$. But now the subsequent measurement, defining the complete event, will not be in the foreign space of position. Instead we measure momentum. The event consists in twice ascertaining momentum: We measure the momentum of a particle whose momentum we have just previously measured and found to be P. We find momentum p. What is the amplitude for this event? The answer is

$$\langle p|P \rangle = \langle n|N \rangle = \delta_{nN} \tag{2-30}$$

The symbol on the right is the Kronecker delta. The bracket $\langle n|N \rangle$ is that function of n given by the Kronecker-δ. It is zero for all n except where $n = N$. There it is unity, the number 1. The lowercase n is the integer state index for p ($= nh/L$), and the uppercase N is that for P ($= Nh/L$).

Here's what this says physically. If the bead is in the state of momentum P, then the amplitude to find any other momentum will be zero: $\langle p \neq P|P \rangle = 0$. Only if $p = P$ is there any possibility of finding the momentum p. And in this case it is a certainty; the probability is unity, and so is the amplitude. If the bead has momentum $p = P$, that's the momentum you find. That's what the equation says.

Next consider a repeated measurement event in position space. The bead is known to be at position X. It's in the state $|X\rangle$. Now measure its position again. What is the amplitude to find position x? The answer is

$$\langle x|X \rangle = \delta(x-X) \tag{2-31}$$

The function of x on the right-hand side is called the Dirac delta-function. It depends on x in this way. It is zero everywhere except in the immediate neighborhood of X. There the function is a continuous single-valued, incredibly tall, sharp, and needlelike pointed spire that centers on $x = X$. It is so narrow at the base that any ordinary function, $f(x)$, does not perceptibly change its value from $f(X)$ over the whole width of the base. Having such a narrow base, it must be very tall ($+\infty$), because the area under the curve is always unity (see Figure 2-9).

Here is the physical picture of this result. If you know the bead is at position X,

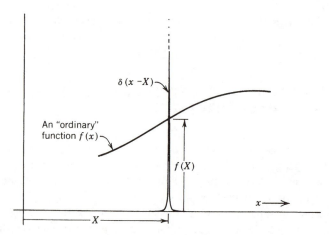

Figure 2.9 The Dirac delta-function $\delta(X - x)$. It guarantees that

$$f(X) = \int_x dx \, \delta(X - x) f(x).$$

then of course you can find it nowhere else. The probability of finding it elsewhere is zero, and so is the amplitude:

$$\delta(x - X \neq 0) = \langle x \neq X | X \rangle = 0.$$

What decides the value of the bracket when the arguments in the bra and ket are equal is the *certainty condition*.

THE CERTAINTY CONDITION: MEASUREMENT YIELDS SOME POINT IN THE OBSERVABLES SPECTRUM

In x-space the certainty condition is the statement that the bead is some-where on the track. If you measure its position, you must find, with certainty (probability 1), that it is between zero and L.

There is a certainty condition for every state space. It is a certainty that if you measure the bead's momentum, you will find some integer value of $n = Lp/h$ between $-\infty$ and $+\infty$. If you measure any observable of the bead, you must find one of the allowed values in the spectrum of this observable. You must find such a result with certainty, no matter in what state the system was prepared. That is the idea of the certainty condition.

Here's that idea applied to our bead in terms of x-space and p-space (n-space):

$$\int_x dx \, P(x|n) = 1 = \sum_n P(n|x) dx \qquad (2\text{-}32)$$

The left-hand equality is the certainty condition in x-space: it is Equation 2-10 rewritten. It says that it is a certainty, for a bead prepared in any state of momentum $|p = nh/L\rangle$, that upon a measurement of its position it will be found somewhere on the track. It will be found within the spectrum of positions.

The right-hand equality is the certainty condition in p-space. It says that for a bead known to be at position x in dx, it is a certainty that upon a momentum measurement, it will be found to have some momentum in the spectrum of momenta available to it. The probability is 1 that it will be found with one of its allowed momenta.

But each individual term in the sum is the probability to find a particular one of the infinity of possible momenta. Each individual term is zero. (See Problem 2.18) That this infinite sum of zeros add up to unity determines the nature of $\langle x|x\rangle$.

Recast the right-hand equality of equation 2-32 in Dirac notation. The process is the subject of Problem 2.17. The result is

$$\langle x|x\rangle\, dx = 1 \tag{2-33}$$

This equation is the verification of (2-21). It says that $\langle x = X|X\rangle$ must be infinity. But it is infinity in such a way that the area under the function $\langle x|X\rangle$ over the differential region $dx\, (=0)$ around $x = X$ is always 1. This is the key property of the Dirac delta-function.

All of the properties of the delta-function are embraced in a single statement. For any single-valued continuous function $f(x)$, $\delta(X-x)$ is the function that makes this integral equation true:

$$\int_x dx\, \delta(X-x)\, f(x) = f(X) \tag{2-34}$$

The range of x in the integration includes the point $x = X$. If this point is not included, then the integral will yield zero, not $f(X)$. This is what defines the Dirac delta-function, $\delta(X-x)$. The behavior of $\delta(X-x)$ issues entirely from this definition (see Problem 2.13).

MEASURE AN OBSERVABLE TWICE:
THE STATE PREPARED IS WHAT YOU FIND

The Dirac δ-function is the continuous space counterpart of the Kronecker-δ. The latter is a function that has meaning only for a discrete variable; the former has meaning only for a continuous variable. The self-space bracket embraces both discrete and continuous spaces. It equals the Kronecter-δ if the variable is discrete (Equation 2-30), and it is the Dirac δ-function if the variable is continuous (Equation 2-31). In either case the self-space bracket represents a state in its own space, the portrayal of the state in its own language.

The idea is best expressed formally in Dirac notation. Suppose Q is one point of the basis set q. Then $\langle q|Q \rangle$ is a self-space bracket. The basic condition that the self-space bracket $\langle Q|q \rangle$ must satisfy, regardless of whether q is discrete or continuous, is

$$\langle Q|\psi \rangle = \sum_q \langle Q|q \rangle \langle q|\psi \rangle \qquad (2\text{-}35)$$

This equation is merely the ket-bra sum $\sum |q\rangle\langle q|$ interposed in $\langle Q|\psi \rangle$. But here Q belongs to the space of q rather than to a foreign space. The ket-bra sum theorem is valid in *any* space, q as well as non-q. It is this validity condition that fixes the meaning of the self-space bracket $\langle Q|q \rangle$. Equation 2-35 is the generalization of Equation 2-34.

Here is what (2-35) says. The bracket $\langle q|\psi \rangle$ is a function of the argument q. The bracket $\langle Q|\psi \rangle$ is exactly the same function. It is the same function, but of a different argument, Q instead of q. The self-space bracket, $\langle Q|q \rangle$, is that function which ensures the validity of (2-35). It is what is needed to guarantee that the equation will be true. For all discrete spectra, the Kronecker-δ does it. For all continuous ones, the Dirac-δ does it.

It's easy to see why. The function $\langle Q|q \rangle$ must pick the value of $\langle q|\psi \rangle$ at the point $q = Q$ out of the sum (or integral). Thus $\langle Q|q \rangle$ must be zero everywhere except where $q = Q$. And there it must be just strong enough to bring out $\langle Q|\psi \rangle$; that is, $dq \langle Q|q \rangle \langle q|\psi \rangle = \langle Q|\psi \rangle$. If q is a discrete variable, this means $dn \langle N|n \rangle \langle n|\psi \rangle = \langle N|\psi \rangle$. Because $dn = 1$, $\langle N|n \rangle$ is a Kronecker-δ. If q is continuous, $\langle Q|Q \rangle$ is infinite, but $dq \langle Q|q=Q \rangle$ is unity.

Here is a summary of the self-space bracket behavior:

$$\langle q \neq Q|Q \rangle = 0 \qquad dq \langle q|q \rangle = 1 \qquad (2\text{-}36)$$

These two statements express the physical idea that if you measure the same observable twice in rapid succession, you will find the same state in both measurements.

MEASUREMENT DESTROYS A STATE, AND MEASUREMENT CREATES A STATE

Suppose you have prepared a particular bead momentum, $p = P$. You have prepared the state $|P\rangle$. Then for every x there is an amplitude that upon measuring position, that particular value of x will be found. The amplitude is $\langle x|P \rangle$.

But now execute the measurement. You will find some particular value $x = X$, say, $X = 6$ Å. What is the state of the bead? It is $|X\rangle$. It is $|6$ Å$\rangle$. The state $|P\rangle$ is destroyed; the state $|X\rangle$ is created. That's what the measurement does.

That before the measurement was made, the bead had momentum P is irrelevant after the measurement. While in $|P\rangle$ it had amplitudes (probabilities) to be at any of the positions x. The state $|P\rangle$ is a superposition of all x-states. But once the measurement is made, the probability distribution collapses: the only nonzero amplitude in x-space is at the position you find in the measurement. To discover a particular position $x = X$

is to know that the bead is in state $|X\rangle$. Upon executing the measurement of position you acquire a precise knowledge of the bead's position. You have erased all probabilities for other positions.

You now know the bead's location precisely; you have destroyed your knowledge of its momentum. After measuring its position you are left with a probability distribution for its momentum. A subsequent measurement might produce any p at all; it need not produce P. There are non-zero amplitudes $\langle p|X\rangle$ for any p: those of (2-19). The act of measurement destroys the state $|P\rangle$ and creates the state $|X\rangle$.

This effect spotlights an essential difference between classical and quantum mechanics. We measure p and then we measure x. In classical physics we would conclude that the bead has momentum p *and* position x. We must not conclude this in quantum mechanics! It has momentum p *or* position x.

Momentum and position measurements are incompatible. The very act of making the second measurement *destroys* the state you first measured. In quantum mechanics you cannot conclude from the pair of measurements that the bead has both momentum p and position x: it has only the property you found in the last measurement. Its state is the last measurement made. Its state is $|x\rangle$. It is only what you measure that you know.

PROBLEMS

2.1 The gunshot is uniformly distributed over the whole area of the target circle: radius $R = 10$ cm. What is the probability per unit length—the radial probability density—of finding gunshot at the radius $r(=$ say, 3 cm)?

Answer:

$$P(3 \text{ cm}) = 0.06 \text{ cm}^{-1}$$

$$dr\, P(r) = \frac{2r}{R^2}\, dr$$

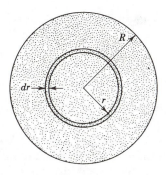

2.2 Calculate the wavelength corresponding to a one-half gram particle (bead) moving at 1 cm/sec. What grating spacing would be necessary to "see" the diffraction of such a particle? What is the smallest grating spacing that can be machined with machine shop tools; that is, how accurately can machinists machine something?

Answer: To the order of 1/1000 of an inch—called 1 mil in machine shop lingo.

To diffract electrons what did Davisson and Germer use for a grating of Angstrom unit spacing—about 10,000 times smaller than can be machined?

2.3 Prove that the allowed momenta for the bead-on-a-ring are indeed those of Equation 2-9. Remember that $e^{2\pi in} = 1$ for any integer, n.

2.4 For a bead with momentum $p = -3h/L$, what is the probability of finding it within the first quadrant of the ring (in the first quarter of the entire circumference)?

$$\int_0^{L/4} dx\, P(x|n=-3) = ?$$

The prescription for the answer is the integral above.

The answer is 1/4.

2.5 From the certainty statement, prove that the constant in (2-8) must indeed have absolute value $1/\sqrt{L}$, as shown. Would the constant $i/\sqrt{L}$ instead of $1/\sqrt{L}$ in (2-8) satisfy (2-10)?

Answer: Yes.

2.6 Equation 2-8 is the state at one moment of time ($t = 0$). An equally legitimate state wave function is the one of (2-8) multiplied by $\exp(-i\omega t)$. This one exhibits the time dependence. The angular frequency ω is the same both for k and for $-k$. Show that positive values of k correspond to waves traveling in the $+x$ direction, and negative k, to negative direction of propagation. Obtain the (phase) velocities for each of these waves. Show that (2-8) multiplied by $\exp(-i\omega t)$ satisfies the certainty condition of (2-10) just as well as (2-8) alone does.

Key: The general traveling wave moving with speed v to the right has the form $f(x - vt)$.

2.7 True or false:

a. $|3\text{ Å}\rangle + |2\text{ Å}\rangle = |5\text{ Å}\rangle$
b. $2\,|3\text{ Å}\rangle = |6\text{ Å}\rangle$

Answer: False. These equations are nonsense. A ket is a mathematical object, but it is not a traditional one: you can't add or multiply the arguments of kets. These equations are false because of the meaning of what a ket is.

2.8 Why is it that for a bead on a ring of circumference, L, no measurement will ever find its momentum to be $3.5\ h/L$?

Answer: No one knows. But the mathematics of quantum theory yields results, like this one, that coincide with experimental findings. That is the reason that we use quantum theory. That quantum theory fits experiment is what validates the theory, but why experiment should give such peculiar results is a mystery. This is the shock to which Bohr referred.

2.9 Suppose there were an infinite repulsive potential barrier at $x = 0$ so that the bead could never be at that position. It is still free to be anywhere between 0 and L. Then the probability distribution shown in Figure 2-2 is one of the important stationary states of the bead. (A stationary state is one that persists in time, and this is one of a discrete set of such states.) The wave function for this state is

$$\psi(x) = \sqrt{2/L}\ \sin\frac{4\pi x}{L}$$

Using the standard procedure for inverting the fourier expansion, find all the coefficients c_n for this wave function.

Answer: $c_2 = -i/\sqrt{2L}$, $c_{-2} = i/\sqrt{2L}$, $c_{\text{other } n} = 0$

2.10 The barrier is removed. The state corresponding to the wave function of Problem 2.9 persists. What are the probabilities to find the following experimental results:

a. that the bead momentum is zero; $|\langle p=0|\psi\rangle|^2 = ?$ is the question put symbolically.

b. that the bead momentum is $3h/L$; $|\langle p=3h/L|\psi\rangle|^2 = ?$

c. that the bead momentum is $2h/L$; $|\langle p=2h/L|\psi\rangle|^2 = ?$

d. that the bead momentum is $2.6h/L$ (What is this question symbolically?).

e. that the bead position is between $L/24\ (1 - .01)$ and $L/24\ (1 + .01)$. Put the question symbolically and give the answer.

f. that the bead position is between $L/24$ and $L/8$.

Answers:

a. zero

b. zero

c. $\dfrac{1}{2}$

d. zero

e. $\dfrac{1}{2400}$

f. $\dfrac{1}{12} + \dfrac{1}{16\pi}$

2.11 Calculate $\langle x|X \rangle$ for the case of the bead-on-a-wire using the structure outlined and demonstrate that the calculation does, in fact, yield the Dirac δ-function, $\delta(x - X)$.

Solution: Applying the ket-bra sum theorem we can write

$$\langle x|X \rangle = \sum_{p} \langle x|p \rangle \langle p|X \rangle$$

Translating this into conventional mathematical language we find

$$\langle x|X \rangle = \sum_{n=-\infty}^{+\infty} \frac{1}{\sqrt{L}} \exp\left(\frac{2\pi inx}{L}\right) \frac{1}{\sqrt{L}} \exp\left(\frac{-2\pi inX}{L}\right) \tag{A}$$

If $x = X$, the summation becomes a particular constant, $1/L$, added to itself an infinite number of times. The summation produces infinity.

If $x \neq X$, the summation is also easily performed. To do so we take note of the identity (valid if $1 \geq |\alpha|$):

$$\frac{1}{1 - \alpha} = 1 + \alpha + \alpha^2 + \alpha^3 + \ldots = \sum_{n=0}^{\infty} \alpha^n = \sum_{n=0}^{-\infty} \left(\frac{1}{\alpha}\right)^n \tag{B}$$

Apply this by substituting for α the quantity $\exp\left[2\pi i(x - X)/L\right]$ where $n \geq 0$ and the reciprocal of this where $n < 0$. A little algebra will then demonstrate that the entire sum, (A), adds precisely to zero when $x - X \neq 0$. (When $x = X$ so that $\alpha \to 1$, the equality still holds: both sides of (A) yield ∞).

This result is the Dirac δ-function: the base length has shrunk to zero, and the height has stretched to ∞. But the area under the curve is unity. You can verify this by integrating Equation A with respect to x over the domain of x, the continuous spectrum of values from zero to L.

To perform the integration you must interchange the order of summation and integration, performing the latter on each term, one at a time, in the series. The result is that in the new series so produced, every single term is just equal to zero except for one, the one with $n = 0$. The integral of this single remaining term yields unity. Thus the integral of the whole sum is unity.

2.12 If $\alpha = 10\text{Å}^{-1}$ what is the period of the function $\sin^3 \alpha\pi x$? What is its wavelength?

Answer: 0.2 Å. It has only a period, not a wavelength.

2.13 The defining property of the Dirac δ-function is that for *any* continuous, single-valued square integrable function, $f(x)$, the following integral yields the value of the function at the point where the argument of the δ-function is zero:

$$\int_{-\infty}^{+\infty} f(x) \, \delta(x - y) \, dx = f(y) \tag{C}$$

Something defined merely by its integral property is called a *distribution*. It is not an ordinary function because many different functions can produce the same effect. There are many representations of the Dirac δ-function. In practice δ(x) is the limit as σ → 0 of any continuous function of x parametrized by σ which has these three properties:

$$1. \, f_{\sigma=0}(x \neq 0) = 0, \quad 2. \, f_{\sigma=0}(x=0) = \infty$$

and   3. $\displaystyle\int_{-\infty}^{\infty} f_{\sigma}(x)dx = 1$

If $f_{\sigma}(x)$ has these three properties then $\delta(x) = \lim f_{\sigma}(x)$ as σ → 0.

That $\delta(-x) = \delta(x)$ is because $\int f(x)\,\delta(x)\,dx = \int f(x)\,\delta(-x)\,dx$ for any $f(x)$. They both yield $f(0)$. In this spirit show that the following are true:

a. $g(x) \equiv \dfrac{d}{dx}\delta(x) = \dfrac{d}{dx}\delta(-x) \left(\text{both yield } \left(-\dfrac{df}{dx}\right)\Big|_{x=0}\right)$ but $g(-x) = -g(x)$

b. $\delta(bx) = \delta(x)/|b| \quad b = $ real number

c. $\displaystyle\int \delta[g(x)]\,dx = \sum_i \left|\dfrac{dg}{dx}\right|_{x=x_i}^{-1}$ all i where $g(x_i)=0$

d. $\delta(x) = \dfrac{1}{2\pi} \displaystyle\int_{-\infty}^{+\infty} e^{ikx}\,dk$ (This is the most important representation)

It's a consequence of the theorem

$$f(x) = \int_{-\infty}^{+\infty} g(k)\,e^{ikx}\,dk$$

$$g(k) = \frac{1}{2\pi} \int_{-\infty}^{+\infty} f(x)\,e^{-ikx}\,dx$$

e. $\delta(x) = \lim_{a\to\infty} \left(\dfrac{\sin ax}{\pi x}\right)$

To prove this use the following fact in conjunction with d.

$$\int_{-\infty}^{+\infty} dk = \lim_{a\to\infty} \int_{-a}^{a} dk$$

f. $\delta(x) = \lim_{\sigma\to 0} \left(\dfrac{\exp(-x^2/4\sigma^2)}{(2\pi\sigma^2)^{1/4}}\right)^2$

Argue the function's shape verbally, and then prove that its integral is unity.

g. $\delta(x) = \lim_{\sigma\to 0} \left(\dfrac{\sigma \sin^2(x/\sigma)}{\pi x^2}\right)$

Note:

$$\int_{-\infty}^{+\infty} \frac{\sin^2 y}{y^2} \, dy = \pi$$

h. $\delta(x) = \dfrac{1}{L} \displaystyle\sum_n e^{2\pi inx/L}$ where $0 \leqslant x < L$, n spans all integers

(see Problem 2.11).

2.14 Consider an atomic bead free to move only on the half-track, $0 < x < L/2$. Because the bead cannot penetrate the wall confining it to the half-track, its wave function must be zero there. Hence its state—whatever it is—must be of the form

$$\psi(x) = \sum_{\nu=1}^{\infty} b_\nu \sqrt{4/L} \sin (2\pi\nu x/L) \qquad 0 \leqslant x \leqslant L/2 \qquad \text{(D)}$$

a. Why?

Answer: This is the most general fourier series expansion possible for a single-valued function of position that goes to zero at its ends, $x = 0, L/2$.

b. Write this in Dirac notation using the ket-bra sum $\sum_\nu |\nu\rangle \langle\nu| = 1$. Show the correspondance between terms in the fourier and Dirac notation equations. Interpret the brackets $\langle x|\psi\rangle$, $\langle x|\nu\rangle$ and $\langle\nu|\psi\rangle$ which arise when you recast Equation D in the form $\langle x|\psi\rangle = \displaystyle\sum_{\nu=1}^{\infty} \langle x|\nu\rangle \langle\nu|\psi\rangle$.

Answer:

$$\langle x|\nu\rangle = \sqrt{\frac{4}{L}} \sin 2\pi\nu x/L$$

$$\langle\nu|\psi\rangle = b_\nu$$

$$\langle x|\psi\rangle = \psi(x)$$

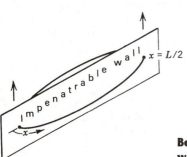

Bead is in state $\nu = 2$ on this side of wall. The wall is suddenly removed.

c. The states, characterized by the value of v, are, as we shall later see, the energy states of the bead. Suppose the bead is in the state $v = 2$ on the right half-track. When the dividing wall is suddenly removed, the bead finds itself on the full track. What is the state—let's call it $|\phi\rangle$—of the bead when the wall is suddenly removed? Describe the state $|\phi\rangle$ by giving its position representation: $\langle x|\phi\rangle = ?$

Answer:
The idea is that the original state is preserved. Although the new state, not being stationary, will change with time, it is unaffected in the first instant by the removal of the wall. So $\langle x|\phi\rangle = \phi(x)$ where

$$\phi(x) = \begin{cases} \sqrt{4/L}\ \sin 4\pi x/L, & 0 \leqslant x \leqslant L/2 \\ \\ 0 & L/2 \leqslant x \leqslant L \end{cases}$$

d. Immediately after removing the wall, what is the probability of finding the bead's momentum to be p? Calculate, from your general answer, the specific probabilities for $p = 2h/L,\ -2h/L,\ 3h/L,\ 3.8h/L$.

Hint: You seek $|\langle p|\phi\rangle|^2$. You find

$$\langle n = pL/h|\phi\rangle = (\delta_{n,2} - \delta_{n,-2})/2i + \left.\frac{1}{\pi(1 - (n/2)^2)}\right|_{\text{when } n = \text{odd}}$$

Thus $\langle 3h/L|\phi\rangle = -4/5\pi$, $\langle 4h/L|\phi\rangle = 0$, $\langle 2L/h|\phi\rangle = 1/2i$ and $\langle 3.8h/L|\phi\rangle = 0$.

2.15 Consider an atomic bead free to move anywhere in one dimension. The domain of position states open to it is all x where $-\infty < x < +\infty$. (This is also a basis space for the system: it is the set of all "words" in one language capable of describing the bead.) For such a bead, a pure state of momentum exists for any value of momentum $p = \hbar k$. Momentum is not quantized; the domain of momentum space is $-\infty < p = \hbar k < \infty$. Prove that

$$\langle x|p\rangle = \frac{1}{\sqrt{h}} \exp\left(\frac{ipx}{\hbar}\right) \tag{E}$$

$$\langle x|k\rangle = \frac{1}{\sqrt{2\pi}} \exp(ikx) \tag{F}$$

Solution (in part): The central point is that according to Fourier's theorem, any function of x over this domain—like $\psi(x)$—can be expanded as an integral over k. Thus any wave function can be represented as (see Problem 2.13d):

$$\psi(x) = \int_{-\infty}^{+\infty} dk \exp(ikx) \phi(k) \tag{G}$$

where $\phi(k)$ is given by

$$\phi(k) = \frac{1}{2\pi} \int_{-\infty}^{+\infty} dx \, \exp(-ikx) \, \psi(x). \tag{H}$$

Writing Equation G in Dirac notation

$$\langle x|\psi\rangle = \sum_k \langle x|k\rangle \langle k|\psi\rangle \tag{I}$$

makes it clear by interpretation of the terms that

$$\langle x|k\rangle = \text{constant } e^{ikx}$$

and

$$\langle k|\psi\rangle = \phi(k)/\text{same constant}$$

The constant is deduced by effecting the guarantee that $\langle k|K\rangle = \delta(k - K)$. Do it with an x-space ket-bra sum. From $\langle p|P\rangle = \delta(p - P)$ we see that the constant for the states labeled by p cannot be the same as that for the states labeled by k.

$$1 = \sum_k |k\rangle \langle k| = \sum_p |p\rangle \langle p| \neq \sum_k |p\rangle \langle p|$$

Getting the constant is called *normalizing the states*.

2.16 A bead-on-a-ring that was known to be in a momentum state with $p = 3h/L$ is found in a position measurement at the origin $x = 0$. What is its momentum?

Answer: It is not possible to say.

Explain this answer. What is it possible to say?

Answer: Only that it is at $x = 0$.

2.17 Prove that Equation 2-33 is just the right-hand side of Equation 2-32 rewritten.

Solution principles: Note that

$$P(n|x) = |\langle n|x\rangle|^2 = \langle x|n\rangle \langle n|x\rangle$$

and keep in mind that

$$\sum_n \langle x|n\rangle \langle n|x\rangle = \langle x|x\rangle$$

2.18

    Event A. Having just measured it, you know the bead's momentum; it's $p = nh/L$, where $n = 2$. What is the probability to find the bead at $x = X$ within $dx = \varepsilon = $ small ($\varepsilon \ll L$)?

Answer: $\varepsilon \, P(X|n) = \varepsilon/L$

Event B. Having just measured it, you know the bead's position to be $x = X$ within $dx = \varepsilon$. What is the probability of finding $n = 2 = pL/h$?

Answer: Zero, $\dfrac{P(n=2|X)}{\langle X|X \rangle} = 0$

That this probability is zero can be understood as follows. Per unit length of the ring no matter where you put the bead it has an equal chance of having any momentum: $P(n|X) = 1/L$. Since there are an infinite number of momenta possible the chance of finding any particular one of them is indeed zero.

That the infinite sum of all such zeros nevertheless adds up to 1 can be comprehended this way. Suppose instead of an infinite number there were only N momenta possible. The probability to find any one of them would be $1/N$. So as N becomes infinite each probability goes to zero. But the sum of all N probabilities, being unity no matter what N is, remains so even as N approaches infinity.

Chapter 3

The Bead on a Track:
Its Measurement Spectra are
Operator Eigenvalues

AN OBSERVABLE IS WHAT YOU MEASURE

You make a measurement. In doing so you are trying to ascertain the value of some property of the system being measured. The property is called an *observable*. The measurement is an observation: ultimately a meter reading. One measures observables.

AN OBSERVABLE HAS A SPECTRUM

The measurement of an observable yields a spectrum of possible values: possible meter readings. You wouldn't be measuring if there weren't more than one possible result! Every observable has a spectrum.

There is a mathematical structure that just parallels this essential feature of the physical world. By its very nature this mathematics is *spectra producing*. The spectrum comes from the mathematical procedure called the *eigenvalue problem*. The eigenvalue problem produces a mathematical spectrum of numbers; every physical observable produces a spectrum of measurement results—numbers. Quantum mechanics is the theory that cements this correspondence.

3.1 OPERATORS OPERATE ON KETS

To enter into this mathematics you need to understand the notion of an *operator*. An operator is is mathematical entity that operates. It operates on a ket. It changes the ket in some way so that something new is produced, a new ket.

To distinguish an operator from what is not an operator we use a *hat;* a circumflex symbol. Thus if the operator $\hat{\Omega}$ operates on the state $|\psi\rangle$ there is a new thing produced called $\hat{\Omega}|\psi\rangle$. The operator and the ket placed together in this order means *this new thing*.

THE OPERATOR'S INSTRUCTION IS ITS REPRESENTATION

That it changes a ket is the first of the two aspects of an operator. The second is that there is a recipe or a prescription for the change. There must be a rule or instruction that tells what is produced when the ket undergoes the operation. An operator is a set of directions for how the ket is to be transformed.

This prescription must be given in some language. The symbol for the instruction for the operator $\hat{\Omega}$ in *x*-language is $\langle x|\hat{\Omega}$. It is also the representation of $\hat{\Omega}$ in the language of *x*. An instruction given in *p*-language is $\langle p|\hat{\Omega}$. The symbol $\langle p|\hat{\Omega}$ denotes the instruction for the operator $\hat{\Omega}$ in *p*-space. It is its representation in *p*-space.

Here's an example. Let's call the translation-in-position-by-length-*b* operator by the symbol $\hat{T}$. The instruction defining $\hat{T}$ is this: in the space of position *x*, it translates any state forward by the length *b*. Thus, if the wave function for the original state is $\psi(x)$, $\hat{T}$ will produce the new function $\psi(x-b)$. In traditional notation this is written:

$$\hat{T}\psi(x) = \psi(x-b) \tag{3-1}$$

Figure 3-1 illustrates the effect of the $\hat{T}$ operator.

The central equality of the three that follow is the prescription in Dirac notation:

$$\hat{T}\psi(x) = \langle x|\hat{T}|\psi\rangle = \langle x-b|\psi\rangle = \psi(x-b) \tag{3-2}$$

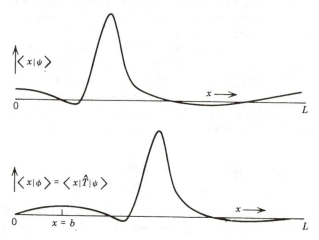

Figure 3.1 The effect of the translation-forward-by-*b*-in-*x*-space operator, $\hat{T}$, on a state in position space.

The left-hand equality defines the meaning of the symbol $\langle x|\hat{T}|\psi\rangle$. It means the effect of the operator $\hat{T}$ on $\psi(x)$. The right-hand equality expresses what $\langle x-b|\psi\rangle$ means. It means that the argument in the function ψ is to be altered: from x to $x-b$. The central equality is the operator equation in Dirac notation form.

Notice that the effect of the operator $\hat{T}$ on $\psi(x)$ is *not* written as $\hat{T}\langle x|\psi\rangle$ but as $\langle x|\hat{T}|\psi\rangle$. This way of writing incorporates a key idea in the mathematical structure: $\langle x|\hat{T}|\psi\rangle$ is the symbol for the effect of the operator $\hat{T}$ on the ket $|\psi\rangle$ in x-space. It implies that the operator $\hat{T}$ can affect $|\psi\rangle$ in other spaces, and indeed, it can. The symbol $\hat{T}\langle x|\psi\rangle$ doesn't carry this critical concept, it is to be strictly shunned. An operator to the left of a bra distorts meaning in the mathematical structure.

The defining equation for $\hat{T}$ is true for any state $|\psi\rangle$. I didn't specify the ψ. Any $\psi(x)$ will do. I explained only how the operator works in x-space. Dirac notation allows us to express the idea that the $\hat{T}$ operation is independent of ψ. It is portrayed like this:

$$\langle x|\hat{T} = \langle x-b| \tag{3-3}$$

If you apply both sides of this equation to any ket at all, the equation will remain valid. This equation is the prescription for $\hat{T}$ in x-space. The prescription is this: Replace x by $x-b$ in the argument. The representation of an operator is carried by the bra on its left, just as the representation of a state is carried by the bra at its left.

	Operator Symbol:
For the change effected:	$\hat{\Omega}|S\rangle$
For the q-space prescription to effect it:	$\langle q|\hat{\Omega}$
Without meaning, a conceptual distortion:	$\hat{\Omega}\langle q|S\rangle$

A second example of an operator is $\hat{D}$. Its prescription in x-space is this: take the derivative with respect to x. Put formally in Dirac notation, this statement is

$$\langle x|\hat{D} = \frac{d}{dx}\langle x| \tag{3-4}$$

No ket need be specified; you may apply this equation to any ket whatsoever. This equation defines the instruction for the $\hat{D}$ operator; it gives its representation in position space.

THE MOST GENERAL THING AN OPERATOR CAN PRODUCE IS A CONSTANT TIMES A KET

To apply the $\hat{D}$ operator to some particular $\psi(x)$ means to take its derivative with respect to x. You get a new function of x; $f(x) = d\psi/dx$. The most general thing this new function of x can be is a new bracket, say $\langle x|\phi\rangle$, multiplied by a number.

If it can be fourier analyzed, any function of x can be viewed as a bracket times a number. You simply ordain that it be so. You do it by associating the ket-bra sum, Equation 2-26 with the fourier expansion (2-27) or (1-6). The general fourier analyzable function is not normalized, brackets are. Their ratio is a number.

Any function of x that has been divided by some number chosen to meet the certainty condition is a bracket. Given $f(x)$, its associated bracket is $\langle x|\phi\rangle$ if $f(x) = c\langle x|\phi\rangle$ where $\sum_x |\langle x|\phi\rangle|^2 = 1$.

To see an example of this idea, apply $\hat{D}$ to the state $|p = 3h/L\rangle$ in x-space. The new state that you get, in this case, turns out to be the old one back again, but multiplied by the number $6\pi i/L$.

This argument holds for any space at all. If the operator $\hat{\Omega}$ operates on $|\psi\rangle$ in q-space, it can produce only some function in q-space. In Dirac notation this function is some bracket $\langle q|\phi\rangle$ times a number. Thus $\langle q|\hat{\Omega}|\psi\rangle = c\langle q|\phi\rangle$. Because it is true for any space, we deduce this profound and important consequence: the most general thing an operator can do is to produce a new ket times some number. Thus

$$\hat{\Omega}|\psi\rangle = c|\phi\rangle \tag{3-5}$$

Notice that by the very meaning of the symbols, multiplicative constants may be placed on either side of a bracket or within it.

$$c\langle q|\psi\rangle = \langle q|c|\psi\rangle = \langle q|\psi\rangle c \tag{3-6}$$

LINEAR OPERATORS ARE WHAT WE USE

Embedded in the mathematics that we use are the conditions of a linear algebra. We are considering only linear operators. All of the operators of quantum mechanics are linear. It appears that nature is adequately described by this class of operators. Linear operators, $\hat{L}$, are those that meet these two conditions:

$$\hat{L}(c|S\rangle) = c(\hat{L}|S\rangle) \tag{3-7}$$

and

$$\hat{L}(|S\rangle + |R\rangle) = \hat{L}|S\rangle + \hat{L}|R\rangle \tag{3-8}$$

The first says that "the effect of the operator on a constant times the ket is the same as that constant multiplied by the effect of the operator on the ket." The second says that "the effect on the sum is the sum of the effects of the operator on a pair of states."

For an example of a nonlinear operator, consider this instruction: in x-space square the amplitude at every point. The result of this operation on $\psi(x)$ is to produce $\psi^2(x)$. Clearly, this operator satisfies neither of the two linearity conditions.

On the other hand both $\hat{T}$ and $\hat{D}$ are linear operators, they each satisfy both of the linearity conditions. You show it in x-space by considering how $\langle x|\hat{T}$ and $\langle x|\hat{D}$ operate in (3-7) and (3-8).

3.2   KET-BRA SUMS TRANSLATE OPERATOR INSTRUCTIONS

The operators $\hat{T}$ and $\hat{D}$ are defined by instructions regarding position space, x. These definitions are useful when the states on which they operate are given in the language of position. But states exist independent of the language in which they are represented; they can be expressed in many languages. So too may operators be expressed in many languages.

The principle governing the instruction for $\hat{T}$ in p-language if it is given in x-language is this: the new ket, representing the effect of $\hat{T}$ on $|\psi\rangle$, must transform in the same way as all kets do, via the ket-bra sum. Here is how this thought is formally implemented.

That we have the instruction for $\hat{T}$ in x-language means that we have the form or rule for $\langle x|\hat{T}$. It is equation 3-3. The form for $\langle x|\hat{D}$ is Equation 3-4. Having this prescription together with the representation of the state $|\psi\rangle$ in x-space, $\langle x|\psi\rangle$, we can calculate $\langle x|\hat{T}|\psi\rangle$. Whatever it comes out to be in detail, it must have the form $c\langle x|\phi\rangle$. The new ket representing the effect of $\hat{T}$ on $|\psi\rangle$ is $|\phi\rangle$; $\langle x|\hat{T}|\psi\rangle$ gives us the representation of $|\phi\rangle$ in x-space.

To get the instruction for $\hat{T}$ in p-language means to deduce the form for $\langle p|\hat{T}$. This key idea determines that form: $\langle p|\hat{T}|\psi\rangle$ had better get us the representation of the *same* state, $|\phi\rangle$, but in p-space. The state $|\phi\rangle$ whose p-representation is produced by $\langle p|\hat{T}|\psi\rangle = c\langle p|\phi\rangle$ must be the same one whose x-representation is produced by $\langle x|\hat{T}|\psi\rangle = c\langle x|\phi\rangle$.

Because the $|\phi\rangle$ are the same, $\langle p|\phi\rangle$ and $\langle x|\phi\rangle$ are related by a ket–bra sum. By inserting an x-space ket-bra sum in $\langle p|\phi\rangle$ we conclude that

$$\langle p|\hat{T}|\psi\rangle = \sum_x \langle p|x\rangle \langle x|\hat{T}|\psi\rangle \tag{3-9}$$

Thus the same ket–bra sum that transforms states also transforms operators. It is as if we split $\langle p|\hat{T}|\psi\rangle$ by inserting the x-space ket–bra sum form of unity at the bra-operator (left hand) vertical.

Because no specific state $|\psi\rangle$ was used in this equation, it can be written as a statement about transforming the language of an operator. The result is good for any state and for any operator and between any two basis spaces. If you have $\langle x|\hat{T}$, here's how you get $\langle p|\hat{T}$.

$$\langle p|\hat{T} = \sum_x \langle p|x\rangle \langle x|\hat{T} \tag{3-10}$$

You may equally well insert a ket–bra sum to replace the vertical in $\langle p|\hat{T}|\psi\rangle$ between $\hat{T}$ and $|\psi\rangle$; at the operator-ket vertical. That's simply because the vertical in $|\psi\rangle$ itself may be so replaced (see Equation 2-24).

Rule: Treat any vertical | as if it were unity. You may replace it with a ket–bra sum.

$$\langle\alpha|\hat{\Omega}|\beta\rangle$$

$$\sum_q |q\rangle\langle q| \underline{\qquad} \underline{\qquad} \sum_r |r\rangle\langle r|$$

IN DIFFERENT LANGUAGES, DIFFERENT INSTRUCTIONS

Let's apply the result to the two operators $\hat{T}$ and $\hat{D}$. Carry out the ket-bra summation in (3-9), in traditional notation. Here are the integrals you must evaluate to carry out this summation for the $\hat{T}$ and $\hat{D}$ operators:

$$\langle p|\hat{T}|\psi\rangle = \int_0^L dx \frac{1}{\sqrt{L}} \exp\left(\frac{-ixp}{\hbar}\right) \psi(x-b) \qquad (3\text{-}11)$$

$$\langle p|\hat{D}|\psi\rangle = \int_0^L dx \frac{1}{\sqrt{L}} \exp\left(\frac{-ixp}{\hbar}\right) d\psi/dx \qquad (3\text{-}12)$$

The calculations lead to the following results:

$$\langle p|\hat{T} = \exp(-ibp/\hbar) \langle p| \qquad (3\text{-}13)$$

$$\langle p|\hat{D} = ip/\hbar \langle p| \qquad (3\text{-}14)$$

To obtain these you must use the periodicity condition for all wave functions, $\psi(x)$ and $\langle x|p\rangle$, that characterize our bead-on-a-track. And you must recognize $\langle p|\psi\rangle$ when you see it. When it is split by an x-space ket–bra sum, it's still $\langle p|\psi\rangle$.

$$\langle p|\psi\rangle = \sum_x \langle p|x\rangle \langle x|\psi\rangle = \int_0^L dx \frac{1}{\sqrt{L}} \exp\left(\frac{-ixp}{\hbar}\right) \psi(x) \qquad (3\text{-}15)$$

The instruction for the $\hat{T}$ operator in p-language is quite different from its instruction in x-language. Equation 3-13 says that the p-space instruction for $\hat{T}$ is this: Multiply by the exponential function $\exp(-ibp/\hbar)$. This is nothing like its x-space instruction.

The prescription for the $\hat{D}$ operator in p-space is this: Multiply by $ip/\hbar$. In p-space there is no differentiation to carry out, as there is in x-space. In p-space you merely multiply the state by a factor.

There is an operator whose prescription is the same in every language! It is the operator $\hat{c}$ whose prescription is this: Multiply by the constant c. The constant may be any number, $3 + 4i$, for example. Carrying out the prescription in x-space yields

$$\langle x|\hat{c}|\psi\rangle = c \langle x|\psi\rangle \tag{3-16}$$

To find the p-space representation of this operator use a ket-bra sum in $\langle p|\hat{c}$. You find for $\langle p|\hat{c}$ exactly the same instruction: multiply by the constant c. Because the instruction is the same in any space, you need not specify the space.

$$\hat{c} = c.$$

SUCCESSIVE OPERATIONS MEAN AN OPERATOR PRODUCT

Two operations done in succession is symbolized by a product of operators. A state $|\psi\rangle$ is operated on first by $\hat{T}$ and then by $\hat{A}$. The first operation produces a new ket $|\phi\rangle$. Thus

$$\hat{T}|\psi\rangle = c|\phi\rangle \tag{3-17}$$

The second is now an operation on a ket also—but on the new ket $|\phi\rangle$. The second operation can produce only another ket. Thus the two operations in succession produce the ket, say, $|\theta\rangle$ where

$$c\hat{A}|\phi\rangle = \hat{A}\hat{T}|\psi\rangle = c'|\theta\rangle \tag{3-18}$$

The order of the $\hat{A}$ and $\hat{T}$ are significant. The operation of $\hat{A}$ first and then $\hat{T}$—represented mathematically as $\hat{T}\hat{A}$—does *not* necessarily produce the same ket $|\theta\rangle$.

These equations are written without specifying a basis. They are true in any basis. For example, the ket $|\theta\rangle$ in the basis x, $\langle x|\theta\rangle$, is the x representation of the result of first $\hat{T}$ and then $\hat{A}$ operating on $|\psi\rangle$; it is $\langle x|\hat{A}\hat{T}|\psi\rangle$ (see Problem 3.5).

3.3 OPERATORS HAVE EIGENVALUES

For each linear operator there is a special set of numbers that characterizes it. This set of characterizing numbers for a particular operator defines the operator just as much as a prescription in a particular language does. The characterizing numbers are even better for the purpose; they define the operator in all languages. These numbers are called the *eigenvalues* of the operator.

There is a certain class of operators whose eigenvalues have this important and remarkable property: they constitute a complete language. They define a *space,* the *home space* of the operator. Every eigenvalue is a point in a complete space generated by the operator. For such operators there are two equivalent mathematical prescriptions for getting the eigenvalues.

IN THE EIGENVALUE PROBLEM, THE FORM IS THE SUBSTANCE

The two distinct statements of the mathematical prescription for getting this spectrum of characterizing numbers are the ket statement and the bra statement.

The ket prescription is embodied in this equation:

$$\hat{\Omega}|q\rangle = \Omega(q)|q\rangle \tag{3-19}$$

It says that for the operator $\hat{\Omega}$ there is a special set of kets, $|q\rangle$ on which the entire effect of $\hat{\Omega}$ is simply an algebraic multiplication. It is the form of this equation that expresses its content. Operators don't usually produce a number times the <u>same</u> ket back again; they generally produce a number times some <u>other</u> ket.

The special kets $|q\rangle$ for which this equation is true are called the *eigenkets* of $\hat{\Omega}$. Each eigenket has an algebraic multiplier, $\Omega(q)$. These are the eigenvalues of the operator $\hat{\Omega}$. They are a set of numbers that depend on q. The task of finding the eigenkets, $|q\rangle$, and the eigenvalues $\Omega(q)$ for the operator $\hat{\Omega}$ is called the *eigenvalue problem*.

The bra prescription defining the eigenvalues of the operator $\hat{\Omega}$ is

$$\langle q|\hat{\Omega} = \Omega(q)\langle q| \tag{3-20}$$

It says that in the special language of q, the effect of the operator $\hat{\Omega}$ is simply multiplication by $\Omega(q)$, no matter on what state it operates.

In the ket statement the q label the eigenkets of $\hat{\Omega}$. In the bra statement the same set of q functions as a language in which a ket can be represented. The q index the spectrum of eigenvalues $\Omega(q)$. If the spectrum is discrete, we must understand by q some integer index. If the spectrum is continuous, then q will be some continuous variable.

HOME SPACE IS p-SPACE FOR $\hat{D}$ AND $\hat{T}$

To exhibit their significance we apply each of these rules to the operators $\hat{D}$ and $\hat{T}$.

Equation 3-4 is the prescription for $\hat{D}$ in x-space, "take the derivative." This prescription does *not* have the form of (3-20); hence x-space is *not* the eigenspace for the $\hat{D}$ operator.

Equation 3-14 is the prescription for $\hat{D}$ in p-space; "multiply by $ip/\hbar$." Hence p-space is the eigenspace for $\hat{D}$; the eigenvalues, $D(p)$, are $ip/\hbar$.

Compare the x-space instruction for $\hat{T}$ (Equation 3-3) with its p-space instruction (Equation 3-13). You will find that only the latter has the form of (3-20). Thus p-space is the eigenspace, or the home space, for $\hat{T}$. Its eigenvalues, $T(p)$, are $\exp(-ibp/\hbar)$. In the language of p the operator is an algebraic multiplier equal to the eigenvalue.

For an example of the ket form of the eigenvalue statement, pick some state in the home space of the operator. Call it $|P\rangle$. This is one of the states in p-space. It is one of the eigenstates of both $\hat{D}$ and $\hat{T}$.

Examine $\hat{D}|P\rangle$. It should produce the eigenvalue $D(P)$ multiplied by the state back again. It should show the form of (3-19) no matter in what space you calculate.

Calculate it in p-space. You will find

$$\langle p|\hat{D}|P\rangle = \frac{ip}{\hbar}\langle p|P\rangle = \frac{iP}{\hbar}\langle p|P\rangle \tag{3-21}$$

The first equality is from (3-14) and the second from $\langle p \neq P|P\rangle = 0$. This equation shows that in p-space, the form of (3-19) is, indeed, met by $\hat{D}|P\rangle$.

Now check the idea in x-space. Calculate $\langle x|\hat{D}|P\rangle$. You will find

$$\langle x|\hat{D}|P\rangle = \frac{d}{dx}\langle x|P\rangle = \frac{iP}{\hbar}\langle x|P\rangle \tag{3-22}$$

The left-hand equality is the meaning of $\hat{D}$ in x-space, Equation 3-4. The right-hand equality is what results from carrying out the differentiation on $\langle x|P\rangle$. The bracket $\langle x|P\rangle$ is given in Equation 2-8. The equality of the extremes confirms the conclusion that p-space kets are eigenkets of $\hat{D}$ with eigenvalues $ip/\hbar$.

AT HOME THE HAT COMES OFF

Consider the operator $-i\hbar\hat{D}$. As it is for $\hat{D}$, the home space for this operator is p-space. The effect of this operator is to produce p itself! The eigenvalues of the $-i\hbar\hat{D}$ operator are the bead momenta; $p = nh/L$. Multiplying equation 3-14 by $-i\hbar$ confirms it. It is reconfirmed in (3-21) and (3-22). The eigenvalue spectrum of this operator is just the spectrum of physical momenta available to the bead. That suggests that we give $-i\hbar\hat{D}$ a name. This operator is the *momentum-generating operator*, or simply the *momentum operator*.

What generates the spectrum of momenta we call the *momentum operator*. We label this operator $\hat{p}$. Thus

$$\hat{p} = -i\hbar\hat{D} \tag{3-23}$$

With this relabeling the two forms of the eigenvalue problem become

$$\hat{p}|p\rangle = p|p\rangle \quad \text{and} \quad \langle p|\hat{p} = p\langle p| \tag{3-24}$$

The key equations of quantum mechanics look like these. The same symbol appears in four places. The symbol corresponds to some observable. In these equations it is the momentum, p. Here is what the equations are meant to convey:

> The operator corresponding to the observable generates, as eigenvalues, the spectrum of the observable. The eigenkets, generated by the operator, are just the pure states of the system, the ones found in measuring that observable. The measurement spectrum serves as a language in which any state can be expressed: the p-spectrum is a complete basis.

All of these profound notions about the momentum observable are contained in a simple notational dictum: *In its home space the hat comes off.* That's the content of the eigenvalue problem: the operator, in its home space, is an algebraic multiplier. Equation 3-24 illustrates this.

THE OBSERVABLE'S SPECTRUM IS THE OPERATOR'S EIGENVALUES

For our bead-on-a-track, position is no less a measurable than is momentum. There must exist, therefore, a position operator $\hat{x}$. This operator must furnish us with position space, just as $\hat{p}$ furnishes us with momentum space. Thus the defining property of the position operator is

$$\hat{x}|x\rangle = x|x\rangle \quad \text{or} \quad \langle x|\hat{x} = x\langle x| \tag{3-25}$$

Because there is such a space, there is such an operator. That in its home space you remove the hat is its critical property. The position operator's eigenvalues is the spectrum of the position observable.

The operator $\hat{x}$ may be represented in p-space. Problem 3.6 is concerned with $\langle p|\hat{x}$ and $\langle x|\hat{p}$. Momentum space is not home for $\hat{x}$, neither is position space home for $\hat{p}$.

MEASUREMENT GENERATES COMPLETE SPACES

In the examples all of the operators are eigen to either p-space or x-space. These both are complete spaces; any bead-on-a-track physical state can be expressed in either of these spaces.

That a space is complete is expressed formally by the ket-bra sum theorem. Because $\Sigma|x\rangle\langle x| = 1$, any state can be completely resolved into all its position possibilities; x-space is complete. The ket-bra sum theorem is the statement of completeness.

The spaces of position and momentum are measurement spaces. These physical measurement spaces are inherently complete: a physical state can always be resolved into the spectrum of measurement results you can get from one of these observables. Given any physical state, you can find its presence distributed among the results of measurement. This is the substance of Chapter 2. It is explored in Problem 3.7.

Quantum Principle 5

3.4 MEASUREMENT SPACES ARE GENERATED BY PHYSICAL OPERATORS

The critical feature of the eigenvalue problem is that the special space of q generated by $\hat{\Omega}$ be complete. Only then does the bra statement of the eigenvalue problem follow from the ket statement. The mathematical space generated by some operator $\hat{\Omega}$ is useful physically only if it is complete.

Among all linear operators there is a special class that is known to produce complete sets of q's. These are called *hermitian operators*. The space of q's created by such operators satisfies the completeness theorem: the q-space ket-bra sum is unity.

All of the operators representing physical observables—like $\hat{x}$ and $\hat{p}$—belong to this class: they are hermitian. The eigenvalue problem for such operators generates the entire word vocabulary of a language. Operators corresponding to physical observables generate physical measurement spaces.

Quantum mechanics is founded on a generalization of this idea. It is this: to every physical observable there corresponds an operator whose eigenvalues are precisely the measurement spectrum of the observable. And the home space of the operator is a complete physical measurement space.

If we tag operators associated with physical observables by the term *physical operators*, then the axiom is simply put: a physical operator generates an observable's measurement space.

In quantum mechanics, nonphysical operators are useful as well as physical ones. But all of the spaces of quantum mechanics are physical ones, those of measurement. Both $\hat{T}$ and $\hat{D}$ are nonphysical operators: there are no observables to which they correspond. The operators $\hat{x}$ and $\hat{p}$ are physical ones. They correspond to measurables—position and momentum. For all four operators the spaces of interest are those of physical measurement, x and p.

When the q-space generated by $\hat{\Omega}$ is complete, it is simple to prove the equivalence of the two statements of the eigenvalue problem.

THE EIGENVALUE STATEMENTS ARE EQUIVALENT

Single out a particular one of the q; we'll call it Q. Then examine the bracketed operator $\langle q|\hat{\Omega}|Q\rangle$. Here is what the two prescriptions imply:

$$\Omega(q) \langle q|Q \rangle = \langle q|\hat{\Omega}|Q \rangle = \Omega(Q) \langle q|Q \rangle \qquad (3\text{-}26)$$

The right-hand equality is a direct application of the ket eigenvalue problem statement, Equation 3-19. The left-hand equality is an equally direct application of the bra statement, Equation 3-20.

The equivalence proof has two parts. (1) It must be demonstrated that this double equation is consistent, that the right-hand equality implies the left, and vice versa. (2) It must be established that the right-hand equality, through apparently a special case of it, is actually completely equivalent to (3-19) and that the left-hand equality is as general as (3-20) is.

That this double equation is consistent hangs only on the fact that $\langle q|Q \rangle$ is a self-space bracket. It is zero unless $q = Q$, because both q and Q are in the same space! By this fact it is easily established that each equality implies the other.

VALID FOR ALL q MEANS VALID FOR ANY STATE

Consider the right-hand equality of (3-26). Because no particular q among those of the spectrum has been specified, the equation is valid for all q. It is, therefore, valid in any basis at all. It must be true, in any foreign basis, say $\langle r|$ that

$$\langle r|\hat{\Omega}|Q \rangle = \Omega(Q) \langle r|Q \rangle \qquad (3\text{-}27)$$

That this equation follows from the right-hand side of (3-26) is merely a matter of a ket-bra sum insertion. We are given a statement about $\hat{\Omega}|Q \rangle$ in the special basis $\langle q|$. We want the statement in the foreign basis $\langle r|$. We must replace the vertical with a ket-bra sum, we apply the theorem $\langle r| = \Sigma \langle r|q \rangle \langle q|$.

To extract (3-27) you must recognize when you can collapse a ket-bra sum to unity. This is as important as the reverse process, replacing unity with a ket-bra sum.

That no specific basis r was chosen in (3-27) means that it's valid in any basis at all; we may drop the bras $\langle r|$ on both sides. That is precisely the content of Equation 3-19. Thus (3-26) $(r.h.s.)$ is as general as (3-19).

A parallel argument shows that the left-hand equality of (3-26), because it holds for all the Q in the spectrum, is just equivalent to (3-20).

Thus the ket statement (3-19) and the bra statement (3-20) are equivalent: they both describe the complete space q and the spectrum of eigenvalues $\Omega(q)$ characterizing the operator $\hat{\Omega}$. Generally one writes for the eigenvalue problem only its ket form; the other form is implied.

THE MOMENTUM OPERATOR GENERATES MOMENTUM

The momentum operator, $\hat{p} = -i\hbar\hat{D}$, generates the momentum measurement space of our bead-on-a-track. It can also get us the representation of $|p \rangle$ in any space.

To get the x-space representation, $\langle x|p \rangle$, cast the eigenvalue problem into x-space:

$$-i\hbar \frac{d}{dx} \langle x|p \rangle = \langle x|\hat{p}|p \rangle = p \langle x|p \rangle \tag{3-28}$$

The right-hand equality is the eigenvalue problem stated in x-space. The left-hand equality displays the instruction for $\hat{p}$ in x-space; it exhibits $\langle x|\hat{p}$. The equality of the extremes amounts to a differential equation for the function of x called $\langle x|p \rangle$. The solution is

$$\langle x|p \rangle = \text{constant} \times \exp\left(\frac{ipx}{\hbar}\right) \tag{3-29}$$

This is the right answer, we obtained it earlier by verbal arguments (see Equation 2-8). From the single valuedness of the $\langle x|p \rangle$ in this equation, we deduce, as before, the allowable spectrum of momenta: $p = nh/L$.

The differential equation yields $\langle x|p \rangle$ to within only an arbitrary constant multiplier. The magnitude of this constant is determined by the certainty condition: having momentum $p = nh/L$, the bead will be found somewhere on the ring. Thus $\langle p=nh/L|p=nh/L \rangle = 1$. The constant $1/\sqrt{L}$ satisfies this condition (see Problem 3.8).

THE HAMILTONIAN OPERATOR GENERATES ENERGY

Energy is an observable. Hence, there must be an energy operator. It produces the spectrum of measurable energies, E. Traditionally the operator that does this is not called $\hat{E}$: it is called $\hat{H}$. That's because it derives from the hamiltonian function of p and x of classical newtonian physics. It is this function written in operator form: the p's and x's in the function wear operator hats.

The hamiltonian operator for our bead-on-a-track is

$$\hat{H} = \frac{\hat{p}^2}{2m} \tag{3-30}$$

There is no potential energy, and so it contains no function of x. In this case it is easy to appreciate why the energy operator must be $\hat{p}^2/2m$. It's because it gives the right energy spectrum, $p^2/2m$.

That $\hat{p}^2/2m$ does, indeed, produce $p^2/2m$ is easily deduced from this simple observation: if the eigenstates of the $\hat{p}$-operator are $|p \rangle$, with eigenvalues p, then the eigenvalues of $\hat{p}^2$ will be p^2. The fact rests only on the meaning of $\hat{p}^2$, that the operator $\hat{p}$ operates twice in succession. It means the effect of $\hat{p}$ on $\hat{p}|p \rangle$: $p^2|p \rangle$ is produced.

The energy spectrum must be related to that of momentum because the classical relationship between the spectra is preserved in quantum mechanics.

Suppose you measure the momentum of the bead. You find p. Now measure its energy. You find E. For a classical bead you know that this energy will be connected to the momentum by $E = p^2/2\,m$. There is no reason to expect that on the quantum scale this relationship is invalid.

Suppose, on the quantum scale, that the energy were different from $p^2/2\,m$. The difference would have to disappear at the macroscopic scale because there we know the relationship between the p-measurement spectrum and the E-measurement spectrum. No evidence from experiment has ever suggested that this relationship shouldn't persist down to the microscopic scale. Barring relativistic effects, it is valid on all scales.

TO MAKE OPERATORS, PUT HATS ON THE OBSERVABLES

The mission of a physical operator is to deliver a physical measurement space; its eigenvalues must correspond to the entire spectrum of measurement results possible for a specific observable in any physical configuration. The operator that does this job is the operator that corresponds to that observable.

For finding the operators that correspond to classical observables the rule that works is this: in algebraic relations between the classical observables, turn all observables into operators: put hats on the observables.

What the hamiltonian function gives us in classical mechanics is an algebraic relation among observables, between p and E in the free-particle case and among x, p, and E in the general case.

View this as a statement about measurement spectra: findings in the spectrum of one observable are related to findings in the spectrum of the other. But the points of each spectrum—the space of each observable—is produced by the operator for that observable. What, in classical physics, is a spectral relationship becomes, in quantum mechanics, an operator relationship.

We have explored the two fundamental basis spaces, x and p. We have all of the matrix elements, $\langle x|p \rangle$, connecting them. We know the instructions for $\hat{x}$ and $\hat{p}$ in two languages and we know their eigenvalues. Being armed with these operators we can investigate any operator built from them. The hamiltonian, $\hat{H} = H(\hat{x},\hat{p})$, is one such. Another example of a familiar physical observable that is an algebraic function of x and p is the velocity; thus $\hat{v} = \hat{p}/m$.

AN OPERATOR FUNCTION IS AN ALGEBRAIC FUNCTION IN ITS HOME SPACE

The kinetic energy operator, $\hat{p}^2/2\,m$, is an example of one operator that is a function of another. For any operator $\hat{q}$ that generates a complete set of eigenstates

$|q\rangle$ and eigenvalues q, we can imagine writing down some function of q in which all the q's wear hats: $f(\hat{q})$. This function of an operator is itself an operator: $\hat{f} = f(\hat{q})$. The hamiltonian $\hat{p}^2/2m$ is a special case: $\hat{H} = H(\hat{p})$.

The meaning of such an operator function comes from its home space instruction: multiply by $f(q)$. The meaning of $f(\hat{q})$ is this: it is that operator that produces $f(q)$ in q-space. It is defined by its home space effect (see Problem 3.9).

$$f(\hat{q})|q\rangle = f(q)|q\rangle \qquad \langle q|f(\hat{q}) = f(q)\langle q| \tag{3-31}$$

3.5 TO SOLVE THE PROBLEM, YOU MAY CHOOSE ANY BASIS

Having the form of the hamiltonian operator, $\hat{H}$, we can now investigate its eigenkets, $|E\rangle$, and eigenvalues, E, directly from the eigenvalue problem that produces them.

$$\hat{H}|E\rangle = E|E\rangle \tag{3-32}$$

a. The x-Space Solution

Casting this equation into x-space yields the amplitudes $\langle x|E\rangle$. They are the solutions of this second-order differential equation.

$$-\frac{\hbar^2}{2m}\frac{d^2}{dx^2}\langle x|E\rangle = E\langle x|E\rangle \tag{3-33}$$

It results from putting the x-space form of the free-particle hamiltonian, (3-30), into the energy eigenvalue problem.

The solution of this equation is

$$\langle x|E\rangle = A\exp\left(ix\sqrt{2m\,E/\hbar^2}\right) + B\exp\left(-ix\sqrt{2m\,E/\hbar^2}\right) \tag{3-34}$$

where A and B are the two arbitrary constants expected from a second-order differential equation.

Like all wave functions, the bracket $\langle x|E\rangle$ must be single valued. Imposing this condition yields

$$L\sqrt{2m\,E/\hbar^2} = 2\pi \times \text{integer} \tag{3-35}$$

This gives us the energy levels of our bead-on-a-track (see Figure 3-2).

$$E = n^2\hbar^2/2m\,L^2 \qquad n = \ldots -2,-1,0,1,2,3,\ldots \tag{3-36}$$

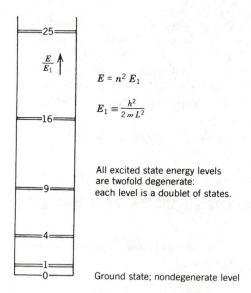

$$E = n^2 E_1$$

$$E_1 \equiv \frac{h^2}{2mL^2}$$

All excited state energy levels
are twofold degenerate:
each level is a doublet of states.

Ground state; nondegenerate level

Figure 3.2 The energy levels of the bead-on-a-track.

b. The p-Space Solution

By casting the eigenvalue problem into p-space we can examine the energy eigenkets in that language.

$$\frac{p^2}{2m} \langle p|E \rangle = E \langle p|E \rangle \tag{3-37}$$

This equation says that *either* the bracket $\langle p|E \rangle$ is zero *or* $E = p^2/2m$. Given a state of energy E, there is no amplitude to find any p other than $\pm\sqrt{2mE}$.

c. The Connection Between Them

The eigenvalue problem gives the same spectrum, whether it is solved in one basis or another. You can choose how you solve an eigenvalue problem. There are many languages in which to do it. Each language has its own insight to offer.

Suppose we insert a p-space ket-bra expansion for the vertical in the bracket $\langle x|E \rangle$. If we agree to mean by P the expression $+\sqrt{2mE}$, then Equation 3-37 tells us that all amplitudes $\langle p \neq \pm P|E \rangle = 0$. In the whole spectrum of p, only two of the amplitudes $\langle p|E \rangle$ are nonzero: the one for $p = -P$ and that for $p = +P$. Hence the p-space ket-bra sum for $\langle x|E \rangle$ collapses; there are only two nonzero terms.

$$\langle x|E \rangle = \langle x|P \rangle\langle P|E \rangle + \langle x|-P \rangle\langle -P|E \rangle \tag{3-38}$$

Compare this equation with (3-34). They are the same! An interpretation completes the isomorphism: that the constant A of (3-34) is the amplitude for an event. The amplitude, "if you know the bead has energy E, to find that it is moving in the forward $(+)$ direction with momentum $p = P$," is $A\sqrt{L}$. The amplitude to find a backward motion of kinetic energy E is $B\sqrt{L}$.

$$A\sqrt{L} = \langle P|E \rangle \qquad B\sqrt{L} = \langle -P|E \rangle \qquad (3\text{-}39)$$

THE MATHEMATICS SIGNALS THE DEGENERACY

The mathematics signals to us that the energy of the bead-on-a-track is *degenerate*. Degeneracy is clear from the conjunction of two facts: (1) The pure momentum states $|p\rangle$, are *simultaneous eigenstates* of both the $\hat{p}$ operator and the $\hat{H}$ operator. (2) Any particular value of energy, E, involves more than one p-state.

Item 1 merely affirms that $|p\rangle$ is eigen to both $\hat{p}$ and $\hat{p}^2$. Item 2 is the content of Equation 3-38 or 3-34. These show us that the energy measurement alone does not define a state; it defines only a linear combination of two states. The states $|P\rangle$ and $|-P\rangle$ have the same energy, E. So does any linear combination of these two states.

To define a specific state one of the amplitudes $\langle P|E \rangle$ or $\langle -P|E \rangle$ must be given; A or B must be given (see Problems 3.10 and 3.11). The label E leaves us without a complete specification of the state: there's an amplitude missing. To specify the state fully requires another number. To get the other number would require a further measurement. That's how we know E is degenerate: to define the state requires further measurement.

The mathematical items 1 and 2 have a direct physical parallel: degeneracy is exposed by *further experiment* (see Chapter 2, "Refer All Matters to Measurement"). It works like this: some states $|E\rangle$, upon measurement of momentum p, will yield $p = +P$, and some will yield $p = -P$, (Item 2). To confirm that these are both states of energy E (Item 1), you must do a second experiment. You must measure the energy of each of the two states $|P\rangle$ and $|-P\rangle$. You will find that they do both have the original energy, E. Thus, by experiment, you have unearthed the degeneracy; two states with the same E.

A PHYSICAL OPERATOR GENERATES A BASIS

That a measurement of energy alone is insufficient to define a state shows that the bare label E doesn't enumerate a complete basis. For the bead-on-a-track the energy spectrum is *not* a complete space. Nevertheless, the energy eigenvalue problem does generate a complete space; it produces all the words of a language: p-language. For each E, the eigenvalue problem solutions produced both $+P$ and $-P$. All of the p-space was produced.

The hamiltonian generated an entire basis, p-space. In addition, it generated, as its eigenvalues, the energy-level spectrum of the system. Both aspects are contained in this single statement of the eigenvalue problem:

$$\hat{H}|p\rangle = E(p)|p\rangle \tag{3-40}$$

To the observable of position corresponds an operator; to that of energy corresponds an energy operator; to that of momentum there corresponds the momentum operator. Each of these operators generates a language that we can speak, the language is a spectrum of measurement results. The observables of nature generate the languages we speak.

Measuring generates measurement space. The measurement results are the language through which reality is perceived. Looking creates reality!

3.6 A DEGENERATE EIGENVALUE IS A STATE'S PARTIAL LABEL

The symbol E alone does not label a state; something else is needed. The ket $|E\rangle$ is not a state: it is an arbitrary combination of two states. The eigenstates of the kinetic energy operator are $|p\rangle$ rather than $|E\rangle$.

Although E is not enough to characterize a state fully, it can partially characterize a state: some additional information is needed to complete the characterization. A ket containing the additional information is a state: the state $|E...\rangle$. It is a state characterized by E *and other things*. The dots mean the other things. For example, the direction of a motion, $+$ or $-$, are other things for our bead-on-a-wire.

By the state $|E...\rangle$ is meant one of a complete set that embraces E. The other labels necessary to define a specific state are not exhibited, but they're understood to be there. The dots remind us of their presence. They show that all the other measurements necessary to define the state have been made.

Here's an example: $|E...\rangle = |+P\rangle$ is a state among those that embrace E. The ket $|E\rangle$ displayed in (3-34) is not a state until either A or B is specified. Specifying A or B defines the dots and thus each of the states $|E...\rangle$ of a complete set. A ket-bra sum over $E...$ is equal to 1. A ket-bra sum over E is *not* equal to 1: the sum over the other things is missing.

The eigenstates of $\hat{H}$ are $|E...\rangle$ so the eigenvalue problem may be written

$$\hat{H}|E...\rangle = E|E...\rangle \tag{3-41}$$

The eigenkets of $\hat{H}$ are $|E\rangle$; its eigenstates are $|E...\rangle$. This notation makes it evident in the eigenvalue problem that E is only part of the complete label for a state. It is useful when we want to focus our attention on a particular property of a state, regardless of its other properties. Thus $|E...\rangle$ embraces all states whose energy is E, regardless of what other properties characterize these states.

NO MEASUREMENT EVER CHANGES
THE VALUE OF A STATE LABEL

Measure $\hat{\lambda}$. You get the state $|\lambda = \Lambda,...\rangle$.

Now measure $\hat{\omega}$. You get state $|\omega...\rangle$.

The state $|\omega...\rangle$ either retains the label $\lambda = \Lambda$ or it sustains no λ-label at all; no state arises as a result of the $\hat{\omega}$ measurement that is characterized by a different λ, say Λ'. This is a fundamental postulate in quantum mechanics, as fundamental as the continuity of the wave functions.

No measurement can ever change the value of a state's label.

If the bead is at position $x = 5$ cm, then you cannot, upon measurement, find it instantaneously subsequent at $x = 27$ cm. The label of the state $|x=5$ cm$\rangle$ cannot change by virtue of a measurement. But the measurement may erase a state index entirely from the ket.

The x-label cannot change, but it may disappear. A measurement of momentum, made on the state $|x=5$ cm$\rangle$, produces the state $|p=P\rangle$. An x-value is forbidden in this ket. No states exists characterized by both position and momentum (see Chapter 2, "Measurement Destroys a State"). The measurement has erased the x-label.

Another example clarifies an important implication of this principle.

You measure the momentum of the bead (operator $\hat{p} = \hat{\lambda}$). You find the value $p = -P = \Lambda$. The bead is left in the state $|p=-P\rangle$.

Now you measure its energy (operator $\hat{H} = \hat{\omega}$). You find the value $E = P^2/2m$. The bead is left in the state $|p=-P\rangle$. The $\lambda = \Lambda$ label is retained, and only the $p = -P$ state is found.

Notice that the bead is *not* left in some undetermined combination of $|-P\rangle$ and $|+P\rangle$ states.

The state $|+P\rangle$ is also an eigenstate of $\hat{H}$ with the same energy, $E = P^2/2m$. Nonetheless, there is no amplitude for its presence. The value $\Lambda' = +P$ can never arise as a result of a measurement on the state $|p=-P\rangle$. Because the index $\lambda = p$ labels the new state as well as the old, its value, $\Lambda = -P$, is preserved.

Here is the quantitative description of the idea.

If you start with a state $|\lambda = \Lambda,...\rangle$ and make the measurement $\hat{\omega}$, the state $|\omega...\rangle$ that you will find in the measurement will be one of these two kinds: $|\omega,\Lambda\rangle$ or $|\omega$,no-λ-label$\rangle$. The new state will be either a simultaneous eigenstate of both $\hat{\omega}$ and $\hat{\lambda}$ or it will be a state in which no λ-label is permitted. If it is a simultaneous eigenstate ($|\omega,\lambda\rangle$), then the following condition will be met:

$$[\hat{\omega}\hat{\lambda} - \hat{\lambda}\hat{\omega}]|\omega...\rangle = 0 \tag{3-42}$$

The proof is Problem 3.16.

THE INEXORABLE ARBITRARY PHASE FACTOR
IS FIXED BY COMPUTATIONAL CONVENIENCE

The pure momentum state wave functions are those of Equation 2-8. They are solutions of the momentum operator eigenvalue problem: Equation 3-28. Consider one of these solutions: the one characterized by $n = 2$; $\langle x|p=2h/L\rangle$. Multiply it by the factor $\exp i\pi/5$.

This new function also solves (3-28). It also meets the conditions of continuity and single valuedness that an acceptable wave function must have. And it meets the certainty condition. This new function ought to qualify as the bracket $\langle x|p=2h/L\rangle$ also. It does!

There are an infinite number of wave functions $\langle x|p=2h/L\rangle$. But any one of them is different from any other only by the factor $\exp i\phi$, where ϕ is a real number.

The wave function $\langle x|p\rangle \exp i\phi$ is just as good as the $\langle x|p\rangle$ listed in (2-8). You may take for each solution, $\langle x|p\rangle$, a different phase factor ϕ_p. The new set, $\langle x|p'\rangle = \langle x|p\rangle \exp i\phi_p$, is as good as the old (see Problem 3.14).

Instead of $1/\sqrt{L}$ as the constant in Equation 3-29, we could just as well choose $e^{i\phi}/\sqrt{L}$. The particular choice $\phi = 0$ is not made on physical grounds; it is made for computational convenience.

IT'S NOT A DIFFERENT STATE
IF IT DIFFERS ONLY IN PHASE

There's no physical statement that determines the phase of a wave function. The phase of a wave function is arbitrary; it is unphysical. It follows that this phase factor must drop out of any physical calculation (see Problem 3.14). If it didn't do so—if it affected a physical calculation—then ϕ would be a physically accessible quantity. Whatever physical process is described by the calculation would fix the experiment to measure ϕ. The fact that the phase factor is, indeed, arbitrary signifies that a bracket is not measurable: $\langle x|p\rangle$ is not measurable; only its absolute square $|\langle x|p\rangle|^2$ is physically accessible. We can't measure the amplitude for an event; we can measure only its probability.

PROBLEMS

3.1 Show that Equation 3-13 is true. Hint: call $x - b \doteq \xi$ in 3-11.

3.2 Prove Equation 3-14.

Answer:

$$\langle p|\hat{D}|\psi\rangle = \sum_x \langle p|x\rangle \qquad \langle x|\hat{D}|\psi\rangle$$

$$= \int_0^L dx\, \frac{\exp(-ipx/\hbar)}{\sqrt{L}} \qquad \frac{d\psi}{dx}$$

$$= \int_0^L dx \, \frac{\exp(-ipx/\hbar)}{\sqrt{L}} \, \frac{ip}{\hbar} \, \psi(x)$$

$$= \frac{ip}{\hbar} \langle p|\psi \rangle$$

The third step comes from an integration by parts.

3.3 The mirror reflection operator $\hat{R}$ produces the reflection about the origin of any function in x-space. It is called the parity operator in physics. Its x-space instruction is $\langle x|\hat{R} = \langle -x|$. What is its p-space instruction?

Note: To investigate an instruction, you must apply it to something. You want $\langle p|\hat{R}|\psi \rangle$. To calculate it, you need this observation:

$$\exp ixp/\hbar = \exp -i(-px/\hbar) = \langle -p|x \rangle \sqrt{L}$$

3.4 The momentum-raising operator, $\hat{\tau}$, may be defined like this: $\hat{\tau}$ produces from any state of pure momentum p a new state of pure momentum $p + Q$. Thus

$$\hat{\tau}|p \rangle = |p+Q \rangle \tag{A}$$

Notice that this operator can produce nonphysical states from our bead-on-a-wire physical states! If $Q = 3.7 \, h/L$, the kets produced by $\hat{\tau}$ are not physical states of the bead-on-a-wire.

Take Q to be one unit of momentum, $Q = h/L$. From the statement on what $\hat{\tau}$ does to a set of pure states, we can find what $\hat{\tau}$ does in the language of these states. Prove that

$$\langle p|\hat{\tau} = \langle p-Q|$$

and

$$\langle x|\hat{\tau} = \exp(iQx/\hbar) \, \langle x|$$

Key notions:

1. $$\langle p|P + h/L \rangle = \langle p - h/L|P \rangle$$
These two self-space brackets express the same idea. They are equivalent.

2. Because $$\langle p|\hat{\tau}|P \rangle = \langle p - h/L|P \rangle$$

is valid for all of the $|P \rangle$, it follows that

$$\langle p|\hat{\tau}|\psi \rangle = \langle p - h/L|\psi \rangle$$

for any $|\psi \rangle$ at all.
Proof:

$$\langle p|\hat{\tau}|\psi \rangle = \langle p|\hat{\tau}| \sum_P |P \rangle \langle P||\psi \rangle = \langle p - h/L|\psi \rangle$$

3. $\langle x|\hat{\tau}|p \rangle = \exp(iQx/\hbar) \, \langle x|p \rangle$ is valid for all of the $|p \rangle$.

3.5 Again taking Q to be one unit of momentum, $Q = h/L$, show that the operator product $\hat{\tau}\hat{D}$ has the following effects on a state of the bead-on-a-track:

$$\langle x|\hat{\tau}\hat{D}|\psi\rangle = \exp\left(\frac{iQx}{\hbar}\right)\frac{d}{dx}\psi(x)$$

and

$$\langle p|\hat{\tau}\hat{D}|\psi\rangle = \frac{i}{\hbar}(p-Q)\langle p-Q|\psi\rangle$$

Show that the two forms are consistent, that the ket $\hat{\tau}\hat{D}|\psi\rangle$ transforms according to the ket-bra sum rule.

Solution: Call by the name $c|S\rangle$ the quantity $\hat{D}|\psi\rangle$. Then

$$\langle x|\hat{\tau}\hat{D}|\psi\rangle = c\langle x|\hat{\tau}|S\rangle$$

$$= c\exp\left(\frac{iQx}{\hbar}\right)\langle x|S\rangle$$

$$= \exp\left(\frac{iQx}{\hbar}\right)\langle x|\hat{D}|\psi\rangle$$

$$= \exp\left(\frac{iQx}{\hbar}\right)\frac{d}{dx}\langle x|\psi\rangle$$

3.6 Show that the position operator $\hat{x}$ of Equation 3-25 must be characterized by this instruction equation in p-space.

$$\langle p|\hat{x} = i\hbar\frac{d}{dp}\langle p|$$

Show that the operator $\hat{p}$ has this x-space instruction:

$$\langle x|\hat{p} = -i\hbar\frac{d}{dx}\langle x|$$

Solution:
What you know is $\hat{x}|x\rangle = x|x\rangle$.
What you want is $\langle p|\hat{x}$, the instruction for $\hat{x}$ in p-space.
You need the quantity "to do only with p" in the formula

$$\langle p|\hat{x}|x\rangle = \text{(to do only with } p)\langle p|x\rangle$$

To get it requires noticing that

$$i\hbar\frac{d}{dp}\exp\left(\frac{-ipx}{\hbar}\right) = x\exp\left(\frac{-ipx}{\hbar}\right)$$

3.7 Suppose that your momentum-measuring instrument for the bead-on-a-track is defective. It fails to respond to some particular momentum, say, $p = 3h/L$. The instrument never displays this momentum, but it responds accurately to all bead momenta except $3h/L$. The instrument's response space is incomplete. It does not respond to the complete space of physical measurement.

How would you deduce the incompleteness of your instrument's measurement space from experiment?

Answer: Notice that all incoming particles are not accounted for among the outgoing ones.

Cast this verbal idea into a mathematical one. Deduce that the physical measurement space of momentum is complete and equals the instrumental measurement space if the instrument is not defective, if it responds to everything.

Answer: If the state of each incoming beam particle is $|\psi\rangle$, then for some $|\psi\rangle$

$$\sum |\langle p|\psi\rangle|^2 < 1 = \langle \psi|\psi\rangle = \sum |\langle p|\psi\rangle|^2$$

all p recorded in the   all physical p
defective apparatus

A defective apparatus does not meet the certainty condition.

3.8 Show that in Equation 3-29, the constant, $L^{-1/2}$, meets the certainty condition for the bead-on-a-track.

3.9 The hamiltonian (Equation 3-30) is a special case of $f(\hat{p})$. By applying $f(\hat{p}) = \hat{p}^2$ show that both Equations 3-37 and 3-33 are true.

3.10 Show that the A and B of Equation 3-34 are related by

$$|A|^2 + |B|^2 = 1/L$$

Hint: It follows from $\langle E|E\rangle = 1$. Interpret the meaning of this result.

3.11 Show that the state given by Equation 3-34 or 3-38 yields the single energy E, regardless of the relative magnitudes of A and B. Thus *any* linear combination of states $|P\rangle$ and $|-P\rangle$ is an eigenstate of $\hat{H}$.

Solution:

$$\hat{H}(A\sqrt{L}|P\rangle + B\sqrt{L}|-P\rangle) = \frac{P^2}{2m}(A\sqrt{L}|P\rangle + B\sqrt{L}|-P\rangle)$$

3.12 Consider the operator $\hat{x}\hat{p}$; its instruction in x-space is

$$\langle x|\hat{x}\hat{p} = -i\hbar x \frac{d}{dx}\langle x|$$

What is the instruction in x-space for the operator $(\hat{x}\hat{p})^2$?

Answer: $\langle x|(\hat{x}\hat{p})^2|\psi\rangle = -\hbar^2(xd/dx + x^2d^2/dx^2)\langle x|\psi\rangle$

What is the instruction in x-space for $\hat{x}^2\hat{p}^2$?

Answer: $-\hbar^2 x^2 d^2/dx^2 \langle x|\psi\rangle$.

The two operations are not the same.
Show the instructions for $\hat{p}\hat{x}$ in both x-space and p-space. Then find $\langle x|\hat{x}\hat{p} - \hat{p}\hat{x}|\psi\rangle$ and $\langle p|\hat{x}\hat{p} - \hat{p}\hat{x}|\psi\rangle$. Show that the first is $i\hbar\langle x|\psi\rangle$ and the second is $i\hbar\langle p|\psi\rangle$. How does this demonstrate that we can write, quite generally, $\hat{x}\hat{p} - \hat{p}\hat{x} = i\hbar$?

3.13 Suppose the point $x = 0$ (and thus also $x = L$) were forbidden to the bead: the bead is boxed between 0 and L. The domain of positions available to this bead-in-a-box are restricted to $0<x<L$.

Being free to be anywhere in this domain, its energy is still determined by the free-particle hamiltonian (Equation 3-30). Solve the eigenvalue problem (Equation 3-32) to obtain the energy levels, E, and wave functions $\langle x|E\rangle$ for this bead-in-a-box. Are the energy levels degenerate?

Answer: The critical issue in the solution is that any wave function must be zero at $x = 0$ and $x = L$. Thus

$$\langle x|E\rangle = \sqrt{\frac{2}{L}} \sin \nu\pi\frac{x}{L}$$

where

$$\nu = 1,2,3 \ldots$$

and

$$E = E_\nu = \nu^2 \frac{h^2}{8\,m\,L^2}$$

3.14 Suppose we change every one of the amplitudes $\langle x|p\rangle$ given in Equation 2-8 by an arbitrary phase factor ϕ_p ($=$real): we define a new set of amplitudes, which we'll call $\langle x|p'\rangle$:

$$\langle x|p'\rangle \equiv exp\; i\phi_p \langle x|p\rangle$$

Because the ϕ_p depends only on p and not on x, this implies a new basis set $|p'\rangle$ where

$$|p'\rangle = exp\; i\phi_p |p\rangle$$

Show that this equation forces $\langle p'|$ to be

$$\langle p'| = exp(-i\phi_p) \langle p|$$

Solution:

$$\langle p'|S \rangle = \langle S|p' \rangle*$$

Show that the ket-bra sum theorem is valid in p'-space

$$\sum_{p'} |p'\rangle \langle p'| = \sum_{p} |p\rangle \langle p|$$

To do it is easy if you procede from meaning and impossible if you don't. The proof requires words rather than algebra: you need to be aware of the values that p' can have. You must appreciate that p'-space is p-space:

$$p' = \frac{nh}{L}$$

Show, in detail, that the expression for $\langle x|E \rangle$ is *exactly* the same when expanded by a p-space (Equation 3-38) or by a p'-space ket-bra sum.

Key: $\langle x| - i\hbar\hat{D}|p' \rangle = p\langle x|p' \rangle$ so $|p'\rangle$ is the solution to the same eigenvalue problem as is $|p\rangle$.

3.15 For a bead-in-a-box (a cut ring), argue that the amplitude $\langle p|x \rangle$ is zero for all p. One can never find the boxed bead in a pure state of momentum. No state $|p\rangle$ can ever be measured for it. How do you know this to be true?

Answer: For no p can $\langle x|p \rangle$ be a wave function, none meets the boundary conditions. For this particular physical configuration—a box—there is no p-spectrum. If there were a momentum that you could find then the bead would be left in a state $|p\rangle$. But that is impossible because no state $|p\rangle$ can satisfy the boundary conditions. For any conceivable $|S\rangle$ $\langle p|S \rangle = 0$ because if it weren't zero the bead could be left in a state $|p\rangle$ by a measurement on the prepared state $|S\rangle$. Hence the expansion $|S\rangle = \sum |p\rangle\langle p|S\rangle$ is not possible since every $\langle p|S \rangle = 0$ so $|p\rangle$ is not a legitimate basis.

The momentum operator is $\hat{p}$, regardless of physical configuration. No state of the bead-in-a-box system can be also an eigenstate of $\hat{p}$.

3.16 An $\hat{\omega}$ measurement on the state $|\lambda = \Lambda...\rangle$ produces either $|\omega,\Lambda\rangle$ or $|\omega,\text{no-}\lambda\text{-label}\rangle$. Show that if it produces $|\omega,\Lambda\rangle$ then Equation 3-42 must be true.

Answer: Because

$$|\omega...\rangle = |\omega,\Lambda\rangle$$

and

$$\hat{\omega}|\omega,\Lambda\rangle = \omega|\omega,\Lambda\rangle$$

and

$$\hat{\lambda}|\omega,\Lambda\rangle = \Lambda|\omega,\Lambda\rangle$$

the equality in Equation 3-42 follows easily.

Chapter 4

The Harmonic Oscillator: Bound Bead in a Symmetric Force Field

THE STUDY OF MODELS
BUILDS PHYSICAL INTUITION

There are no ideal free beads-on-a-track. Nor are there ideal harmonic oscillators. But these two idealizations capture the essential behavior of many real systems. They are computationally tractable models for what is encountered in nature. From such models one acquires physical intuition. That's the goal of this study, to build physical intuition.

4.1 THE HAMILTONIAN WITH HATS
IS THE ENERGY OPERATOR

The traditional ideal harmonic oscillator is shown in Figure 4-1. A mass m is always pulled toward its equilibrium position by a Hooke's law force of strength K. The force is produced by the coil spring in the picture. The displacement of the mass from its equilibrium position is x, and its momentum is p. The coil spring is attached to a stationary wall that is infinitely removed from the equilibrium position. The picture fails to represent this aspect of the idealization. The harmonic oscillator mass we treat may be found at any x between $-\infty$ and $+\infty$; the domain of its position spectrum is all x.

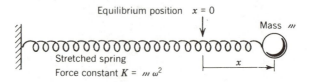

Figure 4.1 The ideal classical harmonic oscillator.

A harmonic oscillator is characterized by a natural oscillation frequency, ω. It arises from the ratio of the force constant to the mass

$$\omega = \sqrt{K/m} \tag{4-1}$$

From classical mechanics, the hamiltonian for such an oscillator is

$$H = \frac{p^2}{2m} + \frac{K x^2}{2} \tag{4-2}$$

This algebraic combination of the dynamical variables x and p produces the energy of the system in classical physics. From that fact we expect the quantum mechanical energies of the system to be generated by the operator $\hat{H}$, where

$$\hat{H} = \frac{\hat{p}^2}{2m} + \frac{K \hat{x}^2}{2} \tag{4-3}$$

The spectrum of energies we seek is given by the eigenvalue problem for E.

$$\hat{H}|E\rangle = E|E\rangle \tag{4-4}$$

As before, the spectrum of positions is generated by $\hat{x}$, and the spectrum of momenta by $\hat{p}$. But now the space of x is no longer finite, and the space of momentum is not quantized (see Problem 2.15).

We shall examine the energy eigenvalue problem and its solutions in three languages: that of position, x, that of momentum, p, and that of the energy index, n ($E = E_n$).

4.2 CAST THE PROBLEM INTO POSITION SPACE

In the space of physical position, the eigenvalue problem becomes a differential equation in x for the system's energy eigenstate wave functions. Using the characteristic frequency, ω, rather than K, in the hamiltonian of Equation 4-3 and casting the eigenvalue problem into x-space, it becomes

$$-\frac{\hbar^2}{2m}\frac{d^2}{dx^2}\langle x|E\rangle + \frac{m\omega^2}{2}x^2\langle x|E\rangle = E\langle x|E\rangle \tag{4-5}$$

This is the equation for the energy eigenstate wave functions: those functions of x called $\langle x|E \rangle$, each of which corresponds to an allowed value, E, of energy.

Only the solutions to this equation that are physically viable can qualify as wave functions. The allowed solutions meet the boundary conditions of physics in general and of this problem in particular. The allowed values of E can be indexed; they form a discrete set in all bound-particle potential well problems. That's because the boundary conditions on the wave function are met only for select values of E. Valid wave functions must be continuous and single valued everywhere in x-space, and they must meet the certainty condition:

$$\int_{-\infty}^{+\infty} dx \, |\langle x|E \rangle|^2 = 1 \tag{4-6}$$

In mathematics this last condition on the wave function is called *square integrability* or *normalizability*. The wave function is normalizable to unity. In physics it says that for any energy state of the system, the probability of finding the mass somewhere between $-\infty$ and $+\infty$ is a certainty; $\langle n|n \rangle = \langle E|E \rangle = 1$.

FOR IMPORTANT DIFFERENTIAL EQUATIONS, THE SOLUTIONS ARE TABULATED

The eigenvalue problem reduces to finding the solutions to the differential Equation 4-5, subject to the three conditions mentioned. This problem is exactly analogous to that of Equation 3-33. The equation there is a special case of the present one: the case $K = 0$.

Consider how the differential Equation 3-33 was solved. There are only two methods for solving differential equations, (1) guessing and (2) looking up the answer. Method 2 is really a variation of Method 1; someone else has done the guessing and published the answers. We used Method 2. Equation 3-34 is the textbook answer.

The solutions to (3-33) involve the special transcendental functions called *exponentials*. Because these are of imaginary argument, the solutions may equally well be written as combinations of other special tabulated functions called *sines* and *cosines*.

Such functions have many representations: there are series expansions, integral forms, lists of tabulated values, graphical depictions, and mathematical relations among the functions by which they are defined. There is no simple algebraic representation for the sine function of θ; we must be content with writing sin θ.

To people who haven't studied them, these functions are forbidding. But to you who have studied them, they are familiar and comfortable to use. You know their properties even though you cannot immediately state their magnitudes for a particular argument: you may not have at your fingertips that sin 2.5 is about 0.6, but you work with sin θ nevertheless. And you know that it is one of the solutions of (3·33). By having written down the solution in (3-34) without deriving it, we are tacitly alluding to the literature of mathematics.

We solve the differential equation (4-5) by the same method: we refer to the literature of mathematics. The solutions to an equation isomorphic (Greek: *iso* = "similar", *morphe* = "form") to (4-5) have been investigated and organized in the course of mathematical history. The man whose name is attached to the investigation is C. Hermite (1822–1905).

The following are six reference books in which you will find sections on Hermite polynomials:

1. W. H. Beyer, editor, *CRC Handbook of Mathematical Science,* CRC Press, West Palm Beach, Florida, 1978.

2. R. C. Weast and S. M. Selby, editors, *Handbook of Tables for Mathematics,* CRC Press, Cleveland, 1975.

3. M. Abramowitz and I. A. Stegun, editors, *Handbook of Mathematical Functions,* National Bureau of Standards, United States, 1964.

4. W. Magnus and F. Oberhettinger, *Formulas and Theorems for the Functions of Mathematical Physics,* Chelsea Publishers, New York, 1954.

5. H. Margenau and G. M. Murphy, *The Mathematics of Physics and Chemistry,* Van Nostrand, New York, 1943.

6. E. T. Copson, *An Introduction to the Theory of Functions of a Complex Variable,* Oxford University Press, London, 1935.

These books, and many others, list a myriad of facts about Hermite polynomials. They list relations among them, series expansions for them, integral forms for them, and so on. One of the important facts about Hermite polynomials is that they play the key role in the solution of Equation 4-5. In treatments of the Hermite polynomials you will always find listed this equation:

$$-\frac{d^2 y_n}{d\xi^2} + \xi^2 y_n = (2n + 1)y_n \tag{4-7}$$

For every value of n (not necessarily an integer) there are two solutions to this second-order differential equation: two $y_n(\xi)$.

Equation 4-7 is isomorphic to our eigenvalue problem, Equation 4-5. These two equations become identical if we make the association

$$x = \xi \sqrt{\frac{\hbar}{\omega m}} \tag{4-8}$$

and

$$E = \left(n + \frac{1}{2}\right)\hbar\omega \tag{4-9}$$

Thus the solutions of (4-7) must solve (4-5).

When mathematicians investigate an equation, they examine all of the solutions, not only those that meet our special requirement of square integrability over the domain

$-\infty < x < +\infty$. Solutions to the equation are possible for any value of n, but only some values of n produce the special class of solutions that meet the square integrability requirement. Along with all the others, this is one of the facts discussed in any reference book treating Hermite polynomials. Because only solutions with nonnegative integer n's are square integrable, only these can be wave functions.

Like the exponential and trigonometric functions, the solutions to (4-7) arise often in mathematics and physics. That's why they are tabulated. As applied to our eigenvalue problem, Equation 4-5, here they are:

$$\langle x|E\rangle = \langle x|n\rangle = i^n \frac{\exp(-m\omega x^2/2\hbar) H_n(x\sqrt{m\omega/\hbar})}{\sqrt{2^n n! \sqrt{\pi\hbar/m\omega}}} \tag{4.10}$$

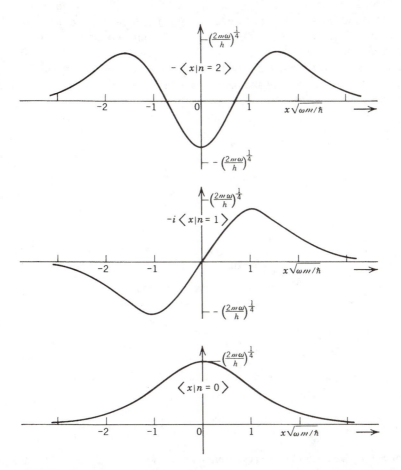

Figure 4.2 The three lowest energy eigenstate wave functions of the harmonic oscillator.

The functions $H_n(\xi)$ are the Hermite polynomials of order n where $n = 0, 1, 2, \ldots$ Here are the first five of them:

$$H_0(\xi) = 1 \tag{4-11}$$

$$H_1(\xi) = 2\xi \tag{4-12}$$

$$H_2(\xi) = 4\xi^2 - 2 \tag{4-13}$$

$$H_3(\xi) = 8\xi^3 - 12\xi \tag{4-14}$$

$$H_4(\xi) = 16\xi^4 - 48\xi^2 + 12 \tag{4-15}$$

The three lowest-energy wave functions, $\langle x|0\rangle$, $\langle x|1\rangle$, and $\langle x|2\rangle$ are sketched in Figure 4-2.

The only physically acceptable solutions to (4-7) are those for nonnegative integer n. But by the association of (4-9), the energy of the oscillator is related to the value of n. Thus the energies are quantized: $E = E_n$. The harmonic oscillator *cannot* have any imaginable positive energy; only a discrete spectrum of positive energies is allowed. The integer, n, indexes the energy levels of the system.

Figure 4-3 shows the energy-level diagram. The wave functions and the potential energy curve are shown on the same diagram. Each wave function is shifted up vertically so as to sit at the level of the energy it represents.

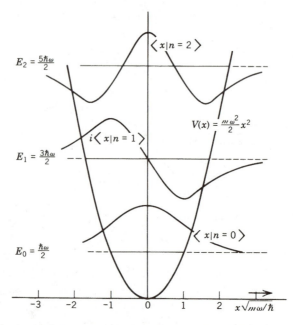

Figure 4.3 The potential energy curve and energy-level diagram of the harmonic oscillator.

HAVING FOUND THE SOLUTIONS, MARK THEIR SIGNIFICANT FEATURES

The harmonic oscillator is a simple mechanical system, albeit a special one; we have explored it in x-space. What is to be learned from the exercise is summarized in these four items:

1. Contained within each wave function of (4-10) is a particular choice for the inexorable arbitrary phase factor. Instead of the factor i^n, I could just as well have chosen 1.

2. The ladder of E values is not degenerate for the harmonic oscillator; they were degenerate for the bead-on-a-ring. The $n + 1/2$ sequence of energies means steps in equal units of $h\omega$ after a zero-point half-step. Because the harmonic oscillator models so many physical systems, its energy spectrum arises repeatedly in the study of nature, and so memorizing it is worthwhile.

3. Note the phenomenon of *tunneling*. The probability density penetrates the potential barrier. Quantum mechanics expressly contradicts classical mechanics in this phenomenon. Classically a particle is constrained always to be within the potential well: classically there is zero probability to be outside it (see Problem 4.1).

Quantum mechanics tells us that the opposite is true: the mass can, in fact, be found where it is classically forbidden to be; outside the well (see Problem 4.2). It *tunnels* there under the barrier. Compare Figure 4-4 with Figure 4-3 to see the idea. The wave function, being finite outside the potential, generates a finite probability outside the potential; the mass can be there.

4. Note the mirror-reflection symmetry of the wave functions: they alternate between even and odd, beginning with an even symmetric ground-state wave function. This is a pervading feature for all symmetric (even) potentials; ones where $V(-x) = V(x)$.

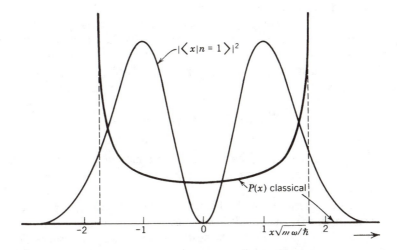

Figure 4.4 The classical and quantum mechanical x-space probability distributions for an oscillator mass with energy $E = 3\hbar\omega/2$.

4.3 THE REFLECTION OPERATOR REFLECTS THE FUNCTION

That all of the wave functions, $\langle x|E\rangle$, have a reflective symmetry property means this: they all are eigenstates of the mirror-reflection operator $\hat{R}$ (see Problem 3.3). This operator produces a reflection about the origin of any function in x-space. Put formally, its x-space instruction is

$$\langle x|\hat{R} = \langle -x| \tag{4-16}$$

As with any operator, when $\hat{R}$ operates on an arbitrary state of the system, $|\psi\rangle$, it produces a new ket; $|\phi\rangle$. The x-space representation of the new ket is $\langle x|\hat{R}|\psi\rangle = \langle -x|\psi\rangle = c\langle x|\phi\rangle$. The new ket is generally not symmetrically connected to the old one. It is not necessarily true that $\langle x|\hat{R}|\psi\rangle$ is $\pm\langle x|\psi\rangle$. The bead-on-a-ring momentum eigenstates serve as examples: $\langle x|\hat{R}|p\rangle \neq \pm\langle x|p\rangle$.

That the same state is returned by the operation of $\hat{R}$ is what characterizes the eigenstates of $\hat{R}$. If we use R to label the eigenvalues of $\hat{R}$, then, as always, we must classify the eigenkets of $\hat{R}$ by the same label. Hence

$$\hat{R}|R\rangle = R|R\rangle \tag{4-17}$$

ODD AND EVEN WAVE FUNCTIONS ARE EIGENSTATES OF $\hat{R}$

To find the eigenvalues is an easy matter. You let $\hat{R}$ operate twice. By virtue of its meaning, $\hat{R}^2$ operating on any state, eigen or not, simply returns the state again. A double reflection equals no reflection. On the other hand, the effect of $\hat{R}^2$ on an eigenstate of $\hat{R}$ must be the same state back again multiplied by R^2. The left-hand equality expresses the first thought, and the right-hand equality expresses the second.

$$\langle x|R\rangle = \langle x|\hat{R}^2|R\rangle = R^2 \langle x|R\rangle \tag{4-18}$$

The validity of each equality implies that the extremes are equal. Thus $R = +1$ and $R = -1$ are the eigenvalues of $\hat{R}$. Eigenstates of $\hat{R}$ have the property that $\langle -x|R\rangle = \pm\langle x|R\rangle$.

The energy eigenstates of the harmonic oscillator have this property: they are eigenstates of $\hat{R}$; they are *simultaneous* eigenstates of both $\hat{H}$ and $\hat{R}$. On the other hand, except for the ground state, the bead-on-a-ring energy eigenstates, $\langle x|p\rangle$, are not simultaneously eigenstates of $\hat{R}$.

A SYMMETRIC POTENTIAL GENERATES EVEN AND ODD ENERGY EIGENSTATES

Whenever the potential energy is symmetric, all of the energy eigenstates can be classified by reflective symmetry, even or odd. Simultaneous eigenstates of $\hat{H}$ and $\hat{R}$ can be found.

The proof is most easily constructed if we consider the meaning of this bracket of a product operator: $\langle -x|\hat{H}\hat{R}|E\rangle$. Cast both the hamiltonian

$$\hat{H} = \frac{\hat{p}^2}{2m} + V(\hat{x}) \tag{4-19}$$

and the ket, $\hat{R}|E\rangle$, on which it operates into $x' = -x$ space. Here is the result (see Problem 4.5).

$$\langle -x|\hat{H}\hat{R}|E\rangle = -\frac{\hbar^2}{2m}\frac{d^2}{d(-x)^2}\langle -x|\hat{R}|E\rangle + V(-x)\langle -x|\hat{R}|E\rangle \tag{4-20}$$

Notice that $\langle -x|\hat{R}|E\rangle$ is just another way of writing a function of $x' = -x$. It's also a function of x; it is the energy wave function $\langle x|E\rangle$. Thus, if the potential $V(-x) = V(x)$, the right-hand side of Equation 4-20 will be exactly equivalent to $\langle x|\hat{H}|E\rangle$ which, in turn, will be equal to $E\langle x|E\rangle$. And because we can rewrite $\langle x|E\rangle$ again as $\langle -x|\hat{R}|E\rangle$, we can conclude that

$$\langle -x|\hat{H}\hat{R}|E\rangle = E\langle -x|\hat{R}|E\rangle \tag{4-21}$$

Because this equation is valid for all x (all $-x$), we may remove the bra $\langle -x|$ to deduce that

$$\hat{H}(\hat{R}|E\rangle) = E(\hat{R}|E\rangle) \tag{4-22}$$

The significance of this equation resides in its form: *The form is the substance in the eigenvalue problem.* The equation says that the new ket $\hat{R}|E\rangle$, generated by $\hat{R}$ operating on $|E\rangle$, is an eigenstate of $\hat{H}$ with energy E. If the energy is not degenerate, there will be only one state with the energy E, the state $|E\rangle$ itself. Hence

$$\hat{R}|E\rangle = constant \times |E\rangle \tag{4-23}$$

But this, too, has the form of the eigenvalue problem. It is the statement that $|E\rangle$ is an eigenstate of $\hat{R}$. The *constant* can be only one of the R's. From the form comes the substance: each nondegenerate energy eigenstate of a particle bound in a symmetric potential is also an eigenstate of $\hat{R}$. It has a mirror-symmetry property, even or odd, by which it can be classified. Each energy level may be assigned a *parity:* plus for even, minus for odd. The first excited state of the harmonic oscillator has odd parity, its ground state has even parity.

4.4   IN POSITION SPACE, THE EIGENVALUE PROBLEM IS A BOUNDARY-VALUE PROBLEM

Example: The Bead in a Partitioned Box

As you might imagine, x-space is a particularly useful space in which to work. Some problems are essentially intractable except in that space. This happens when the potential operator $\hat{V} = V(\hat{x})$ is prohibitive to handle except in its home space where the hat may be removed.

One such particularly instructive potential is that of the split well, shown in Figure 4-5. In the middle of the well is a potential barrier of finite height W and width $2b$. The potential is infinite everywhere outside the box. Thus the particle is in the box; the probability is zero that $|x| > L$. The formal description of $V(x)$ is

$$V(x) = \begin{array}{lll} \infty & \text{when} & L \leq |x| \\ 0 & \text{when} & b < |x| < L \\ W & \text{when} & |x| \leq b \end{array} \qquad (4\text{-}24)$$

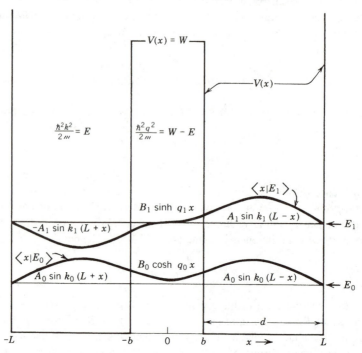

Figure 4.5 The two lowest-energy state wave functions of the bead-in-a-partitioned-box potential.

The energy eigenstates, $|E\rangle$, for a mass in this partitioned box are solutions of

$$\hat{H}|E\rangle = E|E\rangle \tag{4-25}$$

The hamiltonian operator is

$$\hat{H} = \frac{\hat{p}^2}{2m} + V(\hat{x}) \tag{4-26}$$

where the potential energy operator is given in its home space of x by Equation 4-24. Casting the eigenvalue problem into that space produces this differential equation:

$$-\frac{\hbar^2}{2m}\frac{d^2}{dx^2}\langle x|E\rangle + V(x)\langle x|E\rangle = E\langle x|E\rangle \tag{4-27}$$

Extracting each solution is a matter of giving detailed attention to its structure between and at the boundaries. It is a boundary-value problem. The considerations are listed in the following eight items:

1. The potential is a symmetric one: $V(-x) = V(x)$. Hence each solution may have either even symmetry, $\langle -x|E\rangle = \langle x|E\rangle$, or odd symmetry, $\langle -x|E\rangle = -\langle x|E\rangle$. Thus we need consider only half the problem! The region $0 \leqslant x$ suffices. The solution for negative x is either $+$ or $-$ that for positive x.

2. In the region $L \leqslant |x|$ the potential is infinite; $V(x) = \infty$. Therefore the only possible value for the wave function, $\langle x|E\rangle$, in this region is zero. No finite value for $\langle x|E\rangle$ can satisfy the equality in Equation 4-27 if $V(x) = \infty$. Thus, in this region, $\langle x|E\rangle = 0$.

3. To describe $\langle x|E\rangle$ in the region $b < x < L$, it is convenient to invent the parameter $k = k(E)$. Using this function of E instead of E itself is a notational aid in writing the solution. It is defined, as indicated in the figure, by

$$\frac{\hbar^2 k^2}{2m} = E \tag{4-28}$$

In the region $b < x < L$ the solutions to Equation 4-27 are sines and cosines of argument kx.

The continuity of $\langle x|E\rangle$ requires that it be zero at $x = L$. This is an example of the solution meeting the *boundary conditions* of the problem. The only solution to (4-27) in this region that is zero at $x = L$ is $A \sin k(L-x)$ (see Figure 4-5 and Problem 4.7).

4. The case of interest is when the boxed bead does not have enough energy to surmount the barrier. We want to explore those energy levels available to the bead that are less than W: $E < W$. Under this condition the solutions to (4-27) in the region $|x| < b$ are sinh and cosh functions. They are most conveniently expressed via a second parameter, $q = q(E)$. It is defined by

$$\frac{\hbar^2 q^2}{2m} = W - E \tag{4-29}$$

Because the solutions are simultaneously eigenstates of $\hat{R}$, they can be classified by their reflection symmetry property: they each possess an R-ness attribute, ± 1. Because $\cosh qx$ is an even function of x ($R = +1$) and $\sinh qx$ is an odd one ($R = -1$), we know that each is to be associated with a distinctly different energy eigenstate. Therefore we may treat each solution separately. Our knowledge that $\hat{H}$-eigenstates are simultaneously $\hat{R}$-eigenstates spares us the labor of untangling a linear combination of the $\sinh qx$ and the $\cosh qx$ functions. Figure 4-5 illustrates the separation.

5. Because the energy levels may be classified according to reflection symmetry, we need not calculate the solutions in the region $-L < x < -b$. The antisymmetric solution is constructed by taking $-A \sin k(L+x)$ in this region to link up with the odd solution $B \sinh qx$ in the barrier. The symmetric solution is constructed by linking $A \sin k(L+x)$ through $B \cosh qx$ in the barrier to $A \sin k(L-x)$ on the right. Only in this way can the R-property of each solution on the right be preserved on the left (see Figure 4-5).

6. The preceding got us the form of $\langle x|E \rangle$ everywhere in x-space. We still lack the energy levels $E_0, E_1, E_2, \ldots$ and the corresponding magnitudes of $A(A_0, A_1, \ldots)$ and B that go with each energy. These come from matching solutions at the boundary, from the boundary conditions at $x = b$.

In the harmonic oscillator problem, the boundary considerations are embedded in the tabulated solutions: the boundary conditions fix n as a nonnegative integer. In this partitioned box problem we must ourselves find the solutions from the boundary conditions.

Being the amplitude for a physical event, $\langle x|E \rangle$ must be everywhere a continuous function of x. It is therefore continuous at $x = b$. Put formally, this boundary condition is that in the limit, as ε approaches zero:

$$\langle x = b - \varepsilon | E \rangle = \langle x = b + \varepsilon | E \rangle \tag{4-30}$$

The wave function just inside the barrier wall is equal to that just outside the barrier wall.

At this same boundary the potential $V(x=b) = W$ is finite. Consider this fact in conjunction with Equation 4-27, that $\langle x|E \rangle$ must satisfy. You can deduce that the derivative of $\langle x|E \rangle$ is continuous. In the limit, as ε approaches zero

$$\frac{d}{dx} \langle x|E \rangle \big|_{b-\varepsilon} = \frac{d}{dx} \langle x|E \rangle \big|_{b+\varepsilon} \tag{4-31}$$

In fact, the eigenvalue problem, Equation 4-27, requires that the derivative of the wave function be continuous wherever $V(x)$ is finite (see Problem 4.9).

7. The continuity of the wave function and of its derivative through the boundary imposes constraints on A and B (see Problem 4.10). These constraints form a homogeneous set of linear equations. Solutions for such a set don't exist unless the parameters are just right. The allowed values of E are just those that generate the right parameters.

As an example consider the ground state: the state of lowest energy, $E = E_0$. For all symmetrical potentials, this state has even symmetry. An even ground state is shown in the figure. For even states, the two conditions, continuity of $\langle x|E \rangle$ and of its derivative at $x = b$, yield these two relations:

$$B \cosh qb = A \sin k(L-b) \qquad (4\text{-}32)$$

and

$$qB \sinh qb = -kA \cos k(L-b) \qquad (4\text{-}33)$$

For values of $q = q(E)$ and $k = k(E)$ generated by arbitrary energies E there exist no A's and B's that will make this pair of equations simultaneously true. Only for select values of E is this pair solvable. The lowest value of E that does it is E_0.

The mathematical condition that Equations 4-32 and 4-33 be simultaneously solvable for A and B is that the determinant of the coefficients be zero. Thus the energy levels of the system are the values of E that guarantee

$$\det \begin{vmatrix} \cosh qb & -\sin k(L-b) \\ q \sinh qb & k \cos k(L-b) \end{vmatrix} = 0 \qquad (4\text{-}34)$$

In the limit, that the ground-state energy is very much less than the barrier height, the two lowest-state energies shown in the figure are almost degenerate. If $E_0 \ll W$ and $qb > 1$, you will find that $k(L-b) = kd$ is slightly less than π (see Problem 4.12), and so

$$E_0 \approx \frac{h^2}{8\,m\,d^2} \qquad (4\text{-}35)$$

8. Employing any of the allowed energies, both Equation 4-32 and 4-33 tell us the relationship between the coefficients A and B for that energy. However you can deduce only the ratio, not the magnitude of either A or B, from these equations. The magnitudes are a consequence of the certainty condition; that $\langle E|E \rangle = 1$. Inserting an x-space ket-bra sum at the vertical makes it clear how this determines these magnitudes: they must satisfy the integral condition

$$\int_{-\infty}^{+\infty} dx \, |\langle x|E \rangle|^2 = 1 \qquad (4\text{-}36)$$

HAVING FOUND THE SOLUTIONS, MARK THEIR SIGNIFICANT FEATURES

Like the harmonic oscillator, the partitioned box models several physical phenomena: there's something to be learned from its properties. The following are three significant features:

1. From the certainty condition we find $|A|^2$ or $|B|^2$. We are left with a free choice of phase factor: it is arbitrary.

2. There is a discrete sequence of E values, the energy levels. Each energy corresponds to a state of even or odd symmetry. The lowest state is even. For the partitioned box potential, the steps of the energy ladder are not equally spaced, as they are for the harmonic oscillator.

3. According to classical mechanics, for particles with energy $E < W$, the barrier is impenetrable. A classical particle with energy $E_0 < W$ is either on the right or on the left. If it is on the right, it will have zero probability to be on the left (see Problem 4.13).

Quantum mechanics tells us otherwise. The wave function penetrates the barrier; so also does the probability density. The particle can tunnel through the barrier.

Put a particle at $x = +L/2 \pm .01L$. It's to the right of the barrier. Now measure its energy. Suppose you find E_0. It then has a finite probability to be at the left of the barrier. The probability of finding it at $x = -L/2 \pm .01L$ is greater than zero: $.02L|\langle x = -L/2|E_0\rangle|^2 = .02L|A|^2 \sin^2 k_0 L/2 > 0$. The classical result, for exactly the same experimental sequence, is that the probability to find the bead at $x = -L/2 \pm .01L$ is exactly zero! The classical bead doesn't tunnel (see Problem 4.13).

4.5 THE PROBLEM MAY BE CAST INTO MOMENTUM SPACE

In momentum space the harmonic oscillator eigenvalue problem, (4-4), becomes a differential equation in p for the energy eigenstates represented in the p-basis; for $\langle p|E\rangle$. When cast into p-language, (4-4) becomes

$$\frac{1}{2m} p^2 \langle p|E\rangle - \frac{m\hbar^2\omega^2}{2} \frac{d^2}{dp^2} \langle p|E\rangle = E \langle p|E \tag{4-37}$$

Inspection of this equation reveals one of the special features peculiar to the harmonic oscillator: the p-space equation is isomorphic to the x-space equation. The correspondence between this equation and the Hermite polynomial differential equation, (4-7), is brought about if we assign ξ the new meaning:

$$p = \xi \sqrt{m\hbar\omega} \tag{4-38}$$

The correspondence will be completed only if the energy, E, is again related to n, as in (4-9).

As in x-space, so it is in p-space that not all possible solutions to (4-37) are acceptable physically. The only solutions that qualify as physical amplitudes, $\langle p|E\rangle$, are those that meet the physical requirements of continuity, single valuedness, and square integrability. As in x-space, so it is in p-space that this last condition of normalizability selects as acceptable only nonnegative integer values of n. Thus the spectrum of energies generated in the p-space problem are the same as in the x-space one:

$E = (n + 1/2)\hbar\omega$. That the spectrum is independent of the space is the kernel of the eigenvalue problem idea: Equation 4-4 is true in any physical space.

The p-space representation of the harmonic oscillator energy eigenstates is (see Problem 4.16)

$$\langle p|E\rangle = \langle p|n\rangle = \frac{\exp(-p^2/2\,m\,\hbar\omega)\,H_n(p/\sqrt{m\,\hbar\omega})}{\sqrt{2^n n!\,\sqrt{\pi\,m\,\hbar\omega}}} \qquad (4\text{-}39)$$

HAVING ONCE CHOSEN THE PHASE, THERE IS NO MORE FREE CHOICE

In this expression no free choice of the phase factor exists. Here we cannot replace 1 by i^n. The reason is this: we have already used up our free choice of phase. In writing down the x-space representation of the energy eigenstates, a particular phase was chosen. Now, in the p-representation, we are stuck with the choice.

Here's another way of viewing it. What we take for $\langle p|E\rangle$ must do more than satisfy the eigenvalue problem; it must also be derivable from this ket-bra sum expansion.

$$\langle p|E\rangle = \sum_x \langle p|x\rangle\,\langle x|E\rangle \qquad (4\text{-}40)$$

Having chosen a phase for $\langle x|E\rangle$, we cannot now choose one for $\langle p|E\rangle$; this equation forbids it. The phase of $\langle p|E\rangle$ in Equation 4-39 is the proper one relative to the phase chosen in Equation 4-10 for $\langle x|E\rangle$.

The most dramatic view of the harmonic oscillator is in energy-level space, in n-space. To enter this space you need first to absorb some important notions about these two items: the *hermitian conjugate operator* and the *commutator of two operators*.

4.6 A BRACKETED OPERATOR IS A MATRIX ELEMENT

The numbers representing an operator bracketed between pairs of states can be posted in a two-dimensional array. That's why they're called *matrix elements*. A point in the array corresponds to a particular value of the pair. The ket value labels the horizontal position, and the bra the vertical. Each axis runs over all the states of the basis.

Matrix elements are numbers. A number, $\langle\alpha|\hat{\Omega}|\beta\rangle$, is assigned to each point α,β in the array. Figure 4-6 shows an example; the matrix of elements $\langle p|\hat{\tau}|P\rangle$ of the bead-on-a-ring momentum-step-up operator of Problem 3.4.

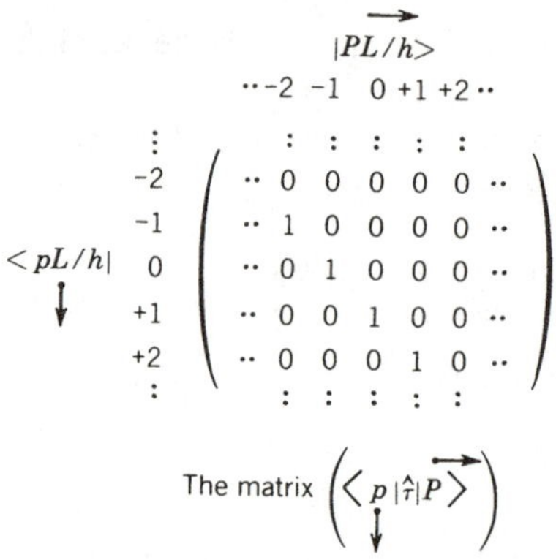

$$|PL/h\rangle$$

$$\cdots -2 \ -1 \ \ 0 \ +1 \ +2 \cdots$$

$$
\langle pL/h|
\begin{array}{c}
\vdots \\
-2 \\
-1 \\
0 \\
\downarrow \ +1 \\
+2 \\
\vdots
\end{array}
\left(
\begin{array}{cccccc}
\vdots & \vdots & \vdots & \vdots & \vdots \\
\cdots & 0 & 0 & 0 & 0 & 0 \cdots \\
\cdots & 1 & 0 & 0 & 0 & 0 \cdots \\
\cdots & 0 & 1 & 0 & 0 & 0 \cdots \\
\cdots & 0 & 0 & 1 & 0 & 0 \cdots \\
\cdots & 0 & 0 & 0 & 1 & 0 \cdots \\
\vdots & \vdots & \vdots & \vdots & \vdots
\end{array}
\right)
$$

The matrix $\left(\langle p \,|\hat{\tau}|\, P \rangle \right)$

Figure 4.6 The matrix of elements for the bead-on-a-track-momentum-step-up operator $\langle p|\hat{\tau}|P\rangle = \langle p|P+h/L\rangle$.

If the space of the states is discrete, it is easy to envision the actual matrix of numbers. The term, *matrix element*, also applies to a continuous space. Because x is a continuous variable, the matrix of elements $\langle x|\hat{\tau}|P\rangle$ is not easily exhibited in a picture: we can't draw it, but we can imagine it. For a particular value of x and P, $\langle x|\hat{\tau}|P\rangle$ is a number in a matrix: it is a matrix element, even though we can't display the matrix pictorially.

4.7 HERMITIAN CONJUGATION IS A MATRIX ELEMENT RULE

Given any operator, physical or not, you can always create from it a new operator called its *hermitian conjugate*. Suppose the original operator is symbolized by $\hat{\Omega}$. Then the new hermitian conjugate operator that you can create from it is $\hat{\Omega}^\dagger$. The dagger symbol, $\dagger$, indicates hermitian conjugation.

The hermitian conjugate operator is defined by stating what its matrix elements are in any space that you want. Suppose $|\alpha\rangle$ and $|\beta\rangle$ are two physical states for which you have the matrix element of $\hat{\Omega}$, the matrix element of $\hat{\Omega}^\dagger$ between these two states is defined in this way:

$$\langle \alpha|\hat{\Omega}^\dagger|\beta\rangle \equiv \langle \beta|\hat{\Omega}|\alpha\rangle^* \qquad (4\text{-}41)$$

This is the instruction for deducing the operator $\hat{\Omega}^\dagger$. It is deduced from what $\hat{\Omega}$ does. The operator $\hat{\Omega}$ bracketed between two particular states α and β amounts merely to a number. The number depends on the values of α and β. The instruction laid down in (4-41) is this: the α,β matrix element of $\hat{\Omega}^\dagger$, $\langle \alpha|\hat{\Omega}^\dagger|\beta\rangle$, is the complex conjugate of the number $\langle \beta|\hat{\Omega}|\alpha\rangle$. If you amass these numbers, $\langle \beta|\hat{\Omega}|\alpha\rangle^*$, for every conceivable pair of states, α and β, then you will have determined the matrix elements of $\hat{\Omega}^\dagger$ for every conceivable basis. Thus $\hat{\Omega}^\dagger$ qualifies as a legitimate operator in its own right. Every one of its matrix elements is accessible by virtue of its defining equation (4-41).

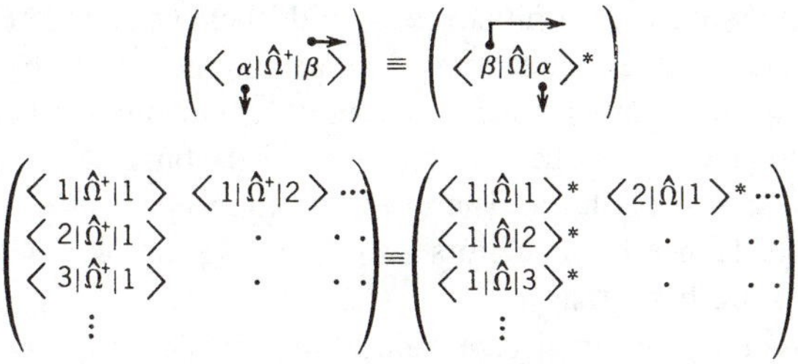

$$\left(\langle \alpha | \hat{\Omega}^\dagger | \overrightarrow{\beta} \rangle \right) \equiv \left(\langle \overrightarrow{\beta | \hat{\Omega} | \alpha} \rangle^* \right)$$

$$\begin{pmatrix} \langle 1|\hat{\Omega}^\dagger|1 \rangle & \langle 1|\hat{\Omega}^\dagger|2 \rangle \cdots \\ \langle 2|\hat{\Omega}^\dagger|1 \rangle & \cdot & \cdot \cdot \\ \langle 3|\hat{\Omega}^\dagger|1 \rangle & \cdot & \cdot \cdot \\ \vdots \end{pmatrix} \equiv \begin{pmatrix} \langle 1|\hat{\Omega}|1 \rangle^* & \langle 2|\hat{\Omega}|1 \rangle^* \cdots \\ \langle 1|\hat{\Omega}|2 \rangle^* & \cdot & \cdot \cdot \\ \langle 1|\hat{\Omega}|3 \rangle^* & \cdot & \cdot \cdot \\ \vdots \end{pmatrix}$$

Figure 4.7 **The elements of the hermitian conjugate matrix.**

The operator $\hat{\Omega}\dagger$ comes to us in the form of a matrix. It is simply the $\hat{\Omega}$ matrix with every element replaced by its complex conjugate and transposed to a new position, its mirror image around the diagonal (see Figure 4-7).

You should be aware of these three facts about hermitian conjugation:

$$(\hat{A} + \hat{B})\dagger = \hat{A}\dagger + \hat{B}\dagger \tag{4-42}$$

$$(c\hat{A})\dagger = c^*\hat{A}\dagger \qquad c = \text{any number} \tag{4-43}$$

$$(\hat{A}\hat{B})\dagger = \hat{B}\dagger\hat{A}\dagger \tag{4-44}$$

The truth of these equations follows easily from the definition in (4-41). For example, the proof of (4-43) runs like this:

$$\langle\alpha|(c\hat{A})\dagger|\beta\rangle = \langle\beta|c\hat{A}|\alpha\rangle^* = c^*\langle\beta|\hat{A}|\alpha\rangle^* = c^* \langle\alpha|\hat{A}\dagger|\beta\rangle = \langle\alpha|c^*\hat{A}\dagger|\beta\rangle \tag{4-45}$$

The first and third equalities are applications of the definition of hermitian conjugation. The second and fourth equalities are examples of (3-16): a constant times a bracket is the bracket of a constant. The equality of the extremes in (4-45) is valid for any pair of states: no particular α and β were mentioned, so you may remove $\langle\alpha|$ and $|\beta\rangle$ in equating the extremes. Doing so produces (4-43) and therefore proves it.

ONLY HERMITIAN OPERATORS
REPRESENT PHYSICAL OBSERVABLES

Among all the linear operators there is a special class which have the property that they are their *own* hermitian conjugates. For these matrices, the array of numbers forming the hermitian conjugate matrix looks exactly like the original matrix array. A self-adjoint or hermitian operator, $\hat{\Lambda}$, is one for which

$$\hat{\Lambda}\dagger = \hat{\Lambda} \qquad (\hat{\Lambda} \text{ hermitian or self-adjoint}) \tag{4-46}$$

The physically significant and important thing about the hermitian property is this: *All operators that represent physical observables must be hermitian ones*.

The reason is simple. Only hermitian operators do these two things: (1) always yield eigenvalues that are real and (2) produce physical spaces, ones in which reverse amplitudes are complex conjugates of forward ones (see Chapter 2, Quantum Principle 3).

Physical observables have spectra of real numbers only. And their measurement can generate only physical states. The operators that represent physical observables must do the same. Hence the operators corresponding to observables are hermitian: physical operators are hermitian.

That the two conditions—real eigenvalues and physical spaces—impose the hermitian property is easily proved. The proof depends on only two items: the meaning of hermitian conjugation and the equivalence of the two eigenvalue problem statements, Equations 3-19 and 3-20. If the eigenspace of $\hat{\Omega}$ is q, then for any ψ and all q (see Problem 4.22),

$$\langle\psi|\hat{\Omega} - \hat{\Omega}^\dagger|q\rangle = \Omega(q)\langle\psi|q\rangle - \Omega^*(q)\langle q|\psi\rangle^* \tag{4-47}$$

The two physical operator conditions—real eigenvalues and amplitude reversal—make the right-hand side zero. The left-hand side can be zero for all q and any ψ only if $\hat{\Omega} = \hat{\Omega}^\dagger$: physical operators are hermitian.

4.8 THE COMMUTATOR IS AN OPERATOR

If two operators operate successively on a state, the total effect will be represented by the product, like the $\hat{A}\hat{T}$ of Equation 3-18. The product $\hat{T}\hat{A}$ may produce an entirely different state. The latter means operate with $\hat{A}$ first and then $\hat{T}$. The former means operate with $\hat{T}$ first and then, on the result you get, operate with $\hat{A}$. Because of its utility, a new operator is defined called the *commutator* of $\hat{A}$ and $\hat{T}$. It is written $[\hat{A},\hat{T}]$ and means

$$[\hat{A},\hat{T}] \equiv \hat{A}\hat{T} - \hat{T}\hat{A} \tag{4-48}$$

Two operators are said to commute if their commutator always produces zero. The pair of operators $\hat{p}$ and $\hat{p}^2$ commute: their commutator is always zero. Asserting that two operators commute is equivalent to saying that the effect of these two operators on any state is independent of the order in which they operate.

COMPATIBLE OBSERVABLES: THEIR OPERATORS COMMUTE

The physical counterpart to an operator product is the successive measurement of two observables. *Compatible* observables are those for which a measurement of one doesn't disturb the measurement result obtained for the other. For such observables the order in which the measurements are made does not affect the results. Operators corresponding to compatible observables commute. For the bead-on-a-ring, the energy and the momentum are compatible observables: for the bead-on-a-ring $[\hat{H},\hat{p}] = 0$.

INCOMPATIBLE OBSERVABLES: NONCOMMUTING OPERATORS

An example of incompatible observables is the pair $\hat{x}$ and $\hat{p}$. They do not commute. The commutator of $\hat{x}$ and $\hat{p}$ is not zero; it is $i\hbar$ (see Problem 4.23).

$$[\hat{x},\hat{p}] = i\hbar \tag{4-49}$$

A measurement of position destroys the result of a previous momentum measurement (see Chapter 2, "Measurement Destroys a State"). Similarly, having measured position, a subsequent measurement of momentum destroys the position state.

This pair of measurements are incompatible because the new result you get invalidates the old one; you can't keep both measurement results to describe the state. The operators corresponding to incompatible observables don't commute.

In contrast with the bead-on-a-wire, the energy and momentum of the harmonic oscillator are incompatible observables; for the harmonic oscillator $[\hat{H},\hat{p}] \neq 0$.

RELATIONSHIPS BETWEEN OBSERVABLES ARE CONNECTIONS BETWEEN OPERATORS

The promised two items—hermitian conjugation and the commutator—have now been unveiled. In the process we encountered the important phenomenon of an algebra of operators. There are relationships between operators in which no language is specified. These connections are valid in any language. Equations 4-49, 4-46, 4-42, 4-43, 4-44 and especially 4-3 and 4-26 are examples. These all are language-independent relationships among operators.

When the operators are physical ones, these language-independent relations are, in fact, relationships among languages! The reason is that to each operator is attached the particular language that the operator creates, the measurement results that the observable generates.

Operator relations represent truths about nature. When, as in Equation 4-3, the operators correspond to observables, the operator equation is a statement about the relationship between observables in nature. "The energy operator is the hamiltonian with hats" is just such a statement.

Not all the operators useful in physics correspond to observables. Some don't correspond to observables at all. Invariably these appear as mathematical tools, computational aids. But an operator that generates results about nature cannot fail to be tagged with an interpretation. Such an operator becomes more than a computational tool; it molds our conceptualization of the physical world even though no measurable is involved. The archetypical case in point arises in considering the harmonic oscillator. Nonphysical ones are the key operators used when working in energy-level space. They are called *creation* and *annihilation operators* or, less dramatically, the *raising* and *lowering operators*.

4.9   THE LOWERING OPERATOR ANNIHILATES ONE UNIT OF ENERGY

The annihilation or lowering operator for the harmonic oscillator is called $\hat{a}$. It may be defined by its relationship to $\hat{p}$ and $\hat{x}$.

$$\hat{a}\,\sqrt{\hbar\omega} \equiv \frac{\hat{p}}{\sqrt{2m}} - i\,\sqrt{\frac{m\omega^2}{2}}\,\hat{x} \tag{4-50}$$

The creation or raising operator is the hermitian conjugate of $\hat{a}$; the creation operator is $\hat{a}^\dagger$.

$$\hat{a}^\dagger\,\sqrt{\hbar\omega} = \frac{\hat{p}}{\sqrt{2m}} + i\,\sqrt{\frac{m\omega^2}{2}}\,\hat{x} \tag{4-51}$$

From these two equations and the harmonic oscillator hamiltonian, (4-3), it is only a matter of algebra to produce these two key relations (see Problem 4.25):

$$[\hat{a},\hat{a}^\dagger] = 1 \tag{4-52}$$

and

$$\hat{H} = \hbar\omega(\hat{a}^\dagger\hat{a} + 1/2) \tag{4-53}$$

The first equation says that the commutator of $\hat{a}$ and $\hat{a}^\dagger$ is unity; it is a consequence of the fact that the commutator of $\hat{x}$ and $\hat{p}$ is $i\hbar$. The second is simply the hamiltonian of (4-3) rewritten in terms of the new operators $\hat{a}$ and $\hat{a}^\dagger$.

These two equations allow us to divorce ourselves from the languages of position and momentum. Using no more than these two alone, we can focus exclusively on the space generated by the operator $\hat{H}$ in the eigenvalue problem

$$\hat{H}|E\rangle = E|E\rangle \tag{4-54}$$

The key attribute of the $\hat{a}$ and $\hat{a}^\dagger$ operators is that they raise and lower energy states. This feature issues from two equations. To derive them is a matter of studying problem

4.26. You apply Equations B and C of that problem to an energy ket $|E\rangle$. It is then only algebra to deduce these two key statements about the properties of $\hat{a}$ and $\hat{a}^\dagger$.

$$\hat{H}\hat{a}|E\rangle \ = \ (E - \hbar\omega)\hat{a}|E\rangle \qquad (4\text{-}55)$$

and

$$\hat{H}\hat{a}^\dagger|E\rangle \ = \ (E + \hbar\omega)\hat{a}^\dagger|E\rangle \qquad (4\text{-}56)$$

It is in their form that their substance lies: both of these equations have the eigenvalue problem form. The first one says that $\hat{a}|E\rangle$ is an eigenstate of $\hat{H}$. It says that the new ket, produced by $\hat{a}$ operating on $|E\rangle$, is itself a state of energy, but with energy less by $\hbar\omega$ than the old one. Hence $\hat{a}$ is the energy-lowering operator: it produces a new state with one unit less of energy. Written mathematically, here is the implication of Equation 4-55 regarding the effect of the $\hat{a}$ operator:

$$\hat{a}|E\rangle \ = \ c_E|E - \hbar\omega\rangle \qquad (4\text{-}57)$$

The idea is profound, but the proof is easy. Substitute Equation 4-57 into Equation 4-55. You get a true statement, the eigenvalue problem for the energy $E' \ = \ E \ - \ \hbar\omega$.

A parallel argument reveals the effect of $\hat{a}^\dagger$: it produces an eigenstate raised in energy by $\hbar\omega$ (see Problem 4.27).

THE HARMONIC OSCILLATOR ENERGY SPECTRUM
IS $\left(n + \dfrac{1}{2}\right)\hbar\omega$

The harmonic oscillator must have a lowest level of energy. You will prove it in Problem 4.28. No energy level of the system may be negative, so it has a lowest energy, E_{gnd}, which is zero or greater.

That there is a ground state energy of the system leads us to this conclusion: that the energy spectrum must be discrete and of the form

$$E \ = \ E_{gnd} + n\hbar\omega \qquad (4\text{-}58)$$

where n is a nonnegative integer.

Here's the argument. Consider the new state produced by the operation of $\hat{a}^\dagger$ on the ground state. Because of what $\hat{a}^\dagger$ does, this new state is an eigenstate of the oscillator with energy $E_{gnd} + \hbar\omega$. But this must be the first excited state of the system: there can be no state with an intermediate energy E' between E_{gnd} and $E_{gnd} + \hbar\omega$. The reason is that the effect of the lowering operator, $\hat{a}$, operating on $|E'\rangle$ would produce an eigenstate with energy less than E_{gnd}. That is impossible if E_{gnd} is the lowest state. Thus E_{gnd} and $E_{gnd} + \hbar\omega$ are the first two states.

Extending this reasoning, the next state is found to be $E_{gnd} + 2\hbar\omega$, and the next, $E_{gnd} + 3\hbar\omega$. Thus the entire spectrum of possible states are, indeed, those listed in Equation 4-58.

To find the value of the system's ground-state energy, we must use the essential property of the ground state: no lower state exists. Thus

$$\hat{a}|E_{gnd}\rangle = 0 \qquad (4\text{-}59)$$

With this property in mind, apply the hamiltonian operator of (4-53) to $|E_{gnd}\rangle$, doing so must produce the ground-state energy. You find from $\hat{H}|E_{gnd}\rangle = E_{gnd}|E_{gnd}\rangle$

$$E_{gnd} = \frac{1}{2}\,\hbar\omega \qquad (4\text{-}60)$$

This is the famous *zero-point energy* of the harmonic oscillator. It is not possible for the harmonic oscillator mass to be at rest with no energy at all. There is always a residual zero-point of energy, given by Equation 4-60.

THE CREATION OPERATOR CREATES AN EXCITATION

Together, Equations 4-60 and 4-58 define the now-familiar spectrum of energies $(n + \frac{1}{2})\hbar\omega$. In view of this spectrum, we can just as well label the energy eigenstates by n as by E; $E = E_n$ and $|E\rangle = |n\rangle$. Thus we may rewrite (4-57) as

$$\hat{a}|n\rangle = c_n|n-1\rangle \qquad (4\text{-}61)$$

Equation 4-61 shows that the operator $\hat{a}$ is the annihilator of an excitation in the system. It leaves the oscillator in a state reduced by one quantum of energy. Even though it represents no physical observable, the operator $\hat{a}$ is a meaningful one conceptually.

From the effect of $\hat{a}$ displayed in (4-61), it follows that the effect of $\hat{a}^\dagger$ must be

$$\hat{a}^\dagger|n\rangle = c^*_{n+1}|n+1\rangle \qquad (4\text{-}62)$$

The proof requires attention to a subtlety that, when once noted, is quite self-evident.

Suppose, in energy index basis space, $|n\rangle$ and $|N\rangle$ characterize two states. Their essential property is that they are in the same basis; a bracket connecting them is a self-space bracket. Then that function of n and N represented by the bracket $\langle n|N-1\rangle$ is precisely the same function as that represented by the bracket $\langle n+1|N\rangle$. They both are zero unless $N = n + 1$, or, equivalently, unless $n = N - 1$. Otherwise, both brackets are unity. That these two brackets are equivalent is the self-evident subtlety.

It follows that

$$c_N\langle n|N-1\rangle = c_{n+1}\langle n+1|N\rangle \qquad (4\text{-}63)$$

The remainder of the proof that (4-62) follows from (4-61) uses only familiar matters. Problem 4.29 is a guide through the proof.

One more question remains: What is c_n?

We can deduce $|c_n|^2$ from this observation (see Problem 4.31):

$$n|n\rangle = \hat{a}^\dagger\hat{a}|n\rangle = |c_n|^2|n\rangle \qquad (4\text{-}64)$$

The right-hand equality registers the successive affects of $\hat{a}$ and $\hat{a}^\dagger$ from (4-61) and (4-62). The left-hand equality is a straightforward combination of the hamiltonian form (4-53), with the energy spectrum $E = (n + \frac{1}{2})\hbar\omega$ in the eigenvalue problem (Equation 4-54).

From the equality of the extremes in Equation 4-64, we learn that the magnitude of c_n is $\sqrt{n}$.

In choosing a phase for each eigenstate $|n\rangle$ when we wrote down the x-space solutions, $\langle x|n\rangle$ in Equation 4-10, we forfeited our free choice of phase factor here. The choice of phase is not arbitrary; we must take $c_n = \sqrt{n}$. In fact, the phase for $\langle x|n\rangle$ was chosen so that here the phase factor would be unity: that $c_n = \sqrt{n}$ is traditional usage in the literature.

MARK THESE HARMONIC OSCILLATOR RESULTS

For the aspiring physicist, these three harmonic oscillator results are key practical tools: they're well worth marking. Keeping in mind that $|E\rangle = |n\rangle$

$$E = \left(n + \frac{1}{2}\right)\hbar\omega \qquad (4\text{-}65)$$

$$\hat{a}|n\rangle = \sqrt{n}|n-1\rangle \qquad (4\text{-}66)$$

$$\hat{a}^\dagger|n\rangle = \sqrt{n+1}|n+1\rangle \qquad (4\text{-}67)$$

YOU MAY CAST INDEX-SPACE STATES INTO POSITION-SPACE LANGUAGE

No reference to x-space or p-space entered the derivations concerning n-space. Everything took place in energy-level space. The only bridge to the physical harmonic oscillator system is the operator equation connecting $\hat{a}$ to $\hat{p}$ and $\hat{x}$, (4-50) or (4-51). To find the x-space representation of the states $|n\rangle$ we use this bridge. Previously we relied on the literature on Hermite polynomials; we now explore another route by which the energy eigenstate wave functions may be obtained. With this route we need know nothing about Hermite polynomials.

We begin with the ground state. Its fundamental property, Equation 4-59, is cast into x-space via Equation 4-50. Thus

$$\langle x|\hat{a}|n=0\rangle = 0 \tag{4-68}$$

becomes

$$\left[\frac{\hbar}{i\sqrt{2m}}\frac{d}{dx} - i\sqrt{\frac{m\omega^2}{2}}\,x\right]\langle x|0\rangle = 0 \tag{4-69}$$

Solve this first-order differential equation for $\langle x|0\rangle$. The answer is just the $n = 0$ solution displayed in Equation 4-10: it agrees with our previous solution.

Having found the ground state, finding any other state is simple. Step up to it one step at a time. The raising operator effects the climb. The first excited state is the ground state *raised* by one step.

$$\langle x|n=1\rangle = \langle x|\hat{a}^\dagger|n=0\rangle \tag{4-70}$$

Thus

$$\langle x|1\rangle = \frac{1}{\sqrt{\hbar\omega}}\left[\frac{\hbar}{i\sqrt{2m}}\frac{d}{dx} + i\sqrt{\frac{m\omega^2}{2}}\,x\right]\langle x|0\rangle \tag{4-71}$$

Solving this for $\langle x|1\rangle$ produces exactly the $n = 1$ solution displayed in (4-10), including the proper phase factor.

Thus we may use the raising and lowering operators in n-space to generate the Hermite polynomials. Working in n-space we solve an x-space differential equation.

PROBLEMS

4.1 Consider the harmonic oscillator of this chapter from the classical newtonian viewpoint. Show first that for an oscillator about which only its total energy, E, is known, the position x varies with time as

$$x = \sqrt{2E/m\omega^2}\,\sin(\omega t + \beta)$$

The β is a constant whose arbitrariness expresses our ignorance of initial conditions.

Now consider the probability $P(x)dx$ of finding the oscillating mass at x in the region dx. Show that $P(x)$ is given by

$$P(x) = \begin{cases} 0 & \sqrt{2E/m\omega^2} \leq |x| \\ [\pi\sqrt{(2E/m\omega^2) - x^2}]^{-1} & |x| < \sqrt{2E/m\omega^2} \end{cases}$$

The key thought generating this result is this: the probability of finding the mass in dx is just the fraction of the time spent by the mass in dx during each period of its oscillation. Mathematically, $dxP(x) = 2dt/T$ where $T = 2\pi/\omega$ and $2dt$ is the time during which the mass is in dx. It is in dx twice during an oscillation.

4.2 According to the previous problem, if the energy were $E = 3\hbar\omega/2$, then there would be no possibility at all of finding the mass anywhere in the region $\sqrt{3\hbar/m\omega} < |x|$ (see Figure 4-4). That is the classical result. What is the quantum mechanical result? What is the probability for the mass to be found where $|x| > \sqrt{3\hbar/m\omega}$ for an oscillator known to have energy $E = 3\hbar\omega/2$?

Answer:

$$\frac{2}{\sqrt{3\hbar/m\omega}} \int_{\sqrt{3\hbar/m\omega}}^{\infty} dx \, |\langle x|n=1\rangle|^2 = \frac{4}{\sqrt{\pi}} \frac{1}{\sqrt{3}} \int_{\sqrt{3}}^{\infty} d\xi \, \xi^2 \, e^{-\xi^2} = 0.1$$

4.3 Consider a mass about that of a proton or neutron bound as an oscillator with a natural frequency of 10 GHz ($\omega/2\pi = 10^{10}$ Hz).

What is the energy of its ground state and first excited state?

In a thermal ensemble of such oscillators, below what temperature could you reasonably expect most of them to be in their ground state? That is, when is kT less than the minimum energy necessary to excite the system: $E_1 - E_0$? How does this temperature compare with room temperature: colder, much colder, hotter?

Answer: $0.5° K = $ very cold.

What is the classical oscillation amplitude—the size of the "1-D orbit"—for this oscillator?

Answer:

$$\sqrt{2E/m\omega^2} = 10 \text{ Å in the ground state.}$$

4.4 The harmonic oscillator is in its first excited state, $n = 1$. What is the probability, that on measuring its position, you will find

a. $x = \sqrt{\hbar/m\omega} \pm 0.01\sqrt{\hbar/m\omega}$

b. $x = 0 \pm 0.01\sqrt{\hbar/m\omega}$

Answers:

a. $0.02\sqrt{\hbar/m\omega} \, |\langle x=\sqrt{\hbar/m\omega} \, |n=1\rangle|^2 = .008$

b. $0.02\sqrt{\hbar/m\omega} \, |\langle x=0|n=1\rangle|^2 = 0$

4.5 Proceeding from the meaning of the symbols, construct a proof that Equation 4-20 is valid. Exhibit the x-space differential form of these four matrix elements where $|\psi\rangle$ is an arbitrary state of the system.

a. $\langle x|\hat{H}|\psi\rangle$ Answer: $\left[-\dfrac{\hbar^2}{2m}\dfrac{d^2}{dx^2} + V(x) \right]\langle x|\psi\rangle$

b. $\langle -x|\hat{H}|\psi\rangle$ Answer: $\left[-\dfrac{\hbar^2}{2m}\dfrac{d^2}{dx^2} + V(-x) \right]\langle -x|\psi\rangle$

c. $\langle x|\hat{R}\hat{H}|\psi\rangle$ Answer: same as for b.

d. $\langle -x|\hat{R}\hat{H}|\psi\rangle$ Answer: same as for a.

Notice, by comparing d with (4-20), that only if the potential is symmetrical is $\hat{R}\hat{H}$ equal to $\hat{H}\hat{R}$.

4.6 The absence of any potential at all is certainly an example of a symmetric potential. Why, therefore, do the energy eigenstates of the bead-on-a-ring, $\langle x|p \neq 0\rangle$, *not* exhibit even and odd symmetry? What element of the proof that $\langle x|E\rangle$ is a simultaneous eigenstate of $\hat{R}$ and $\hat{H}$ is *not* met by the bead-on-a-ring?

When E is degenerate, the eigenstates need not have reflection symmetry; nevertheless, eigenstates of energy *may always be found* that do have such symmetry. Obtain these alternative excited energy eigenstates of the bead-on-a-ring; the ones that are simultaneously eigenstates of $\hat{R}$.

Answer:

$$\langle x|E\rangle = \langle x|v, R = +1\rangle = \sqrt{\frac{2}{L}} \cos \frac{2\pi v x}{L}$$

and

$$\langle x|E\rangle = \langle x|v, R = -1\rangle = \sqrt{\frac{2}{L}} \sin \frac{2\pi v x}{L}$$

where

$$v = 1,2,3 \ldots$$

and

$$E_v = \frac{h^2 v^2}{2mL^2}$$

Although they are eigenstates of energy, these are not simultaneous eigenstates of momentum. The states $|n = Lp/h\rangle$ are simultaneous eigenstates of $\hat{p}$ and $\hat{H}$. The states $|v\rangle$ are simultaneous eigenstates of $\hat{R}$ and $\hat{H}$.

4.7 Prove that if $[a \sin kx + b \cos kx]$ must always be zero at $x = L$, then the combination can be written as $A \sin k(L-x)$.

  Prove also that if $[\alpha \exp(ikx) + \beta \exp(-ikx)]$ is to be zero at $x = L$, then it too will reduce to $A \sin k(L-x)$.

4.8 Prove that the functions given in Figure 4-5 for each individual region are, indeed, solutions of (4-27) for that region.

4.9 Prove Equation 4-31, and then prove that $d\langle x|E\rangle/dx$ must be continuous in all regions of space where $V(x)$ is finite.

  Hint: Integrate $\langle x|\hat{H}|E\rangle$ over any region of space centered at $x = X$ from $X - \varepsilon$ to $X + \varepsilon$. From the result in the limit of vanishing ε, you conclude that $\dfrac{d}{dx}\langle x|E\rangle$ cannot change over the region 2ε. You find

$$\frac{d}{dx}\langle x|E\rangle|_{x-\varepsilon} = \frac{d}{dx}\langle x|E\rangle|_{x+\varepsilon}$$

4.10 Show, for the even solution to (4-27), that the two boundary conditions at $x = b$ produce the pair of constraints given in the text as Equations 4-32 and 4-33.
  Find the pair of constraints generated by the odd solutions.

4.11 Exhibit the determinant equation that determines the energies of the odd symmetry states of the bead-in-a-partitioned-box. The one that determines the even states is (4-34).

4.12 Even when $qb > 1$, Equation 4-34 remains a transcendental equation; it is an equation that relates $\tan kd$ to k itself. You must solve it graphically. Do so in the case $k/q = \sqrt{E_0/W} \ll 1$ and thus confirm (4-35). Interpret this result by comparing it with the limiting case $W \to \infty$.

The equation to be solved is

$$k = -q \tan kd$$

4.13 Calculate the probability densities $P(x)$ for a classical particle inside the partitioned box whose energy is $E < W$. The principle on which the calculation rests is exactly the same as that used in Problem 4.1. The probability of finding the mass in dx is $P(x)dx$: this probability is just the fraction of time dt/T that a particle spends in dx. The time, T, is that of one complete cycle of the bead in its travels. Thus

$$\frac{dt}{T} = \frac{dx}{L - b}$$

if $b < x < L$. Notice that if the particle is in the right-hand well, it will spend no time at all in the left one, and so there $P(x < b) = 0$. For each energy there are two distributions possible: left one or right one.

4.14 Noting that the operator $\hat{x}$ represented in p-space is given by

$$\langle p|\hat{x} = i\hbar \frac{d}{dp} \langle p|$$

prove that (4-37) follows from (4-4).

4.15 a. Show that the substitutions (4-8) and (4-9) are the *only* ones that can cast (4-5) into the form (4-7). b. Show that the substitutions (4-9) and (4-38) are the *only* ones that can recast (4-37) into the form of (4-7).

4.16 Show that (4-39) is indeed a solution of the differential equation, (4-37), and that it fits every p-space boundary condition. Base your proof on the Hermite polynomial properties mentioned in the text.

4.17 Produce the hermitian conjugate matrix of $\hat{\tau}$: the one whose elements are $\langle p|\hat{\tau}\dagger|P\rangle$. The operator $\hat{\tau}$ is that of Figure 4-6.

Solution:

$$\langle n|\hat{\tau}\dagger|N\rangle = \langle N|\hat{\tau}|n\rangle* = \langle N|n+1\rangle* = \langle n+1|N\rangle \text{ or } \langle n|N-1\rangle$$

where $n = \dfrac{Lp}{h}$ and $N = \dfrac{LP}{h}$.

4.18 The differentiation-in-x-space operator $\hat{D}$, was defined by Equation 3-4. Show that the hermitian conjugate operator $\hat{D}\dagger$ has the property

$$\langle x|\hat{D}\dagger = -\frac{d}{dx} \langle x|$$

Answer: By Equation 3-14, $\langle p|\hat{D} = \dfrac{ip}{\hbar} \langle p|$. Hence

$$\langle x|\hat{D}\dagger|p\rangle = \langle p|\hat{D}|x\rangle* = \left(\frac{ip}{\hbar} \langle p|x\rangle \right)* = -\frac{ip}{\hbar} \langle x|p\rangle = -\frac{d}{dx} \langle x|p\rangle$$

The equality of the extremes is valid for all $|p\rangle$, thus for any $|\psi\rangle$.

4.19 Prove that the operator $\hat{x}$ is hermitian and the translation operator $\hat{T}$ of Equation 3-3 is not. Find $\hat{T}\dagger$.

Answers:

$$\langle x|\hat{x}\dagger|\psi\rangle = \langle \psi|\hat{x}|x\rangle* = (x\langle \psi|x\rangle)* = x\langle x|\psi\rangle = \langle x|\hat{x}|\psi\rangle$$

$$\langle x|\hat{T}\dagger|p\rangle = \langle p|\hat{T}|x\rangle* = \text{via (3-13)} \left[\exp\left(\frac{-ibp}{\hbar}\right)\langle p|x\rangle \right]* = \exp\left(\frac{ibp}{\hbar}\right)\langle x|p\rangle$$

thus

$$\hat{T}^\dagger = \exp\left(\frac{ib\hat{p}}{\hbar}\right)$$

4.20 If two operators are equal to each other in one basis space, can they be different from each other in another basis space?

Answer: No. They are equal to each other in all bases.

$$\text{If } \langle q|\hat{A} = \langle q|\hat{B}, \text{ then } \langle r|\hat{A} = \langle r|\hat{B}, \text{ since } \langle r| = \Sigma \langle r|q\rangle \langle q|$$

That the equality transcends all spaces is why we can write operator equations like $\hat{A} = \hat{B}$, without referring to any specific space.

4.21 Prove Equations 4-42 and 4-44.

Solution key: For any pair of physical states $|\psi\rangle$ and $|\phi\rangle$,

$$\text{if } \hat{B}|\psi\rangle = c|\phi\rangle \text{ then } \langle\psi|\hat{B}^\dagger = c^*\langle\phi|$$

Prove this first, and use it to verify (4-44).

4.22 Show the truth of Equation 4-47, which proves that all physical operators are hermitian.

4.23 Prove that the commutator of $\hat{x}$ and $\hat{p}$ is $i\hbar$ (See Problem 3.12). To prove operator theorems, you may work in your favorite basis because any one will do. To prove $[\hat{x},\hat{p}] = i\hbar$, it suffices to prove that $\langle x|[\hat{x},\hat{p}]|p\rangle = i\hbar\langle x|p\rangle$. Show that this result automatically demonstrates that $\langle\alpha|[\hat{x},\hat{p}]|\beta\rangle = i\hbar\langle\alpha|\beta\rangle$ where α and β are arbitrary states. Thus you may remove $\langle\alpha|$ and $|\beta\rangle$ or $\langle x|$ and $|p\rangle$.

4.24 Prove that the expression 4-51 given for the hermitian conjugate of $\hat{a}$ is, indeed, its hermitian conjugate.

Solution: Dagger the operator $\hat{a}$. Note that physical operators are hermitian.

$$(\hat{a})^\dagger = \left(\frac{\hat{p}}{\sqrt{2m\hbar\omega}} - i\hat{x}\sqrt{\frac{m\omega}{2\hbar}}\right)^\dagger = \frac{\hat{p}^\dagger}{\sqrt{2m\hbar\omega}} + i\hat{x}^\dagger\sqrt{\frac{m\omega}{2\hbar}}$$

$$= \frac{\hat{p}}{\sqrt{2m\hbar\omega}} + i\hat{x}\sqrt{\frac{m\omega}{2\hbar}}$$

4.25 Do the operator algebra necessary to get from the definition of $\hat{a}$ to the statements, Equations 4-52 and 4-53, about its commutator with $\hat{a}^\dagger$ and its relation to $\hat{H}$.

Hint: First show that

$$\hbar\omega\hat{a}\,\hat{a}^\dagger = \frac{\hat{p}^2}{2m} + \frac{m\omega^2}{2}\hat{x}^2 - \frac{i\omega}{2}[\hat{x},\hat{p}]$$

and

$$\hbar\omega\hat{a}^\dagger\,\hat{a} = \frac{\hat{p}^2}{2m} + \frac{m\omega^2}{2}\,\hat{x}^2 + \frac{i\omega}{2}\,[\hat{x},\hat{p}]$$

4.26 Prove these five fundamental facts about commutators:

$$[\hat{A},\hat{A}] = 0$$
$$[c,\hat{A}] = 0 \qquad c = \text{constant, number}$$
$$[\hat{B},\hat{A}] = -\,[\hat{A},\hat{B}]$$
$$[\hat{A},\hat{B}+\hat{C}] = [\hat{A},\hat{B}] + [\hat{A},\hat{C}]$$
$$[\hat{A},\hat{B}\hat{C}] = \hat{B}[\hat{A},\hat{C}] + [\hat{A},\hat{B}]\hat{C} \tag{A}$$

The first four equations are easy to remember. It is worth memorizing also the last one, (A). It is a powerful calculational tool. For example, to show that

$$[\hat{a},\hat{H}] = \hbar\omega\hat{a} \tag{B}$$

use the commutator $[\hat{a},\hat{a}^\dagger\hat{a}]$. It is easily done via Equation (A). Use (A) to show that

$$[\hat{a}^\dagger,\hat{H}] = -\hbar\omega\hat{a}^\dagger \tag{C}$$

Show that (C) follows from (B) by taking the hermitian conjugate of the latter; that is, use $[\hat{a},\hat{H}]^\dagger = [\hat{H},\hat{a}^\dagger]$.

4.27 Prove, using (4-56), that

$$\hat{a}^\dagger|E\rangle = \text{const.} \times |E+\hbar\omega\rangle$$

4.28 Prove that no energy level of the harmonic oscillator can be negative.

Hint: Calculate $\langle E|\hat{H}|E\rangle$ to find

$$E\langle E|E\rangle = \sum_p \frac{p^2}{2m}|\langle p|E\rangle|^2 + \sum_x \frac{Kx^2}{2}|\langle x|E\rangle|^2$$

From this finding, make the proof.

4.29 Show, by means of (4-61) and (4-63), that the matrix element $\langle n|\hat{a}|N\rangle$ may be written in two ways, as follows:

$$\langle n|\hat{a}|N\rangle = c_N\,\langle n|N-1\rangle = c_{n+1}\,\langle n+1|N\rangle$$

Using this, calculate the matrix element $\langle N|\hat{a}^\dagger|n\rangle$. The result, plus an argument noting that N represents any arbitrary state of the basis, leads to Equation 4-62.

4.30 Use the double equality shown in the previous problem to deduce the instruction for the $\hat{a}$ operator in n-space:

$$\langle n|\hat{a} = c_{n+1}\,\langle n+1|$$

4.31 A recasting of (4-54) so that the label n is exhibited looks like this:

$$\hat{H}|n\rangle = E_n|n\rangle$$

where the function of n that is E_n is

$$E_n = \frac{1}{2}\hbar\omega + n\hbar\omega$$

Show that these plus the hamiltonian of (4-53) lead directly to the left-hand equality in (4-64).

4.32 Draw the matrix whose elements are $\langle n|\hat{a}^\dagger|N\rangle$: the $\hat{a}^\dagger$ matrix in the basis of energy levels. Draw the $\hat{a}$ matrix also.

4.33 Draw the matrix whose elements are $\langle n|\hat{x}|N\rangle$.

Hint: $\hat{x}$ is related to $\hat{a}$ and $\hat{a}^\dagger$.

4.34 The harmonic oscillator mass is in the potential well $V(x) = Kx^2/2$, so pure momentum states $|p\rangle$ are not eigenstates of energy. Can the mass ever be in a pure momentum state?

Answer: Yes. The position state $|x=.05$ cm$\rangle$ is not an energy eigenstate, but the mass can be found at $x = .05$ cm. The momentum state $|p=3.6$ gm-cm/sec$\rangle$ is not an energy eigenstate, but a momentum measurement may uncover this momentum, in which case the bead has been found in the pure momentum state $|p=3.6$ gm-cm/sec$\rangle$. Only if you measure energy, do you find energy eigenstates, but such states are not the only ones to be found in nature.

Chapter 5

The Bead in a Spherical Shell:
Two Dimensions
with Angular Momentum

5.1 Multidimensional Systems

To locate a point in space, three positions must be specified, not merely one. In the previous chapters only one-dimensional systems (1-D) were contemplated. Now we shall consider multidimensional systems.

The leap from one dimension to two dimensions involves matters of substance. Going from two dimensions to three or more is a mere matter of formalities. In this chapter we make the grand leap: we explore 2-D systems.

TWO DIMENSIONS REQUIRE
A BIVARIABLE STATE LABEL

Consider again our familiar point-mass ideal bead. Let it now be free to move anywhere in the x-y plane. Its position measurement space is two dimensional. To locate the bead, two numbers must be specified, the value of x and the value of y. Together, the pair constitute a nondegenerate position measurement: there is only one state per 2-D point.

The domain of this 2-D space is the entire x-y plane: $-\infty < x < +\infty$ and $-\infty < y < +\infty$. Each of the points in this continuous 2-D domain is a possible position measurement result. All of the points together comprise the spectrum of possible position measurement results; they are a basis space.

BEAD IS AT POSITION x, y: IT'S DENOTED $|x,y\rangle$

It requires two numbers to define a single point in this basis space. You need a value for both x and for y. Thus a state of the bead is labeled by a bivariable in the ket. A position state of the bead must be $|x,y\rangle$.

Values for x are measurable; x is an observable. Hence there must be an operator $\hat{x}$. It produces x as its eigenvalues. For our 2-D bead there is also a value of y; it, too, is an observable. Hence there must be an operator $\hat{y}$ to produce the spectrum of y values. The possible experimental findings corresponding to these two operators are the spectrum of values x,y. This spectrum comprises the simultaneous eigenvalues of $\hat{x}$ and $\hat{y}$. Thus

$$\hat{x}|x,y = x|x,y\rangle \qquad \text{and} \qquad \langle x,y|\hat{x} = x\langle x,y| \qquad (5\text{-}1)$$

$$\hat{y}|x,y\rangle = y|x,y\rangle \qquad \text{and} \qquad \langle x,y|\hat{y} = y\langle x,y| \qquad (5\text{-}2)$$

BEAD HAS MOMENTUM p_x, p_y: IT'S DENOTED $|p_x, p_y\rangle$

Suppose we choose not to measure position. We measure other things instead. There must be operators corresponding to those other observables. By making these measurements new basis spaces are created.

Instead of x and y we might measure the components of the momentum of the bead, p_x and p_y. In so doing we create the state $|p_x, p_y\rangle$ labeled by the values of the momenta obtained in the measurement. There must, of course, be operators $\hat{p}_x$ and $\hat{p}_y$ whose eigenvalues are p_x and p_y. The state $|p_x, p_y\rangle$ is an eigenstate simultaneously of the two operators $\hat{p}_x$ and $\hat{p}_y$. Thus

$$\hat{p}_x|p_x, p_y\rangle = p_x|p_x, p_y\rangle \qquad \text{and} \qquad \langle p_x, p_y|\hat{p}_x = p_x\langle p_x, p_y| \qquad (5\text{-}3)$$

$$\hat{p}_y|p_x, p_y\rangle = p_y|p_x, p_y\rangle \qquad \text{and} \qquad \langle p_x, p_y|\hat{p}_y = p_y\langle p_x, p_y| \qquad (5\text{-}4)$$

2-D AMPLITUDES YIELD PROBABILITIES

As in 1-D, so in 2-D there are amplitudes for events. The amplitude, $\langle x,y|p_x, p_y\rangle$, is for this event: to find a bead, known to have momentum $\mathbf{p}$ whose x-component is p_x and whose y-component is p_y, at the position $\mathbf{r} = \mathbf{e}_x x + \mathbf{e}_y y$.

By $\mathbf{e}_x$ is meant the unit vector in the x-direction; by $\mathbf{e}_y$ that for the y-direction.

The key property of this amplitude is that its absolute square yields the probability density for the event. The probability density $P(xy|p_x p_y)$ is a function of x and y or of $\mathbf{r}$. Thus $\langle x,y|p_x, p_y\rangle$ is another way of writing something that is a function of $\mathbf{r} = x\mathbf{e}_x + y\mathbf{e}_y$. It is that function of $\mathbf{r}$ whose absolute value squared is the probability density $P(\mathbf{r}|\mathbf{p})$.

$$|\langle x,y|p_x,p_y\rangle|^2 = P(\mathbf{r}|\mathbf{p}) \qquad (5\text{-}5)$$

Thus Quantum Principle 1, embodied in Equation 2-1 for a 1-D system, is met in 2-D.

A 2-D VOLUME IS AN AREA

A probability density function in a four-dimensional space describes our 2-D system. The four dimensions are $x, y, p_x,$ and p_y. The density $P(\mathbf{r}|\mathbf{p})$, is a probability per unit *volume* of **p**-space and per unit *volume* of **r**-space.

The probability of finding the bead known to have momentum **p** in $d^V p \equiv dp_x\, dp_y$ at a point **r** in $d^V r \equiv dx\, dy$ is $dx\, dy\, P(\mathbf{r}|\mathbf{p})\, dp_x\, dp_y$. Hence, the elemental volume of x, y space is $d^V r \equiv dx\, dy;$ that of **p**-space, $dp_x dp_y$. An elemental 1-D volume is a *length;* elemental 2-D volumes are *areas,* and, of course, 3-D volumes are *volumes.*

The 2-D Pure Momentum State Wave Function

The bracket $\langle x,y|p_x,p_y\rangle$ is the representative of the pure momentum state $|\mathbf{p}\rangle$ in the language of position, **r**. We can deduce the pure momentum state wave function $\langle x,y|p_x,p_y\rangle$ for our bead-in-a-plane. It must be this function of x and y:

$$\langle x,y|p_x,p_y\rangle = \frac{1}{h}\exp i(xp_x + yp_y)/\hbar \qquad (5\text{-}6)$$

It must be this function for reasons just parallel to those advanced for the 1-D bead-on-a-ring, Equation 2-8. Here's the way to see it.

THE SAME EVENT, THE SAME AMPLITUDE

1. A bead in a pure momentum state moves in some fixed particular direction with momentum p. Call the coordinate along which the motion takes place ξ. The bead's motion is in the ξ-direction: it behaves like a 1-D bead described by ξ or p. The one-dimensional pure momentum state, $\langle \xi|p\rangle$, for such a bead is

$$\langle \xi|p\rangle = \text{constant} \times \exp i\,\xi\, p/\hbar \qquad (5\text{-}7)$$

2. Now imagine a coordinate perpendicular to the ξ-axis (see Figure 5-1) Call distances along this axis η.

A pure momentum state defines a unidirectional momentum no matter in how many dimensions you care to view it. We can describe the 1-D bead in two dimensions even though it has no motion in the second dimension. We view the bead as having

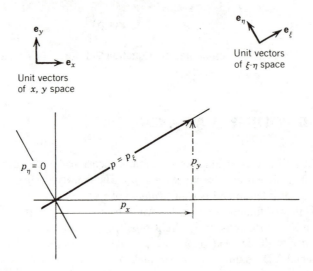

***Figure 5.1* The 2-D view of the 1-D bead. Its pure momentum state wave function is $\langle\xi,\eta|p,0\rangle$. It has $p_\xi = p$ and $p_\eta = 0$.**

momentum $p_\xi = p$ and $p_\eta = 0$ in the space of ξ and η. The event to which $\langle\xi,\eta|p_\xi=p,p_\eta=0\rangle$ refers is just the event to which $\langle\xi|p\rangle$ refers: that a bead with momentum $\mathbf{p} = \mathbf{e}_\xi p$ be found at ξ. Thus

$$\langle\xi,\eta|p,0\rangle = \text{constant} \times \exp i\,\xi\,p/\hbar \tag{5-8}$$

3. Viewing this same bead in the x-y coordinate system does not change its physical nature. Using x and y, the position vector is $\mathbf{r}$ where

$$\mathbf{e}_\xi\xi + \mathbf{e}_\eta\eta = \mathbf{r} = \mathbf{e}_x x + \mathbf{e}_y y \tag{5-9}$$

Its momentum is given by

$$\mathbf{e}_\xi p = \mathbf{p} = \mathbf{e}_x p_x + \mathbf{e}_y p_y \tag{5-10}$$

The critical quantity that appears in the pure momentum state wave function is ξp. Viewed in the x-y coordinate system, this is merely the vector dot product $\mathbf{r}\cdot\mathbf{p}$.

$$\xi p = \mathbf{r}\cdot\mathbf{p} = xp_x + yp_y \tag{5-11}$$

4. In ξ-η space the wave function for this bead is written $\langle\xi,\eta|p,0\rangle$. In x-y space it is written $\langle x,y|p_x,p_y\rangle$. These must be equal because they refer, in two different languages, to precisely the same event. The event is that a bead with momentum $\mathbf{p}$ be found at position ξ,η (or x,y). Thus

$$\langle \xi, \eta | p, 0 \rangle = \langle x, y | p_x, p_y \rangle \qquad (5\text{-}12)$$

Using Equation 5-11 in Equation 5-8 to implement this equality establishes the form of $\langle x, y | p_x, p_y \rangle$ put forth in Equation 5-6.

5. That the wave function in Equation 5-6 is properly normalized is a matter of the certainty condition: that $\langle p_x, p_y | p_x, p_y \rangle \, dp_x \, dp_y = 1$. From the substance and consequences of Problem 2.15 comes the proof. It amounts to calculating the bracket $\langle p_x, p_y | P_x, P_y \rangle$ via an x, y-space ket-bra sum realizing that it is equal to the 2-D Dirac δ-function $\delta^2(\mathbf{p} - \mathbf{P}) = \delta(p_x - P_x)\delta(p_y - P_y)$ (see Problem 5.1).

To perform the calculation needs only the natural generalization, from 1-D, of the meaning of a sum over a continuous variable (see Chapter 2, "A Continuous Variable Sum Means An Integral"). For our 2-D plane the sum over x, y means

$$\sum_{x,y} \equiv \int_{-\infty}^{+\infty} dy \int_{-\infty}^{+\infty} dx \qquad (5\text{-}13)$$

The foregoing five items conclude the argument for the validity of the expression, given in Equation 5-6, for the 2-D pure momentum state wave function $\langle x, y | p_x, p_y \rangle$.

MULTIDIMENSIONAL AMPLITUDES SATISFY ALL QUANTUM PRINCIPLES

The wave function $\langle x, y | p_x, p_y \rangle$ of Equation 5-6 satisfies all of the Quantum Principles 1 through 5 that governed the 1-D case.

From the amplitude a probability can be calculated, Principle 1.

The pure momentum state exhibits the de Broglie association between particle momentum and wave vector, Principle 2. In fact it exposes the natural generalization of Equation 2-6 (see Problem 5.1).

$$\mathbf{p} = \hbar \mathbf{k} \qquad (5\text{-}14)$$

From the amplitude reversal theorem, Principle 3, we are able to deduce the reverse event amplitude

$$\langle p_x, p_y | x, y \rangle = \langle \mathbf{p} | \mathbf{r} \rangle = \frac{1}{h} \exp(-i\mathbf{r} \cdot \mathbf{p}/\hbar) \qquad (5\text{-}15)$$

The ket-bra sum theorem on the superposition of states is valid, Principle 4. But now there are two variables over which to sum, not just one. And because the elemental volume of position space is $d^V r = dx \, dy$, the ket-bra sum rule becomes (see Problem 5.1):

$$\int_{x,y} dx\, dy\, |x,y\rangle \langle x,y| \;=\; \sum_{x,y} |x,y\rangle \langle x,y| \;=\; 1 \qquad (5\text{-}16)$$

Recognizing the elemental volume of momentum space as $d^v p = dp_x\, dp_y$, the p-space ket-bra sum derives from this meaning attached to sums over the continuous variables of p-space.

$$\sum_{p_x \cdot p_y} \equiv \int_{-\infty}^{+\infty} dp_y \int_{-\infty}^{+\infty} dp_x \qquad (5\text{-}17)$$

Quantum Principle 5 assures us that there are operators that generate the spaces of measurement. Hence operators $\hat{x}$ and $\hat{y}$ exist that generate the space x,y. And operators $\hat{p}_x$ and $\hat{p}_y$ exist that generate, as their eigenvalues, the spectrum of measurable momenta p_x and p_y.

THE EIGENVALUE PROBLEM YIELDS POSITION-SPACE INSTRUCTIONS FOR MOMENTUM OPERATORS

We can deduce the r-space representation of the operators $\hat{p}_x$ and $\hat{p}_y$. To be commensurate with the r-space representation of a pure state of momentum, $\langle \mathbf{r}|\mathbf{p}\rangle$, these operators must be the same in x,y space, as each of them is in x- and y-space separately.

$$\langle x,y|\hat{p}_x \;=\; -i\hbar\, \frac{\partial}{\partial x}\, \langle x,y| \qquad (5\text{-}18)$$

and

$$\langle x,y|\hat{p}_y \;=\; -i\hbar\, \frac{\partial}{\partial y}\, \langle x,y| \qquad (5\text{-}19)$$

Proof: Each yields its eigenvalue p_x or p_y, as it must, when applied to a pure momentum state $|p_x,p_y\rangle$ (see Problem 5.2).

5.2 VECTOR OPERATOR: ONE WITH OPERATOR COMPONENTS

These two equations may be collapsed into a single one if vector notation is used. To do it, I must define what I mean by a vector operator: it is a vector whose components are operators.

The traditional handwritten notation for a normal vector is an arrow over the symbol; $\vec{r} = \mathbf{r}$. A convenient handwritten notation for a vector operator is an arrowed hat, a bent arrow. In print, boldface with a hat is used. The momentum vector operator is $\vec{\hat{p}} = \hat{\mathbf{p}}$. Its meaning is

$$\vec{\hat{p}} = \hat{\mathbf{p}} = \mathbf{e}_x \hat{p}_x + \mathbf{e}_y \hat{p}_y \tag{5-20}$$

With this notation the pair of equations 5-18 and 5-19 can be written as a single one:

$$\langle \vec{r} | \vec{\hat{p}} = \langle \mathbf{r} | \hat{\mathbf{p}} = -i\hbar \nabla \langle \mathbf{r} | \tag{5-21}$$

By the symbol ∇ is meant the *del operator* for differentiation which in *x-y* space is

$$\nabla = \mathbf{e}_x \frac{\partial}{\partial x} + \mathbf{e}_y \frac{\partial}{\partial y} \tag{5-22}$$

Using these results, it is a straightforward exercise to establish all of the six commutator relations among the four operators $\hat{x}$, $\hat{y}$, $\hat{p}_x$, and $\hat{p}_y$ (see Problem 5.3). They are

$$[\hat{x},\hat{y}] = [\hat{x},\hat{p}_y] = [\hat{y},\hat{p}_x] = [\hat{p}_x,\hat{p}_y] = 0 \tag{5-23}$$

and

$$[\hat{x},\hat{p}_x] = [\hat{y},\hat{p}_y] = i\hbar \tag{5-24}$$

BRACKET PRODUCTS, YES

The 2-D pure momentum state wave function given in Equation 5-6 has the special property that it may be disjoined into a product of two independent 1-D wave functions. Thus

$$\langle x,y|p_x,p_y \rangle = \frac{1}{h} \exp i(xp_x + yp_y)/\hbar = \langle x|p_x \rangle \langle y|p_y \rangle \tag{5-25}$$

This decomposition into a product is sometimes useful. However, it is not always possible to form such a product.

Resolution into a product is not a general property of 2-D brackets. A simple example where the resolution is not possible is this one: $\langle \rho,\phi|p_x,p_y \rangle$. This is the same pure momentum state but expressed in polar coordinates ρ,ϕ instead of the rectinear

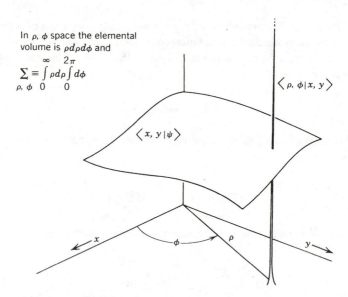

In ρ, ϕ space the elemental volume is $\rho d\rho d\phi$ and

$$\sum_{\rho, \phi} \equiv \int_{0}^{\infty} \rho d\rho \int_{0}^{2\pi} d\phi$$

$\langle \rho, \phi | x, y \rangle$

$\langle x, y | \psi \rangle$

Figure 5.2 **The same state $|\psi\rangle$ in two different spaces. The 2-D delta function in x,y space, $\langle \rho,\phi|x,y\rangle = \delta(x-\rho \cos \phi)\,\delta(y-\rho \sin \phi)$, is a selector function; it selects the value of $\psi(x,y)$ at the point $x = \rho \cos \phi$, $y = \rho \sin \phi$.**

$$\psi(\rho \cos \phi, \rho \sin \phi) = \langle \rho,\phi|\psi\rangle = \int\int_{-\infty}^{+\infty} dx\, dy\, \langle \rho,\phi|x,y\rangle\, \psi(x,y)$$

ones, x and y. The polar coordinates are related to x and y by $x = \rho \cos \phi$ and $y = \rho \sin \phi$ (see Figure 5-2). The bracket $\langle \rho,\phi|p_x,p_y\rangle$ is given by

$$\langle \rho,\phi|p_x,p_y\rangle = \frac{1}{h} \exp i(p_x \cos \phi + p_y \sin \phi)\rho/\hbar \neq \langle \rho|p_x\rangle \langle \phi|p_y\rangle \qquad (5\text{-}26)$$

BRA OR KET PRODUCTS, NO

Multidimensional *wave functions* can sometimes be resolved into products. Multidimensional *states* can not be resolved into products: you may *not* replace the 2-D state $|p_x,p_y\rangle$ with the symbol $|p_x\rangle|p_y\rangle$. Such a ket product symbol is meaningless.

The reason is this: whether or not a resolution into a product may be made depends on the basis space. The product ket symbol implies that a product form is possible in any basis, that the 2-D state, independent of basis, can be resolved into a product. It's not true.

The example demonstrates it. For a pure momentum state the product form is possible in the x,y basis but not in the ρ,ϕ basis. Only the representation in some special bases may be resolved into a product. Thus it is a conceptual distortion to write a product of kets or of bras: only products of brackets make sense.

In the literature of quantum mechanics the product of two kets does appear. In those cases a symbol, such as $|\alpha\rangle|\beta\rangle$, indeed, means that no bases exist which can cause confusion, that the product form is possible in any basis. The spin-state spatial-state products of Chapter 6 (equation 6-9) and of Chapter 9 (equation 9-58) are examples.

SIMULTANEOUS EIGENSTATES SURVIVE THE MEASUREMENT PROCESS

An essential concept concerning multidimensional systems is that of a *simultaneous eigenstate*. The idea is this: that the state you are measuring must survive the measurement process. The state must be simultaneously eigen to *all* the observables measured.

In a one-dimensional system, only one measurement is needed to define the state of the system. In two dimensions, two are needed. We must be sure that the second measurement does not destroy the results of the first. A wrong second measurement won't define the state: it will destroy it.

Consider this example. We measure the x-component of position of our bead-in-a-plane. In so doing we create a state $|x,?\rangle$. Next we choose to measure the energy of the system. We find energy E.

Have we measured a system in the state $|x,E\rangle$? We have not. Can we say that the state of the system is $|x,E\rangle$? No, we cannot. No such state as $|x,E\rangle$ exists. It is impossible to have a state labeled by x and E. The operators $\hat{x}$ and $\hat{H}$ do not have simultaneous eigenstates. A measurement of energy destroys the knowledge of position.

THERE ARE IMPOSSIBLE STATE LABELS

After the energy measurement the state is, indeed, one of energy E. But it can no longer be said that the bead is at position x. The new state is $|E,?\rangle$ not $|x,E\rangle$. The old state of position is destroyed by the energy measurement. The state label $|x,E\rangle$ is an impossible one. There exists no state that survives both a position and an energy measurement.

The proof is a matter of further experiment. Having measured the energy you now measure position again. You do not find the same value of $x!$ The position has not survived the measurement process.

The incompatibility is evident for a measurement pair in one dimension. You measure the momentum of a position state. The state labeled $|x\rangle$ is destroyed (see Chapter 2, "Measurement Destroys a State"). A state labeled $|p\rangle$, not $|x,p\rangle$, is created. The proof is a matter of repeating the position measurement. In a state of momentum $|p\rangle$ the bead has an amplitude to be at any position at all! It is no longer fixed at its original position. Thus states $|p\rangle$ and $|x\rangle$ are possible but $|x,p\rangle$ is impossible.

A 1-D state is completely defined by the one label, x or p, alone. But in two dimensions the matter is more subtle. There are *necessarily* two labels.

COMPATIBLE OPERATORS
HAVE SIMULTANEOUS EIGENSTATES

In two or more dimensions the only way to specify a state is by a series of *compatible* measurements. Compatible measurements are those that don't interfere with one another. A later one doesn't destroy the reliability of data taken in an earlier one. The essential feature of compatible observables is that both measurement results be *simultaneously valid*.

The mathematical statement of this notion is this: the two eigenvalues of the corresponding operators must simultaneously define a state, a simultaneous eigenstate. Such a state survives the series of measurements; later measurements don't invalidate earlier data.

For example, a measurement of position x yields $x = b$. A y-position measurement follows. It yields $y = 2b$. The legitimate state $|x=b, y=2b\rangle$ is defined by these two measurements. The two operators $\hat{x}$ and $\hat{y}$ are compatible. The state carries both the x and the y label. It is simultaneously an eigenstate of both $\hat{x}$ and $\hat{y}$.

The proof resides in repeating the $\hat{x}$ measurement. You get $x = b$ again with certainty.

SIMULTANEOUS MEASUREMENTS:
THERE'S NO SUCH THING

That we must consider which observables are compatible reveals a fundamental feature of the quantum view of nature: we may not assume that two measurements can be made at the same time. All measurements are made sequentially. Quantum mechanics forbids the assumption that two observables may be measured simultaneously. They are measured one after the other. So the measurements had better be compatible if the result is to be meaningful.

It is not really measurement simultaneity that is the important consideration. What is important about a pair of measurements is that you can make them in either order and get the *same result:* that they commute. That's what makes a compatible measurement pair.

To be able to do them in either order is to be able to do them simultaneously. But if you can't reverse the order of measurement, if they are noncommuting measurements, then you certainly cannot expect to perform them simultaneously. Commutability is the essential consideration in thinking about compatible measurements.

FURTHER EXPERIMENT TESTS COMPATIBILITY

The physical test of compatibility resides in further experiment. Suppose you make a measurement. You get result Λ. This is necessarily an eigenvalue of some operator, say $\hat{\lambda}$. The operator, $\hat{\lambda}$, corresponds to the observable measured. Having measured the result $\lambda = \Lambda$, you conclude that the state of the system is $|\Lambda \ldots \rangle$. The dots represent other measurements that must be made to define the state fully: $\lambda = \Lambda$ is only a partial characterization of the state (see Chapter 3, "A State's Partial Label").

To fill out the characterization, you make a second measurement. You measure an observable that corresponds to some other operator, $\hat{\omega}$. You find the result ω.

The compatibility of the two measurements resides in the answer to this question: Is the system characterized by the state $|\Lambda, \omega \rangle$? If the system is such a simultaneous eigenstate of $\hat{\lambda}$ and $\hat{\omega}$, then the second measurement did not destroy the results of the first one; rather, it complemented the first measurement by further defining the state.

To answer the question, you do another experiment. You make a $\hat{\lambda}$ measurement again. If $\hat{\lambda}$ and $\hat{\omega}$ are compatible, you must get $\lambda = \Lambda$ again—with certainty. If you get $\lambda = \Lambda'$, some other eigenvalue of $\hat{\lambda}$, then measurement $\hat{\omega}$ must have destroyed the original state $|\Lambda \ldots \rangle$. It produced a superposition state with many λ's possible, one of which was detected in the second $\hat{\lambda}$ measurement. Only if the second $\hat{\lambda}$ measurement yields the original $\lambda = \Lambda$ with certainty can you deduce that $\hat{\lambda}$ and $\hat{\omega}$ are compatible. The further experiment tests for compatibility.

5.3 COMPATIBLE OPERATORS ALWAYS COMMUTE

Here is the mathematical rendering of the compatibility test. Make the three measurements $\hat{\lambda}$, $\hat{\omega}$, and then $\hat{\lambda}$ again. Do them repeatedly so as to quarantee that the same value of λ is obtained twice: the third measurement must yield the same value of λ as the first, for all λ.

The first measurement yields $|\Lambda \ldots \rangle$. The $\hat{\omega}$ measurement yields the state $|\omega \ldots \rangle$. If this state sustains a λ-value, it must be the same value $\lambda = \Lambda$ (see Chapter 3, "No Measurement Ever Changes the Value of a State Label"). Hence measurement compatibility means that the state is $|\omega \ldots \rangle = |\omega, \Lambda \rangle$.

On this state, a $\hat{\lambda}$ measurement must produce $\lambda = \Lambda$, the physical test result for compatibility.

The set of all $|\omega, \lambda \rangle$ covers the entire physical measurement space. They are a complete set. If the system is, indeed, found in such a state, then the operators $\hat{\omega}$ and $\hat{\lambda}$, which produce the pair of labels ω and λ, will commute.

$$[\hat{\omega}, \hat{\lambda}] = 0 \tag{5-27}$$

Proof:
That (see Problem 3.16)

$$[\hat{\omega}, \hat{\lambda}]|\omega, \lambda \rangle = 0 \tag{5-28}$$

is true for all ω, λ implies that for any $|\psi \rangle$

$$[\hat{\omega}, \hat{\lambda}]|\psi \rangle = 0 \tag{5-29}$$

That this equation is true for any $|\psi \rangle$ is precisely the meaning of Equation 5-27.

EVERY DIMENSION HAS AT LEAST ONE STATE LABEL

The bead-in-a-plane has two degrees of freedom; it is a 2-D system. A state of the bead requires at least two labels in the ket. More than two labels may be used, but the ket must have at least two.

More than two labels are due either to *degeneracy* or to *redundancy*. If an eigenvalue label is degenerate, then a second label must clarify which, among the degenerate states, is under consideration (see Chapter 3, "A Degenerate Eigenvalue Is a State's Partial Label").

The other reason for extra labels is redundancy. The state $|p\rangle$ is evidently equivalent to the state $|p,p^2\rangle$. The value of p being given, to specify p^2 is redundant; the operators $\hat{p}$ and $\hat{p}^2$ are not independent of each other.

Redundant labels in a ket are not useful, but they don't invalidate the state. Redundant state labels are not impossible ones; they are merely extra baggage.

To define completely the state of a 2-D system, there must be at least two labels in the ket. There must be at least N labels in the ket of an N-dimensional system.

DEGENERACY MEANS AN EXTRA LABEL

Suppose the operator $\hat{\lambda}$ produces degenerate eigenvalues. If the eigenvalues are twofold degenerate, there are two states, $|\lambda,1\rangle$ and $|\lambda,2\rangle$, with the same λ. An example is the energy of the 1-D bead-on-a-ring; two different states, $|E,+\rangle = |p=\sqrt{2mE}\rangle$ and $|E,-\rangle = |p=-\sqrt{2mE}\rangle$, have the same energy.

Degeneracy means that there is an extra label needed to define the state: a supplementary measurement is necessary (see Chapter 3, "A Degenerate Eigenvalue Is a State's Partial Label"). The bead-on-a-ring illustrates the idea. Take for $\hat{\lambda}$ the hamiltonian, $\hat{H}$. An energy measurement is not enough to define a state. After the energy measurement, an auxilliary measurement must be made to determine the direction of the motion.

If you insist on characterizing a state by the degenerate eigenvalue, say $\lambda_1 = \lambda_2 = \Lambda$, from the measurement $\hat{\lambda}$, then you must make an auxillary measurement to define the state fully. You make the measurement $\hat{\gamma}$. The extra label, γ, breaks the degeneracy; it enumerates the states with the same λ-value. For example, $\lambda = \Lambda$, $\gamma = 1$ cites the first case of $\lambda = \Lambda$ and $\lambda = \Lambda$, $\gamma = 2$ cites the second. The label λ is degenerate; the label λ,γ is not.

ANOTHER LABEL, ANOTHER OPERATOR

A 2-D state with degenerate λ may sustain three labels, $|\lambda,\gamma,\omega\rangle$.
But if you need three measurements to define a state, all three must be compatible.

The three-label state you want to measure can only be one for which no later measurement destroys earlier data.

The only way that all three measurements can be compatible is if all of the three operators representing these measurements commute with one another. The proof is as follows:

1. Because $\hat{\lambda}$ yields degenerate eigenvalues, another measurement must be made in order to select one from among the several states $|\lambda...\rangle$. Call the operator corresponding to this measurement $\hat{\gamma}$: it produces eigenvalues γ that characterize states $|\lambda,\gamma...\rangle$. Simultaneous eigenstates of both $\hat{\lambda}$ and $\hat{\gamma}$ are necessarily formed because $\hat{\gamma}$ selects one from among the several degenerate states of $\hat{\lambda}$.

2. That simultaneous eigenstates of $\hat{\lambda}$ and $\hat{\gamma}$ exist implies that these operators commute.

3. If a simultaneous eigenstate, $|\lambda,\gamma,\omega\rangle$, exists, then $\hat{\omega}$ commutes both with $\hat{\lambda}$ and with $\hat{\gamma}$. The proof is Problem 3.6.

Thus all three operators will be compatible if they commute. *All of the operators whose eigenvalues label a state must commute.*

FURTHER MEASUREMENT BREAKS THE DEGENERACY

The key feature of degeneracy is that there are auxilliary labels. These arise because degenerate results demand further experiment to define a state (see Chapter 2, "Refer All Matters to Measurement"). The result of a further measurement is catalogued as an auxilliary label in the state ket.

A COMPLETE COMMUTING SET OF OPERATORS: *N* FOR *N* DEGREES OF FREEDOM

If you choose to measure nondegenerate observables, a system with N degrees of freedom will have an N-dimensional ket. Thus, a point in the basis space is defined by N numbers. The basis space is N dimensional. Each dimension is the spectrum of an operator's eigenvalues. A basis point consists of the eigenvalues of N-independent operators, each of which commutes with all the others. A complete ket is labeled by the eigenvalues of N-independent commuting operators. Such a set of operators is called a *complete commuting set of operators*. An N-D state is specified by N eigenvalues, one from each operator belonging to a complete commuting set.

The physical side of the thought is this: N measurements suffice to define an N-dimensional system. A system with N degrees of freedom can be characterized by N measurements, one each of N compatible and independent observables.

DIMENSIONALITY IS CONSERVED

If the original state space is two dimensional, then all other state spaces to which it transforms will also be two dimensional. There exists no situation in which a 3-D state arises from a 2-D one. Two-ness is conserved.

So it is in N-dimensions. Dimensionality is conserved. If it takes N labels to specify a state minimally in one basis, then in any other basis the ket must also be N-dimensional.

PROCEED FROM MEANING

To specify the position of the bead in the 2-D plane requires two measurements. You may measure the x and y position coordinates; you may equally well measure the ρ and ϕ position coordinates.

To describe the momentum fully also demands two measurements. We have discussed the pair p_x and p_y. In classical physics one may equally well imagine measuring p_ρ and p_ϕ, the components of momentum in the ρ and ϕ directions. In quantum mechanics, however, to write the symbol $|p_\rho,p_\phi\rangle$ merely announces confusion; it does not denote a state. This ket is an impossible one.

The confusion will be dissolved if you pay careful attention to meaning. By the ket $|p_\rho,p_\phi\rangle$ one must mean the ket $|p_\rho,p_\phi,\phi\rangle$. If p_ρ and p_ϕ were measured, so necessarily must have been the angular position of the bead; else how could the $\mathbf{e}_\rho$ and $\mathbf{e}_\phi$ directions have been known? The momentum pair p_ρ and p_ϕ have meaning only if ϕ is first specified. Without having ϕ, neither p_ρ nor p_ϕ makes sense. To write the bivariable ket $|p_\rho,p_\phi\rangle$ is to practice self-deception, a third label belonging in the ket has been omitted.

But no such three-label ket exists. Three nondegenerate measurements are too many to label a 2-D state. The pair ϕ and p_ϕ do not commute. The ket $|\phi,p_\rho\rangle$ does represent a state, but not $|p_\rho,p_\phi\rangle$, nor $|\rho,p_\rho\rangle$, nor $|\rho,p_\phi\rangle$. These last three are antinomies, they express only an internal contradiction. The angle ϕ has been omitted. It belongs in these kets but, being there, destroys them. By contrast, the kets $|x,p_y\rangle$, $|p_x,p_y\rangle$, and $|y,p_x\rangle$ are legitimate states: the operators corresponding to each of the label-pairs are independent; they commute (Equation 5-23); and there are no hidden labels (see Problem 5.7).

The essential difference between these two cases is in the coupling between measurement spaces. The characterization of the momentum $\mathbf{p}$ by p_x and p_y takes place entirely in the space of momentum. It is independent of position space. The characterization of $\mathbf{p}$ by its components p_ρ and p_ϕ depends intimately on the position-space location of the bead. It is not a characterization purely in momentum space, it is not independent of position (see Figure 5-3).

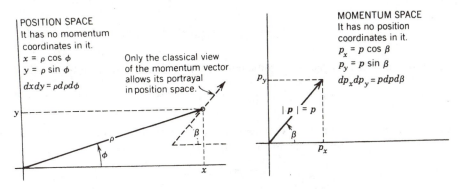

Figure 5.3 **The decoupled spaces of position and momentum.**

MOMENTUM STATES ARE IN MOMENTUM SPACE

To characterize a momentum measurement properly, one must enter a *pure momentum space*, it should be decoupled from position space. In the figure the angle between the momentum vector **p** and the fixed horizontal axis is called β. And the magnitude of the momentum vector is $|\mathbf{p}| = p$. Like p_x and p_y, the measurements of β and p are independent of what value ϕ, the position angle, takes. The state $|p,\beta\rangle$ is a valid one, the measurements have no hidden dependence on position.

The ket $|p,\beta\rangle$ represents a pure momentum state, one of fixed momentum **p**. It represents the same pure momentum state as $|p_x,p_y\rangle$ where $p_x = p \cos\beta$ and $p_y = p \sin\beta$. Thus the amplitude $\langle\rho,\phi|p,\beta\rangle$ refers, in a different language, to precisely the same event as does $\langle x,y|p_x,p_y\rangle$. For a given event the amplitude is a certain quantity, no matter in what language you describe it; the two brackets are, therefore, equal. Using the result in Equation 5-6 for $\langle x,y|p_x,p_y\rangle$ we deduce that

$$\langle\rho,\phi|p,\beta\rangle = \frac{1}{h} \exp\left[i\,\frac{\rho p}{\hbar} \cos(\beta - \phi)\right] \tag{5-30}$$

Both the language p_x,p_y and the language (basis) p,β are suitable for describing the state in momentum space. There is a third way. It is the language p,L_z.

MOMENTUM ANGLE OR ANGULAR MOMENTUM: EITHER FORMS A BASIS

The operator $\hat{L}_z$ produces the eigenvalues L_z. It is related to the operator $\hat{\beta}$, which has as its eigenvalues the angular orientation, β, of the momentum vector. The p,β-space instruction for $\hat{L}_z$ is

$$\langle p,\beta| \; \hat{L}_z \; = \; -i\hbar \frac{\partial}{\partial\beta} \langle p,\beta| \tag{5-31}$$

It follows immediately that its commutator with $\hat{\beta}$ is given by (see Problem 5.8):

$$[\hat{\beta},\hat{L}_z] = i\hbar \tag{5-32}$$

Like $\hat{\beta}$ and $\hat{p}$, the operator $\hat{L}_z$ is a physical, one: it corresponds to an observable that one measures, the angular momentum of the 2-D bead.

5.4 WHAT PRODUCES ANGULAR MOMENTUM IS THE ANGULAR MOMENTUM OPERATOR

Let $\hat{L}_z$ operate on a position state of the bead locating it at ρ,ϕ. In momentum space $\hat{L}_z$, acting on this state, produces $\rho p \sin(\beta - \phi)$ (see Problem 5.9):

$$\langle p,\beta| \; \hat{L}_z \; |\rho,\phi\rangle = \rho p \sin(\beta - \phi) \langle p,\beta|\rho,\phi\rangle \tag{5-33}$$

But $\rho p \sin(\beta - \phi)$ is just the angular momentum, $(\mathbf{r} \times \mathbf{p})_z$, in the x-y plane. Because $\hat{L}_z$ produces $(\mathbf{r} \times \mathbf{p})_z = xp_y - yp_x$ in one language, it produces it in all of them. The operator $\hat{L}_z$ has exactly the same effect as does the operator $\hat{x}\hat{p}_y - \hat{y}\hat{p}_x$ in any space at all. Thus $\hat{L}_z$ qualifies as the angular momentum operator (see Problems 5.10 and 5.11).

OPERATOR RELATIONS ARE BASIS INDEPENDENT

Because we can recast Equation 5-33 as

$$\langle p,\beta| \; \hat{L}_z|\rho,\phi\rangle = \langle p,\beta \; |\hat{x}\hat{p}_y - \hat{y}\hat{p}_x|\rho,\phi\rangle \tag{5-34}$$

it follows that

$$\hat{L}_z = \hat{x}\hat{p}_y - \hat{y}\hat{p}_x \tag{5-35}$$

The proof is a matter of meaning. An operator relation means that the equality holds between any pair of states; Equation 5-35 means that for any pair of states Ψ and S

$$\langle\Psi| \; \hat{L}_z|S\rangle = \langle\Psi| \; \hat{x}\hat{p}_y - \hat{y}\hat{p}_x \; | \; S\rangle \tag{5-36}$$

But use a p,β ket-bra sum to describe $\langle\Psi|$ and a ρ,ϕ ket-bra sum to describe $|S\rangle$. That (5-34) is true makes (5-36) true, and hence (5-35) is true.

Given an operator instruction in any one language, recast it so that it has purely an operator form, as in (5-34). Then the operator relationship will be true in all languages. This application illustrates the profound significance of the results in Problem 4.20 and of the section in Chapter 4, "Relationships between Observables Are Connections between Operators." Operator relationships are language independent.

THE ARCHETYPICAL MECHANICAL ROTATOR IS A BEAD-IN-A-SPHERICAL-SHELL

Reconsider the bead-on-a-wire of Chapters 2 and 3. Suppose the circumference of the track were not constrained to be very large, and suppose the track circle were free to orient its plane in any direction. This amounts to a new system: a bead-in-a-spherical-shell. The bead can now move anywhere in a spherical surface at a fixed distance, R, from the origin.

For the bead-on-a-wire the track radius, $R = L/2\pi$, was kept large so as to avoid angular momentum considerations. Here we want expressly to focus on angular momentum. The present bead has only angular momentum.

In the classical picture the bead is at position $\mathbf{r}$ moving with momentum $\mathbf{p}$ in some direction. The constraint that the bead is confined to the shell may be put mathematically as

$$\mathbf{r} \cdot \mathbf{r} = R^2 \tag{5-37}$$

Position space is the two-dimensional one labeled by θ and ϕ. The coordinate r is not a variable; $r = R$ (see Figure 5-4).

The classical bead has a vector angular momentum about the origin of amount

$$\mathbf{L} = \mathbf{r} \times \mathbf{p} \tag{5-38}$$

The component of this vector in the z-direction is

$$L_z = (\mathbf{r} \times \mathbf{p})_z = xp_y - yp_x \tag{5-39}$$

The components in the other directions are, of course, just the two cyclical permutations of (5-39). Replacing, say x by y, y by z, and z by x yields the formula for L_x.

There is no potential energy in this system, only kinetic energy. But there is a momentum constraint; the bead can't move off the shell. The momentum vector $\mathbf{p}$ is always perpendicular to $\mathbf{r}$. Thus

$$\mathbf{r} \cdot \mathbf{p} = 0 \tag{5-40}$$

This constraint on the momentum means that only two of the three components are independent. The momentum space for this system is intrinsically two dimensional.

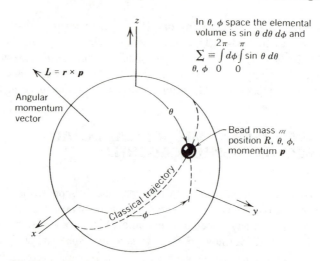

Figure 5.4 The bead in a spherical shell. It's confined to positions at a fixed distance R from the origin.

In view of 5-37 and (5-40), the magnitude of the angular momentum, $|\mathbf{L}| = L$, is directly related to the total momentum magnitude by

$$L^2 = (\mathbf{r} \times \mathbf{p}) \cdot (\mathbf{r} \times \mathbf{p}) = R^2 p^2 \qquad (5\text{-}41)$$

It follows that the hamiltonian for the bead-in-a-shell is

$$H = \frac{p^2}{2m} = \frac{L^2}{2mR^2} = \frac{L^2}{2I} \qquad (5\text{-}42)$$

The energy is purely rotational. The bead-in-a-spherical-shell is the archetypical mechanical rotator. It has moment of inertia $I = mR^2$.

The classical picture is that the bead may wander anywhere on the sphere surface. At each instant of time it is at a position characterized by two variables: θ and ϕ. Simultaneously it has an angular momentum vector $\mathbf{L}$. Its energy resides entirely in the magnitude of its angular momentum. Classically the bead may have any energy and any angular momentum in any direction.

2-D STATES COME FROM TWO SIMULTANEOUS EIGENVALUE PROBLEMS

The quantum view follows from these considerations. In an experiment to locate the position of the bead the spectrum of possible results we could get defines a basis space: the 2-D space of θ and ϕ. A position state of the bead is $|\theta,\phi\rangle$.

There must be two operators, $\hat{\theta}$ and $\hat{\phi}$, whose eigenvalues are the spectrum of measured θ, ϕ values. The possible experimental findings for θ and ϕ are the eigenvalues of these operators. A state $|\theta, \phi\rangle$ is a simultaneous solution of the two eigenvalue problems:

$$\hat{\theta}|\theta, \phi\rangle = \theta|\theta, \phi\rangle \tag{5-43}$$

and

$$\hat{\phi}|\theta, \phi\rangle = \phi|\theta, \phi\rangle \tag{5-44}$$

5.5 Angular Momentum: Its Operators and Eigenvalues

Another space for this system is that of angular momentum. Its components are observables. They can be measured. Thus there must be angular momentum state labels and angular momentum operators.

The angular momentum operators are the classical angular momenta turned into operators. Thus, in addition to

$$\hat{L}_z = \hat{x}\hat{p}_y - \hat{y}\hat{p}_x \tag{5-45}$$

we also fabricate from $\mathbf{L} = \mathbf{r} \times \mathbf{p}$ the operators

$$\hat{L}_x = \hat{y}\hat{p}_z - \hat{z}\hat{p}_y \quad \text{and} \quad \hat{L}_y = \hat{z}\hat{p}_x - \hat{x}\hat{p}_z \tag{5-46}$$

Because of its physical significance and its mathematical convenience we create another operator; $\hat{L}^2$. It comes from the three, $\hat{L}_x$, $\hat{L}_y$, and $\hat{L}_z$, already manufactured. The definition of $\hat{L}^2$ is

$$\hat{L}^2 \equiv \hat{L}_x^2 + \hat{L}_y^2 + \hat{L}_z^2 \tag{5-47}$$

The eigenvalues of $\hat{L}^2$ constitute the measurement spectrum of the square of the total angular momentum.

Because dimensionality is conserved, the space created by the angular momentum operators must be a two-dimensional one. No matter in what language the system is described, its number of coordinates cannot change. The spatial state, $|\theta, \phi\rangle$ is two dimensional; so must be the angular momentum space.

This means that it is impossible to form a state that is simultaneously eigen to more than two of the angular momentum operators. Of the four operators, $\hat{L}_x$, $\hat{L}_y$, $\hat{L}_z$ and $\hat{L}^2$, only two can characterize a state. All four cannot commute.

In fact, no two of the three components commute with one another. From the definitions in (5-45) and (5-46), it is a straightforward matter to show that (see Problem 5.16)

$$[\hat{L}_x, \hat{L}_y] = i\hbar\hat{L}_z \quad \text{(all cyclical permutations)} \tag{5-48}$$

So no simultaneous eigenstate of $\hat{L}_x$, and $\hat{L}_y$, exists; they don't commute.

But all three components do commute with $\hat{L}^2$ (see Problem 5.17).

$$[\hat{L}_x, \hat{L}^2] = [\hat{L}_y, \hat{L}^2] = [\hat{L}_z, \hat{L}^2] = 0 \qquad (5\text{-}49)$$

Thus, of the two needed to characterize a state, one must be $\hat{L}^2$. A simultaneous eigenstate of $\hat{L}^2$ and $\hat{L}_z$ exists. It is this basis that we will use; that of the simultaneous eigenstates of $\hat{L}^2$ and $\hat{L}_z$.

THERE ARE RAISING AND LOWERING OPERATORS FOR ANGULAR MOMENTUM EIGENSTATES

We seek the eigenvalues of $\hat{L}^2$ and $\hat{L}_z$. Call these $\lambda\hbar^2$ and $m\hbar$ respectively. We don't assume here that m is an integer index like what appeared for the bead-in-a-plane. Because $\hbar$ has the dimensions of angular momentum, this way of writing the eigenvalues merely assures us that λ and m are dimensionless quantities.

The object is to find the λ's and m's where

$$\hat{L}^2|\lambda,m\rangle = \lambda\hbar^2|\lambda,m\rangle \qquad (5\text{-}50)$$

and

$$\hat{L}_z|\lambda,m\rangle = m\hbar|\lambda,m\rangle \qquad (5\text{-}51)$$

Two very important operators, useful in finding λ and m, are the raising and lowering operators, $\hat{L}_+$ and $\hat{L}_-$. They are not physical operators. They are the analogue of the creation and annihilation operators of the harmonic oscillator. They are defined in terms of $\hat{L}_x$ and $\hat{L}_y$.

$$\hat{L}_- = \hat{L}_x - i\hat{L}_y \qquad (5\text{-}52)$$

and

$$(\hat{L}_-)^\dagger = \hat{L}_+ = \hat{L}_x + i\hat{L}_y \qquad (5\text{-}53)$$

The key relations among the angular momentum operators are those of Equations 5-48 and 5-49. We may frame these using the set $\hat{L}^2$, $\hat{L}_z$, $\hat{L}_+$, and $\hat{L}_-$ instead of the set $\hat{L}^2$, $\hat{L}_z$, $\hat{L}_x$, and $\hat{L}_y$. We discover that

$$[\hat{L}_-, \hat{L}_z] = \hbar\hat{L}_- \qquad (5\text{-}54)$$

and

$$[\hat{L}_z, \hat{L}_+] = \hbar\hat{L}_+ \qquad (5\text{-}55)$$

Of course because it commutes with $\hat{L}_x$ and $\hat{L}_y$, $\hat{L}^2$ also commutes with $\hat{L}_+$ and $\hat{L}_-$ (see Problem 5.17).

If we operate with both sides of (5-54) on the state $|\lambda,m\rangle$, we can see that $\hat{L}_-$ is, indeed, the lowering operator. The operation yields

$$\hat{L}_z\,(\hat{L}_-|\lambda,m\rangle) = (m-1)\hbar\,(\hat{L}_-|\lambda,m\rangle) \qquad (5\text{-}56)$$

The form of this equation is its substance: it has the eigenvalue problem form. It says that the new state, $\hat{L}_-|\lambda,m\rangle$, is itself an eigenstate of $\hat{L}_z$, one with eigenvalue $(m-1)\hbar$. And λ is not changed. Because $\hat{L}_-$ and $\hat{L}^2$ commute, they have simultaneous eigenstates. $\hat{L}_-$ does not change an eigenstate, $|\lambda \dots\rangle$, of $\hat{L}^2$. Thus the ket that is $\hat{L}_-|\lambda,m\rangle$ must be labeled by λ and by $m-1$.

Put mathematically, the implication of (5-56) is

$$\hat{L}_-|\lambda,m\rangle = c_m\hbar|\lambda,m-1\rangle \qquad (5\text{-}57)$$

This equation says that $\hat{L}_-$ is an m-lowering operator. When it operates on a state, $|\lambda,m\rangle$, it produces a new state of the same λ but with the m eigenvalue lowered by one, from m to $m-1$. The dimensionless multiplier c_m remains to be determined.

Operating with (5-55) on the state $|\lambda,m\rangle$ and following the same line of reasoning easily demonstrates that $\hat{L}_+$ raises m by 1. When it operates on a state $|\lambda,m\rangle$, a new state $|\lambda,m+1\rangle$ is produced. Because the operator $\hat{L}_+$ is the hermitian conjugate of $\hat{L}_-$, the constant that arises is related to c_m. In Problem 5.19 you will find that

$$\hat{L}_+|\lambda,m\rangle = c^*_{m+1}\hbar|\lambda,m+1\rangle \qquad (5\text{-}58)$$

THE ANGULAR MOMENTUM HAS INDICES

We are now ready to discover the m and the λ. The process takes three steps:

1. There must be a maximum m. We'll call it $\ell = m_{max}$. There must also be an m_{min}. The reason is that $\lambda\hbar^2$ represents the eigenvalue, held fixed, of $\hat{L}^2$. This limits how large an eigenvalue of $\hat{L}_z$ can be.

Two compatible measurements determine the state. The measurement corresponding to $\hat{L}^2$ has yielded a value $\lambda\hbar^2$. Now the measurement corresponding to $\hat{L}_z$ is made. A value $m\hbar$ is found. And the state is a simultaneous one: the second measurement has not destroyed the state left by the first measurement; the value of λ persists. We cannot have a larger value for a mere component of the angular momentum than for the total angular momentum. Rather, we must have that $(m\hbar)^2 \leq \lambda\hbar^2$. Thus, for fixed λ, m is bounded; there is an m_{max} and m_{min}.

Formally put, the effect of $\hat{L}_x^2 + \hat{L}_y^2$ can never be negative:

$$0 \leqslant \langle \lambda,m | \hat{L}_x^2 + \hat{L}_y^2 | \lambda,m \rangle = \langle \lambda,m | \hat{L}^2 - \hat{L}_z^2 | \lambda,m \rangle = \hbar^2(\lambda - m^2)$$

2. The eigenvalues, m, are discrete. They consist of a ladder of unit steps. The highest rung of the ladder is $m_{max} = \ell$; the next lowest eigenvalue is $m = \ell - 1$. The next lowest is $m = \ell - 2$, and this continues on down to $m = m_{min}$.

$$m_{max} = \ell = m_{min} + \text{some integer } N \tag{5-59}$$

Here's the argument. It's just parallel to that given in Chapter 4, "The Harmonic Oscillator Energy Spectrum Is $(n + 1/2)\hbar\omega$."

Consider the new state produced by the lowering operator, $\hat{L}_-$, on the highest state $|\lambda,\ell\rangle$. This new state has eigenvalue $m = \ell - 1$ and belongs to the spectrum of eigenstates of $\hat{L}_z$. That's the content of Equation 5-57.

But this must be the next lowest state of the operator $\hat{L}_z$. There can be no state with an eigenvalue m' between ℓ and $\ell - 1$. The reason is this: the effect of the raising operator, $\hat{L}_+$, operating on $|\lambda,m'\rangle$, would produce an eigenstate of the system with m greater than $m_{max} = \ell$! This is impossible; so no state with m' between ℓ and $\ell - 1$ exists.

Thus the highest two states have m-values of ℓ and $\ell - 1$. By extension of this reasoning, the next lower state is found to have $m = \ell - 2$, and the next $\ell - 3$, and so forth.

The lowering cannot go on indefinitely: m cannot go to minus infinity. The process stops at some $m = m_{min}$. This minimum m must lie an integral number, N, of steps below the maximum m. That's what Equation 5-59 says.

3. We still need to find out what ℓ is and how it relates to λ. These are two unknowns. They require two equations, two physical statements. These are (a) that the minimum state cannot be lowered further and (b) that the maximum state cannot be raised further. Thus

$$\hat{L}_-|\lambda,m_{min}\rangle = 0 \tag{5-60}$$

and

$$\hat{L}_+|\lambda,m_{max}\rangle = 0 \tag{5-61}$$

To extract conclusions from these statements we need to be aware of two alternative formulas for the operator $\hat{L}^2$. Equation 5-47 defines $\hat{L}^2$. It can be recast in terms of $\hat{L}_+$ and $\hat{L}_-$ instead of $\hat{L}_x$ and $\hat{L}_y$. There are two ways to do it (see Problem 5.20):

$$\hat{L}^2 = \hat{L}_+\hat{L}_- - \hbar\hat{L}_z + \hat{L}_z^2 \tag{5-62}$$

and

$$\hat{L}^2 = \hat{L}_-\hat{L}_+ + \hbar\hat{L}_z + \hat{L}_z^2 \tag{5-63}$$

Now apply (5-62) to the ket $|\lambda,m_{min}\rangle$, keeping (5-60) in mind. Apply (5-63) to the ket $|\lambda,m_{max}\rangle$, taking note of (5-61). Using (5-59), the result is (see Problem 5.21)

$$-(\ell - N) + (\ell - N)^2 = \lambda = \ell + \ell^2 \qquad (5\text{-}64)$$

There are two conclusions to be drawn from these equations.

The first is that the eigenvalues of the $\hat{L}^2$ operator are related to ℓ, the maximum m-value, by $\lambda = \ell(\ell + 1)$. We no longer need λ. The label for an eigenstate of $\hat{L}^2$ is more conveniently taken as ℓ. It is the angular momentum index: the square of the angular momentum is restricted to the values $\ell(\ell + 1)\hbar^2$.

The second conclusion is that ℓ must be one of the nonnegative half-integers

$$\ell = 0, \tfrac{1}{2}, 1, \tfrac{3}{2}, \ldots \qquad (5\text{-}65)$$

These comprise the allowed spectrum of ℓ-values. This fact is a straightforward deduction from (5-64). Equating the extremes proves it because N is a nonnegative integer.

That concludes the three steps to discover λ and m. Here is a summary of the results:

$$\hat{L}^2|\ell,m\rangle = \ell(\ell+1)\hbar^2|\ell,m\rangle \qquad (5\text{-}66)$$

$$\hat{L}_z|\ell,m\rangle = m\hbar|\ell,m\rangle \qquad (5\text{-}67)$$

where 2ℓ is a nonnegative integer and m runs from minus ℓ to plus ℓ in integer steps

$$m = -\ell, -\ell+1, -\ell+2, \ldots \ell-1, \ell \qquad (5\text{-}68)$$

For example, if $\ell = 2$, then the possible values of m are $-2, -1, 0, +1$, and $+2$. If $\ell = 1/2$, the possible values of m are $-1/2$ and $+1/2$. Only two eigenstates of $\hat{L}_z$ are possible for the $\ell = 1/2$ case. There are five for the $\ell = 2$ case.

THE VECTOR MODEL OF ANGULAR MOMENTUM: A PICTORIAL DEVICE

There is a pictorial device for visualizing the results. In the scheme called the Vector Model of Angular Momentum, we represent the angular momentum by a vector of length $\sqrt{\ell(\ell + 1)}$. That the z-component is quantized means that there are only select angles that this vector can make with the z-axis. It may point only in directions such that its projection onto the z-axis falls into integer steps. We envision the angular momentum vector as sweeping out all possible directions in the x-y plane. It does so keeping a fixed z-component.

This scheme is a mnemonic device. It's the visual display of Equations 5-65, 5-66, 5-67, and 5-68. Figure 5-5 pictures the state $|\ell=2, m=-1\rangle$ as an angular momentum of magnitude $\sqrt{2(2 + 1)}\,\hbar$ whose z-component, $-\hbar$, is one of the five possible ones shown.

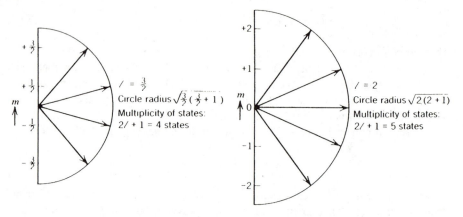

Figure 5.5 Two vector model diagrams.

This model illustrates why m is bounded: the z-component is never greater in magnitude than the total angular momentum.

5.6 Raising and Lowering Operator Instructions

We need only one more quantity to complete our entry into angular momentum space. We need the c_m of Equations 5-57 and 5-58. It's an easy matter to get $|c_m|^2$. Apply (5-62) to a state $|\ell,m\rangle$ noting the effect of the operators $\hat{L}$ (Equation 5-57), $\hat{L}_1$ (Equation 5-58), $\hat{L}^2$ (Equation 5-66), and $\hat{L}_z$ (Equation 5-67). The result is

$$\ell(\ell+1) = |c_m|^2 + m(m-1) \tag{5-69}$$

Choosing c_m to be real is the traditional and most convenient choice for the arbitrary phase factor. With this choice the effects of the lowering and raising operators are

$$\hat{L}\,|\ell,m\rangle = \hbar\sqrt{\ell(\ell+1) - m(m-1)}\,|\ell,m-1\rangle \tag{5-70}$$

and

$$\hat{L}_1|\ell,m\rangle = \hbar\sqrt{\ell(\ell+1) - m(m+1)}\,|\ell,m+1\rangle \tag{5-71}$$

HAVING SOLVED THE PROBLEM, MARK THE RESULTS

The special relations that characterize the angular momentum operators are these (Equations 5-48 and 5-49):

$$[\hat{L}_x, \hat{L}_y] = i\hbar\hat{L}_z \qquad \text{(all cyclical permutations)} \tag{5-72}$$

and

$$[\hat{L}_x, \hat{L}^2] = [\hat{L}_y, \hat{L}^2] = [\hat{L}_z, \hat{L}^2] = 0 \qquad (5\text{-}73)$$

Eigenstates, $|\ell,m\rangle$, can be found that are simultaneous for the operator pair $\hat{L}^2$ and $\hat{L}_z$ where

$$\hat{L}^2|\ell,m\rangle = \ell(\ell+1)\hbar^2|\ell,m\rangle \qquad (5\text{-}74)$$

and

$$\hat{L}_z|\ell,m\rangle = m\hbar|\ell,m\rangle \qquad (5\text{-}75)$$

and m runs in integer steps from $-\ell$ to $+\ell$. For these eigenstates there are raising and lowering operators $\hat{L}_+$ and $\hat{L}_-$ where

$$\hat{L}_\pm |\ell,m\rangle = \hbar\sqrt{\ell(\ell+1) - m(m\pm1)} \, |\ell,m\pm1\rangle \qquad (5\text{-}76)$$

5.7 ORBITAL ANGULAR MOMENTUM STATES HAVE A POSITION-SPACE REPRESENTATION

A position state of our bead-in-a-spherical-shell is $|\theta,\phi\rangle$. An angular momentum state is $|\ell,m\rangle$. We want to discover the transformation matrix elements, $\langle\theta,\phi|\ell,m\rangle$, between these. A general state $|\Psi\rangle$ can be discussed in either language. The transformation matrix elements enable us to speak both languages.

The dictionary entries $\langle\theta,\phi|\ell,m\rangle$, themselves have physical significance. Each is the amplitude for an event, to find the bead at position θ,ϕ when it is known to be in a state of angular momentum indexed by ℓ and m.

CAST THE EIGENVALUE PROBLEM IN THE LANGUAGE OF YOUR CHOICE

The generic procedure for getting transformation matrix elements is to cast the eigenvalue problem into the language you wish to speak. In the present case it works like this. Cast (5-74) and (5-75) in the language of θ and ϕ. Thus

$$\langle\theta,\phi|\hat{L}^2|\ell,m\rangle = \ell(\ell+1)\hbar^2 \langle\theta,\phi|\ell,m\rangle \qquad (5\text{-}77)$$

$$\langle\theta,\phi|\hat{L}_z|\ell,m\rangle = m\hbar \langle\theta,\phi|\ell,m\rangle \qquad (5\text{-}78)$$

To execute the operator instructions, we need the forms of the operators $\hat{L}_z$ and $\hat{L}^2$ in the language θ,ϕ. We need expressions for $\langle\theta,\phi|\hat{L}^2$ and $\langle\theta,\phi|\hat{L}_z$. Together with those for $\langle\theta,\phi|\hat{L}_+$ and $\langle\theta,\phi|\hat{L}_-$, they are

$$\langle\theta,\phi|\hat{L}^2 = -\frac{\hbar^2}{\sin^2\theta}\left[\sin\theta\frac{\partial}{\partial\theta}\left(\sin\theta\frac{\partial}{\partial\theta}\right)+\frac{\partial^2}{\partial\phi^2}\right]\langle\theta,\phi| \tag{5-79}$$

$$\langle\theta,\phi|\hat{L}_z = -i\hbar\frac{\partial}{\partial\phi}\langle\theta,\phi| \tag{5-80}$$

$$\langle\theta,\phi|\hat{L}_- = \hbar e^{-i\phi}\left[-\frac{\partial}{\partial\theta}+i\,ctn\,\theta\frac{\partial}{\partial\phi}\right]\langle\theta,\phi| \tag{5-81}$$

$$\langle\theta,\phi|\hat{L}_+ = \hbar e^{i\phi}\left[\frac{\partial}{\partial\theta}+i\,ctn\,\theta\frac{\partial}{\partial\phi}\right]\langle\theta,\phi| \tag{5-82}$$

These expressions come from a coordinate transformation. The execution is tedious, but the idea is simple. The four operators all are well defined in x, y, z space through Equations 5-45 and 5-46. To get them in the new space is merely a matter of coordinate transformation, from x, y, z, to $r = R$, θ, and ϕ. The procedure is precisely parallel to that in Problem 5.12. You find the expressions for x, $\partial/\partial x$, and so forth, in terms of θ, $\partial/\partial\theta$, and the like, and with them, cast the operators $\hat{x}$, $\hat{p}_x$ and so on, and thus $\hat{L}_z$, and so on, into r, θ, ϕ space. Problems 5.22 and 5.23 are guides through the process (see Figure 5-6).

Notice that the variable r does not appear, only the two dimensions θ and ϕ. Angular momentum is not related to r.

One differential equation that governs the Dirac bracket $\langle\theta,\phi|\ell,m\rangle$ comes from using the instruction (5-80) in (5-78). It is

$$-i\frac{\partial}{\partial\phi}\langle\theta,\phi|\ell,m\rangle = m\langle\theta,\phi|\ell,m\rangle \tag{5-83}$$

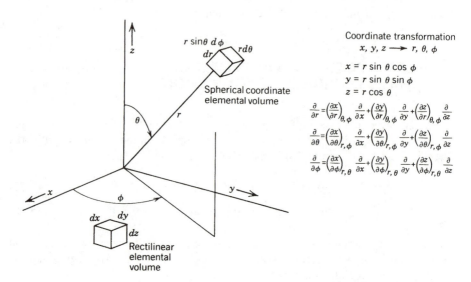

Figure 5.6 Rectilinear–spherical coordinate transformation.

The other differential equation that governs this same bracket comes from the explicit representation of (5-77) via (5-79). It is

$$-\frac{1}{\sin\theta}\frac{\partial}{\partial\theta}\left(\sin\theta\,\frac{\partial\langle\theta,\phi|\ell,m\rangle}{\partial\theta}\right) - \frac{1}{\sin^2\theta}\frac{\partial^2\langle\theta,\phi|\ell,m\rangle}{\partial\phi^2} = \ell(\ell+1)\,\langle\theta,\phi|\ell,m\rangle \quad (5\text{-}84)$$

These two differential equations are the specific implementation of (5-77) and (5-78). They are the eigenvalue problems cast into θ and ϕ space. The transformation matrix elements, $\langle\theta,\phi|\ell,m\rangle$, are the simultaneous solutions of these two equations.

PHYSICAL POSITION SOLUTIONS, $\langle\theta,\phi|\ell,m\rangle$ ALLOW ONLY INTEGER ℓ AND m

The bracket $\langle\theta,\phi|\ell,m\rangle$ is the amplitude to find the bead at physical position θ, ϕ. Thus, the bracket must be continuous, single valued, and square integrable in θ, ϕ space.

The condition of single valuedness is imposed via the geometry of the shell pictured in Figure 5-4. The coordinates θ, $2\pi + \phi$ refer to the same point in physical space as do θ,ϕ. Similarly, the position $-\theta$, ϕ is the same as θ, $\phi + \pi$.

Hence the three brackets $\langle-\theta,\phi+\pi|\ell,m\rangle$, $\langle\theta,\phi+2\pi|\ell,m\rangle$ and $\langle\theta,\phi|\ell,m\rangle$ all must have the same value: they refer to the same physical event.

Now consider the solution to (5-83). It is

$$\langle\theta,\phi|\ell,m\rangle = e^{im\phi} \times \text{some function of } \theta \quad (5\text{-}85)$$

For this to be a physical solution m must be an integer. That's the only way that the value of the bracket at $\phi + 2\pi$ is indistinguishable from its value at ϕ.

Integer m's can come only from integer ℓ's (see Equation 5-68). Thus only integer ℓ's characterize physical position solutions for angular momentum. Half-integer ℓ's cannot represent the angular momentum of an orbiting particle. Only integer ℓ's describe *orbital* angular momentum.

The energy of the ideal mechanical rotator is governed by its orbital angular momentum. Equation 5-42 shows that the eigenvalues of the hamiltonian are effectively those of $\hat{L}^2$. Because orbital angular momentum is labeled only by integer ℓ values, the energy levels of the ideal mechanical rotator are those shown in Figure 5-7.

The next step is to find the yet unknown "function of θ." Using the ϕ dependence deduced in (5-85), the differential equation of (5-84) can be recast in terms of only the one variable θ.

$$-\frac{1}{\sin\theta}\frac{\partial}{\partial\theta}\left(\sin\theta\,\frac{\partial}{\partial\theta}\langle\theta,\phi|\ell,m\rangle\right) + \frac{m^2}{\sin^2\theta}\,\langle\theta,\phi|\ell,m\rangle = \ell(\ell+1)\,\langle\theta,\phi|\ell,m\rangle \quad (5\text{-}86)$$

This equation is a familiar one in the literature of mathematics. It is satisfied by the associated Legendre function, $P_\ell^m\,(\cos\theta)$. This function, the equations it satisfies, and

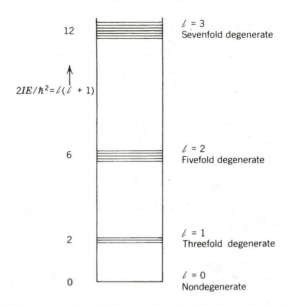

Figure 5.7 The energy-level diagram for the ideal mechanical rotator.

its other properties are discussed in the six books listed in Chapter 4. Legendre polynomials are like the trigonometric functions, exponentials, and the Hermite polynomials: functions tabulated because the differential equations that they solve arise so often in mathematics, physics, and engineering.

A Legendre polynomial is the "function of θ" in (5-85) that solves (5-86). Using P_ℓ^m (cos θ) to construct a properly normalized simultaneous solution to (5-83) and (5-84) results in yet other tabulated functions, $Y_{\ell,m}(\theta,\phi)$. These are called the *spherical harmonic functions*. They are simply normalized products of exp $im\phi$ with P_ℓ^m (cos θ).

These functions characterize the normal mode standing waves of vibration of a sphere shell: the fundamental and all the harmonics of a spherical drumhead. They are the 2-D analogue on a sphere of 1-D sines and cosines. The latter are the *harmonic functions* for a line. Spherical harmonic functions are those for a sphere shell.

$$\langle \theta,\phi|\ell,m\rangle = Y_{\ell,m}(\theta,\phi) \tag{5-87}$$

The first few of the $Y_{\ell,m}(\theta,\phi)$ are these:

$$\langle \theta,\phi|\ell=0,m=0\rangle = Y_{0,0}(\theta,\phi) = 1/\sqrt{4\pi} \tag{5-88}$$

$$\langle \theta,\phi|1,0\rangle = Y_{1,0}(\theta,\phi) = \sqrt{3/4\pi}\ \cos\theta \tag{5-89}$$

$$\langle \theta,\phi|1,\pm1\rangle = Y_{1,\pm1} = \mp\sqrt{3/8\pi}\ \sin\theta\ \exp(\pm i\phi) \tag{5-90}$$

$$\langle \theta,\phi|2,0\rangle = Y_{2,0} = \sqrt{5/16\pi}\ (3\cos^2\theta - 1) \tag{5-91}$$

$$\langle \theta, \phi | 2, \pm 1 \rangle = Y_{2, \pm 1} = \mp \sqrt{15/8\pi} \, \cos \theta \, \sin \theta \, \exp(\pm i\phi) \qquad (5\text{-}92)$$

$$\langle \theta, \phi | 2, \pm 2 \rangle = Y_{2, \pm 2} = \sqrt{15/32\pi} \, \sin^2 \theta \, \exp(\pm 2i\phi) \qquad (5\text{-}93)$$

ELEMENTAL VOLUMES ARE AREAS
IN 2-D BASIS SPACES

The elemental volume of θ, ϕ space is $\sin \theta \, d\theta \, d\phi$. Hence, the meaning assigned to the *sum over θ, ϕ space* is

$$\sum_{\theta, \, \phi} \equiv \int_0^{2\pi} d\phi \int_0^{\pi} d\theta \, \sin \theta \qquad (5\text{-}94)$$

Because the differential area on a sphere of radius R is $R^2 \sin \theta \, d\theta \, d\phi$, the solid angle part, $d\Omega = \sin \theta \, d\theta \, d\phi$, carries the elemental volume of θ, ϕ space; it furnishes a complete description for spheres of any size (see this chapter, "A 2-D Volume Is an Area").

The elemental volume in angular momentum space is $d\ell \, dm = 1$ and

$$\sum_{\ell, m} \equiv \sum_{\ell=0}^{\infty} \sum_{m=-\ell}^{\ell}$$

THE CERTAINTY CONDITION
GOVERNS THE NORMALIZATION

The bracket, $\langle \theta, \phi | \ell, m \rangle$, has a precise physical interpretation. Its absolute value squared multiplied by an elemental volume in θ, ϕ space is a probability; that, at θ and ϕ within the elemental volume of θ, ϕ space, a bead with angular momentum indices ℓ and m will be found. That a bead with these indices will be found *somewhere* is a certainty. Thus

$$1 = \langle \ell, m | \ell m \rangle = \int_0^{2\pi} \int_0^{\pi} \sin \theta \, d\theta \, d\phi \, |\langle \theta, \phi | \ell, m \rangle|^2 \qquad (5\text{-}95)$$

It is this condition that generates the nomalization factors shown in (5-88) through (5-93).

5.8 RAISING AND LOWERING OPERATORS GENERATE THE SPHERICAL HARMONICS

One way to get the wave functions, $\langle\theta,\phi|\ell,m\rangle = Y_{\ell,m}(\theta,\phi)$ is by solving the differential Equation 5-86. There is another way to get them. The raising and lowering operators generate them.

For a given value of ℓ, there is a highest value of m. This is when m equals ℓ itself. The state, $|\ell,\ell\rangle$, has the property that it cannot be raised any further by the raising operator $\hat{L}_+$. Implementing this statement in θ,ϕ space produces

$$0 = \langle\theta,\phi|\hat{L}_+|\ell,\ell\rangle = \hbar e^{i\phi}\left[\frac{\partial}{\partial\theta} + i\,\text{ctn}\theta\,\frac{\partial}{\partial\phi}\right]\langle\theta,\phi|\ell,\ell\rangle \qquad (5\text{-}96)$$

The left-hand equality says that a highest state cannot be further raised. The right-hand equality is just the θ,ϕ language instruction for $\hat{L}_+$. The equality of the extremes yields the form of the wave function, $\langle\theta,\phi|\ell,\ell,\rangle$. Solve it using the known ϕ-dependence, $e^{i\ell\phi}$. The solution is

$$\langle\theta,\phi|\ell,\ell\rangle = \text{constant} \times e^{i\ell\phi}\sin^\ell\theta \qquad (5\text{-}97)$$

The constant comes from the normalization condition that $\langle\ell,\ell|\ell,\ell\rangle = 1$. (see Problem 5.26).

For the other wave functions where $m \neq \ell$, the process is easy. Just lower $|\ell,\ell\rangle$ with the lowering operator.

$$\hbar\,e^{-i\phi}\left[-\frac{\partial}{\partial\theta} - \ell\,\text{ctn}\,\theta\right]\langle\theta,\phi|\ell,\ell\rangle = \langle\theta,\phi|\hat{L}_-|\ell,\ell\rangle = \hbar\sqrt{2\ell}\,\langle\theta,\phi|\ell,\ell-1\rangle$$

The left-hand equality expresses the θ,ϕ representation of $\hat{L}_-$ (5-81). The right-hand equality expresses the lowering property of $\hat{L}_-$ (5-70). The equality of the extremes gives us a formula for $\langle\theta,\phi|\ell,\ell-1\rangle$ in terms of $\langle\theta,\phi|\ell,\ell\rangle$.

Another lowering produces $\langle\theta,\phi|\ell,\ell-2\rangle$ from $\langle\theta,\phi|\ell,\ell-1\rangle$. Continuing the process generates all of the brackets $\langle\theta,\phi|\ell,m\rangle$ by a series of lowerings from $\langle\theta,\phi|\ell,\ell\rangle$. Thus we can get all of the spherical harmonics via the raising and lowering operators; working through ℓ,m space we can solve a θ,ϕ space differential equation.

PROBLEMS

5.1 Any single valued continuous function of x and y can be expanded as an integral over $\mathbf{k}$, where $\mathbf{k}\cdot\mathbf{r} = xk_x + yk_y$, as

$$\psi(x,y) = \int_{-\infty}^{+\infty}\int_{-\infty}^{+\infty} dk_x\,dk_y\,\exp(i\mathbf{k}\cdot\mathbf{r})\,\phi(k_x,k_y)$$

By associating $\psi(x,y)$ with a wave function $\langle xy|\psi \rangle$ (see Problem 2.15), use this expansion together with the 2-D ket-bra sum theorem

$$\sum_k |k\rangle\langle k| = \sum_{k_x,k_y} |k_x,k_y\rangle \langle k_x,k_y| = 1$$

and the summation-over-a-continuous-variable rule

$$\int d^2k = \int \int dk_x, dk_y = \sum_k$$

and the 2-D self-space bracket δ-functions

$$\delta(\mathbf{r}, \mathbf{R}) = \delta(x-X)\delta(y-Y) = \langle \mathbf{r}|\mathbf{R}\rangle$$

$$\delta(\mathbf{k}, \mathbf{K}) = \delta(k_x-K_x)\delta(k_y-K_y) = \langle \mathbf{k}|\mathbf{K}\rangle$$

to deduce that

$$\phi(k_x,k_y) = \left(\frac{1}{2\pi}\right)^2 \int_{-\infty}^{+\infty} \int_{-\infty}^{+\infty} dx\,dy \exp(-i\mathbf{k}\cdot\mathbf{r})\,\psi(x,y)$$

Solution: The two key ingredients are

1. that you associate $\langle k_x,k_y|\psi \rangle$ with $\phi(k_x,k_y)$:
$$\phi(k_x,k_y) = \text{constant} \times \langle k_x,k_y|\psi \rangle$$
2. that you calculate $\langle \mathbf{k}|\mathbf{K}\rangle$ realizing that it is equal to $\delta(k_x-K_x)\,\delta(k_y-K_y)$. Use the formula given in Problem 2.13(d).

5.2 Prove (5-18) and (5-19) by applying them to (5-6). Why does this constitute proof of their general validity? See problem 4.23 and page 62.

5.3 Establish the six commutator relations between $\hat{x},\hat{y},\hat{p}_x$ and $\hat{p}_y$ in Equations 5-23 and 5-24.

Answer: For all x, y, and any $|\psi\rangle$

$$\langle x,y|[\hat{x},\hat{p}_y]|\psi\rangle = -i\hbar \left[x\frac{\partial\psi}{\partial y} - \frac{\partial(x\psi)}{\partial y} \right] = 0$$

5.4 Prove this δ-function equality (see Figure 5-2):

$$\frac{1}{\rho}\delta\left(\rho-\sqrt{x^2+y^2}\right)\delta(\phi-\tan^{-1}y/x) = \delta(x-\rho\cos\phi)\,\delta(y-\rho\sin\phi)$$

It allows us to attach meaning to integrals that otherwise look forbidding. For example, the function of ρ and ϕ that is the integral:

$$\int_{-\infty}^{+\infty} \int_{-\infty}^{+\infty} dx\,dy\,\delta(\rho-\sqrt{x^2+y^2})\,\delta(\phi-\tan^{-1}y/x)\,\frac{1}{\sqrt{x^2+y^2}} \exp\left[-(\alpha x^2+\beta y^2)\right]$$

is this one

$$\exp[-\rho^2 (\alpha \cos^2 \phi + \beta \sin^2 \phi)]$$

Solution: The proof requires that you insert an x,y-space ket-bra sum in $\langle \rho,\phi|\psi\rangle$ showing that

$$\psi(\rho \cos \phi, \rho \sin \phi) = \int_{-\infty}^{+\infty} \int_{-\infty}^{+\infty} dx \, dy \, \langle \rho,\phi|x,y\rangle \, \psi(x,y)$$

and insert a ρ,ϕ-space ket-bra sum in $\langle x,y|\psi\rangle$ showing that

$$\psi(x,y) = \int_{0}^{\infty} \int_{0}^{2\pi} d\phi \, \rho d\rho \, \langle x,y|\rho,\phi\rangle \, \psi(\rho \cos \phi, \rho \sin \phi)$$

By its role as a function of x and y in the first ket-bra sum, the bracket $\langle \rho,\phi|x,y\rangle$ must be interpreted as $\delta(x - \rho \cos \phi) \, \delta(y - \rho \sin \phi)$. By its role as a function of ρ and ϕ, the bracket, $\langle x,y|\rho\phi\rangle$, in the second ket-bra sum means $\frac{1}{\rho} \delta(\rho - \sqrt{x^2 + y^2}) \, \delta(\phi - \tan^{-1} y/x)$. By the event reversal theorem:

$$\langle \rho,\phi|x,y\rangle = \langle x,y|\rho,\phi\rangle^*$$

Notice that if $\langle x,y|\psi\rangle = \psi(x,y)$ then $\langle \rho,\phi|\psi\rangle = \psi(\rho \cos \phi, \rho \sin \phi) \neq \psi(\rho,\phi)$.

An alternative proof needs only some logical deductions from this change-of-variable observation:

$$\int_{-\infty}^{+\infty} \int_{-\infty}^{+\infty} dxdy \, \gamma(\rho,\phi) \, f(x,y) = \int_{0}^{\infty} \rho d\rho \int_{0}^{2\pi} d\phi \, \gamma(\rho,\phi) \, f(\rho \cos \phi, \rho \sin \phi)$$

5.5 You make the following three instantaneously sequential measurements of the bead-on-a-ring-of-circumference-L.

Measurement 1: Its position x is found to be $L/6$ within $dx = .01L$.

Measurement 2: Its momentum p is then found to be $6h/L$.

Measurement 3: Its energy E is then found to be $18h^2/mL^2$.

a. What is the probability of finding $x = L/6$ within $dx = .01L$ in the next (fourth) measurement?

b. What is the probability of finding $p = 6h/L$ if the fourth measurement is of momentum instead of position?

Answers: a. 0.01 b. 1.0 Notice that the energy measurement does not disturb the momentum measurement result: the amplitude to find $p = -6h/L$ is zero. Had Measurement 2 not been made, this amplitude would not be zero; it would be $1/\sqrt{2}$.

5.6 Prove that if the complete set of states $|\lambda,\gamma,\omega\rangle$ exists, then all three operators $\hat{\lambda}$, $\hat{\gamma}$, and $\hat{\omega}$ will commute.

Keys:

$$\hat{\lambda}|\lambda,\gamma,\omega\rangle = \lambda|\lambda,\gamma,\omega\rangle$$

$$\hat{\gamma}|\lambda,\gamma,\omega\rangle = \gamma|\lambda,\gamma,\omega\rangle$$

$$\hat{\lambda}|\lambda,\gamma,\omega\rangle = \omega|\lambda,\gamma,\omega\rangle$$

and

$$|\psi\rangle = \sum |\lambda,\gamma,\omega\rangle \langle\lambda,\gamma,\omega|\psi\rangle$$

5.7 Which of the following free bead-in-a-plane 2-D kets are impossible?

$|p_x,p_y,E\rangle$
$|x,p_y,E\rangle$
$|y,p_x\rangle$
$|y,p_y\rangle$
$|p_x,E\rangle$
$|x\rangle$
$|y,E\rangle$
$|x,y,p_x\rangle$
$|x,y,\rho\rangle$
$|x,y,\phi\rangle$
$|x,y,\rho,\phi\rangle$

Answers: 2nd, 4th, 7th, and 8th; the 6th is incomplete.

5.8 Demonstrate that the p,β-space instruction for $\hat{L}_z$, given in (5-31) implies the commutator rule of (5-32).

Answer: For any $|\psi\rangle$ and all p,β

$$\langle p,\beta|[\hat{\beta},\hat{L}_z]|\psi\rangle = -i\hbar\left[\frac{\beta\partial\langle p,\beta|\psi\rangle}{\partial\beta} - \frac{\partial(\beta\langle p,\beta|\psi\rangle)}{\partial\beta}\right] = i\hbar\langle p,\beta|\psi\rangle$$

5.9 Calculate $\langle p,\beta|\hat{L}_z|\rho,\phi\rangle$ to demonstrate the validity of (5-33).

Key:

$$\langle p,\beta|\hat{L}_z|\rho,\phi\rangle = -i\hbar\frac{\partial}{\partial\beta}\left\{\frac{1}{h}\exp\left[\frac{-i\rho p \cos(\beta-\phi)}{\hbar}\right]\right\}$$

5.10 The two momentum measurement spaces p_x, p_y and p, β are related by

$$p_x = p \cos \beta$$

$$p_y = p \sin \beta$$

It follows that

$$
\begin{pmatrix} \dfrac{\partial}{\partial p_x} \\[2ex] \dfrac{\partial}{\partial p_y} \end{pmatrix} = \begin{pmatrix} \cos \beta & -\dfrac{1}{p}\sin \beta \\[2ex] \sin \beta & \dfrac{1}{p}\cos \beta \end{pmatrix} \begin{pmatrix} \dfrac{\partial}{\partial p} \\[2ex] \dfrac{\partial}{\partial \beta} \end{pmatrix}
$$

Demonstrate the truth of the two simultaneous equations represented by this matrix relation.

Solution Key: Calculate $\partial \beta(p_x,p_y)/\partial p_x$, and so forth and use them in the chain rule for differentiation.

$$
\frac{\partial}{\partial p_x} = \left(\frac{\partial p}{\partial p_x}\right) \frac{\partial}{\partial p} + \left(\frac{\partial \beta}{\partial p_x}\right) \frac{\partial}{\partial \beta}
$$

5.11 You must get the same result for **p** whether you measure p_x and p_y or p and β. From this notion, together with the equations in Problem 5.10, deduce the p,β-space operator instructors for $\hat{p}_x$, $\hat{p}_y$, $\hat{x}$, $\hat{y}$, and $\hat{x}\hat{p}_y - \hat{y}\hat{p}_x$. Thus establish that $\langle p,\beta|\hat{L}_z = \langle p,\beta|(\hat{x}\hat{p}_y - \hat{y}\hat{p}_x)$

Partial Answer:

$$
\langle p,\beta|\hat{p}_x = p \cos \beta \langle p,\beta|
$$

$$
\langle p,\beta|\hat{x} = i\hbar \left[\cos \beta \frac{\partial}{\partial p} - \left(\frac{\sin \beta}{p}\right) \frac{\partial}{\partial \beta} \right] \langle p,\beta|
$$

5.12 Show that the instruction for $\hat{L}_z$ in $\langle \rho,\phi|$ space must be

$$
\langle \rho,\phi|\hat{L}_z = -i\hbar \frac{\partial}{\partial \phi} \langle \rho,\phi|
$$

Answer: Demonstrate that

$$
\langle \rho,\phi|(\hat{x}\,\hat{p}_y - \hat{y}\,\hat{p}_x) = -i\hbar \frac{\partial}{\partial \phi} \langle \rho,\phi|
$$

5.13 Both p and P are momentum vector magnitudes. Show that each 2-D transformation matrix element $\langle p,\beta|P,L_z\rangle$ given by

$$
\langle p,\beta|P,L_z\rangle = \frac{1}{\sqrt{2\pi}} \exp\left(\frac{i\beta L_z}{\hbar}\right) \langle p|P\rangle
$$

is a simultaneous solution of two eigenvalue problems: one for $\hat{p}$ and one for $\hat{L}_z$ when cast into the language $\langle p,\beta|$.

Answer: Calculate $\langle p,\beta|\hat{p}|P,L_z\rangle$. See whether it equals $P\langle p,\beta|P,L_z\rangle$. See whether $\langle p,\beta|\hat{L}_z|P,L_z\rangle = -i\hbar\partial\langle p,\beta|P,L_z\rangle/\partial\beta$ equals $L_z\langle p,\beta|P,L_z\rangle$.

5.14 Show that the wave function given in Problem 5.13 implies that L_z must be quantized

$$L_z = m\hbar \text{ where } m = \ldots -1, 0, +1, +2 \ldots$$

So m indexes the states of L_z.

Key: $\langle p, \beta + 2\pi | P, L_z \rangle$ and $\langle p, \beta | P, L_z \rangle$ represent the same event.

5.15 Using the form shown in Problem 5.13 for the 2-D transformation matrix elements $\langle p, \beta | P, L_z \rangle$, together with the form for $\langle \rho, \phi | p, \beta \rangle$ given in (5-30), prove that a state of momentum magnitude p and angular momentum $L_z = m\hbar$ represented in ρ, ϕ space may be written

$$\langle \rho, \phi | p, m \rangle = \int_0^{2\pi} d\beta \, \frac{1}{\sqrt{2\pi\hbar^2}} \exp\left[i\left(\frac{\rho p}{\hbar}\right) \cos(\beta - \phi) + im\beta\right]$$

Solution:

$$\langle \rho\phi | p, m \rangle = \sum_{P,\beta} \langle \rho\phi | P, \beta \rangle \langle P, \beta | p, m \rangle$$

and from (2-32)

$$\sum_P f(P) \langle P | p \rangle = f(p)$$

5.16 Prove the truth of (5-48). It is merely a matter of establishing the identity

$$[\hat{y}\hat{p}_z - \hat{z}\hat{p}_y, \hat{z}\hat{p}_x - \hat{x}\hat{p}_z] = i\hbar(\hat{x}\hat{p}_y - \hat{y}\hat{p}_x)$$

Use the results in Problem 4.26 to do it.

5.17 Show that $\hat{L}^2$ commutes with $\hat{L}_x$, $\hat{L}_y$, $\hat{L}_z$, $\hat{L}_-$, and $\hat{L}_+$. Use (5-48) and the rules of Problem 4.26, Equation (A) to do it.

5.18 Operate with the commutator $[\hat{L}_z, \hat{L}_+] = \hbar\hat{L}_+$ on the state $|\lambda, m\rangle$ to produce the eigenvalue equation that (5-58) solves.

5.19 Deduce equation 5-58. The argument exactly parallels that leading to Equation 4-63. Note that

$$c_M \langle \lambda, M-1 | \lambda, m \rangle = c_{m+1} \langle \lambda, M | \lambda, m+1 \rangle$$

5.20 Prove the commutation relation

$$[\hat{L}_+, \hat{L}_-] = 2\hbar\hat{L}_z.$$

Use this commutator result and (5-47) to demonstrate the truth of (5-62) and (5-63).

5.21 Derive Equation 5-64.

> *Solution:* The left-hand equality is the combination of (5-60) and (5-62) using (5-59). The right-hand equality is (5-61) with (5-63).

5.22 Show that this matrix equation captures all of the information given in Figure 5-6 relating derivatives in r,θ,ϕ space to those in x,y,z space. It expresses the chain rule for differentiation, as in Problem 5.10.

$$\begin{pmatrix} \dfrac{\partial}{\partial r} \\[2mm] \dfrac{\partial}{\partial \theta} \\[2mm] \dfrac{\partial}{\partial \phi} \end{pmatrix} = \begin{pmatrix} \sin\theta\cos\phi & \sin\theta\sin\phi & \cos\theta \\[2mm] r\cos\theta\cos\phi & r\cos\theta\sin\phi & -r\sin\theta \\[2mm] -r\sin\theta\sin\phi & r\sin\theta\cos\phi & 0 \end{pmatrix} \begin{pmatrix} \dfrac{\partial}{\partial x} \\[2mm] \dfrac{\partial}{\partial y} \\[2mm] \dfrac{\partial}{\partial z} \end{pmatrix}$$

If we agree to mean by the symbol ρ, $\sqrt{x^2+y^2}$ and by the symbol r, $\sqrt{x^2+y^2+z^2}$, then show that the 3×3 matrix may be written in x,y,z coordinates as

$$\begin{pmatrix} x/r & y/r & z/r \\ xz/\rho & yz/\rho & -\rho \\ -y & x & 0 \end{pmatrix}$$

How could you get the inverse matrix, the one relating derivatives in x,y,z,-space to those in r,θ,ϕ space? Here it is:

$$\begin{pmatrix} \dfrac{\partial}{\partial x} \\[2mm] \dfrac{\partial}{\partial y} \\[2mm] \dfrac{\partial}{\partial z} \end{pmatrix} = \begin{pmatrix} x/r & xz/\rho r^2 & -y/\rho^2 \\[2mm] y/r & yz/\rho r^2 & x/\rho^2 \\[2mm] z/r & -\rho/r^2 & 0 \end{pmatrix} \begin{pmatrix} \dfrac{\partial}{\partial r} \\[2mm] \dfrac{\partial}{\partial \theta} \\[2mm] \dfrac{\partial}{\partial \phi} \end{pmatrix}$$

where the 3×3 matrix may also be written

$$\begin{pmatrix} \sin\theta\cos\phi & \cos\theta\cos\phi/r & -\sin\phi/r\sin\theta \\ \sin\theta\sin\phi & \cos\theta\sin\phi/r & \cos\phi/r\sin\theta \\ \cos\theta & -\sin\theta/r & 0 \end{pmatrix}$$

5.23 Using the results of Problem 5.22, work out the combination $x\partial/\partial y - y\partial/\partial x$ in terms of r, θ, ϕ, $\partial/\partial r$, $\partial/\partial\theta$, and $\partial/\partial\phi$. Do the same for $y\partial/\partial z - z\partial/\partial y$ and $z\partial/\partial x - x\partial/\partial z$. From these deduce the representations of $\hat{L}^2$, $\hat{L}_z$, and so on, in r,θ,ϕ space and thus verify Equations 5-79 through 5-82.

5.24 Show that each of the six spherical harmonics exhibited in the text does, indeed, satisfy the single-valuedness condition that

$$\langle -\theta,\phi|\ell,m\rangle = \langle\theta,\phi+\pi|\ell,m\rangle$$

5.25 The bead is in angular momentum state $\ell = 2$, $m = -1$. What is the probability of finding it at the position $\theta = \pi/6$, $\phi = \pi/4$ within $d\theta = 0.01$ radians and $d\phi = 0.02$ radians? Using Equation 5-92 the answer is $45/128\pi \ 10^{-4}$.

5.26 Calculate the constant of Equation 5-97 for the cases $\ell = 0$, 1, and 2.

$$\text{Answers: } \sqrt{1/4\pi}, \ \sqrt{3/8\pi}, \ \sqrt{15/32\pi},$$

5.27 A position measurement has just determined the bead to be at $\theta = \pi/6$ and $\phi = \pi/2$ in $d\theta = .01$ radians and $d\phi = .01$ radians. An instantaneous subsequent measurement of the square of the total angular momentum ($\hat{L}^2$) is made. The quantum number $\ell = 1$ is found. What is the probability of finding $m = -1$?

Answer: Since m may have any of three values the probability of finding the particular value $m = -1$ is

$$\frac{|\langle \ell = 1, m = -1|\pi/6, \pi/2\rangle|^2}{\sum\limits_{-1}^{+1} |\langle \ell = 1, m|\pi/6, \pi/2\rangle|^2} = 1/8$$

Chapter 6

Spin, Matrices, and The Structure of Quantum Mechanics

6.1 MAGNETIC MOMENT SIGNALS ANGULAR MOMENTUM

Angular momentum was presumed to be an observable. How is it observed? It is deduced from magnetic moment measurements.

The angular momentum of a circulating charge produces a magnetic dipole moment. An elementary magnetic dipole is nothing but charge with angular momentum. Outside the orbit the magnetic field of a circulating electrical charge is indistinguishable from that due to a magnetic dipole.

To calculate its magnetic dipole moment is to find the dipole to which the circulating charge is equivalent. You can always find a magnetic pole pair (a dipole) with which you can replace the current loop. The circulating charge $+e$ also has mass m. Then, because a circulating charge is like a current loop, associated with its orbital angular momentum is a magnetic moment.

The magnetic moment is the product of the current and the area traced out by the current. In Figure 6-1 you see that the charge e moves the distance $\mathbf{v}\, dt$ in the time dt. That a charge $e = I dt$ is transported in time dt defines the current I. In this same time this current traces out the area shown as d(**area**).

The angular momentum of the circulating particle has exactly the same $\mathbf{r} \times \mathbf{v}$ vector attribute as does the magnetic moment. From the figure it is clear that the relationship between orbital angular momentum, $\mathbf{L}$, and magnetic moment, $\mathbf{M}$, is

$$\mathbf{M} = \frac{e}{2m}\mathbf{L} \qquad (6\text{-}1)$$

The ratio of the two is a scalar. It depends only on an inherent property of the particle, its charge-to-mass ratio. This proportionality constant is traditionally measured in units

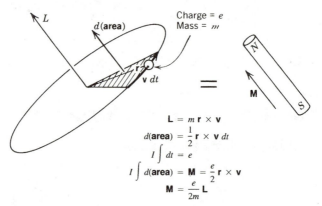

$$\mathbf{L} = m\,\mathbf{r} \times \mathbf{v}$$
$$d(\text{area}) = \frac{1}{2}\,\mathbf{r} \times \mathbf{v}\,dt$$
$$I\int dt = e$$
$$I\int d(\text{area}) = \mathbf{M} = \frac{e}{2}\,\mathbf{r} \times \mathbf{v}$$
$$\mathbf{M} = \frac{e}{2m}\,\mathbf{L}$$

Figure 6.1 The angular momentum of a circulating electrical charge produces a magnetic dipole moment.

of Bohr magnetons. If e is the charge of the electron and m is the electronic mass, then

$$e\hbar/2\,m = 1 \text{ Bohr magneton} = 58 \text{ micro-eV/tesla} \qquad (6\text{-}2)$$

Making $\hbar$ part of the proportionality constant is a computational convenience, it carries the angular momentum units for us. Magnetic dipole moments have the dimensions of Bohr magnetons.

THE STERN–GERLACH APPARATUS DISPLAYS THE MAGNETIC MOMENT SPECTRUM

We want to know the angular momentum of our bead-in-a-shell. It's an atom-sized system; the shell is a few angstrom units in diameter. For the conceptual convenience of dealing with electrically neutral pseudoatoms, we imagine that the circulating charge is just balanced by an equal and opposite charge located at the origin of each shell. Because it contains circulating electrical charge, the bead-in-a shell system is a magnetic dipole. It has a magnetic dipole moment. To measure its magnetic dipole moment is to measure its angular momentum.

To measure its magnetic moment, we use the wonderful apparatus of Stern and Gerlach. We prepare a beam of identical dipoles and send the beam through the apparatus pictured in Figure 6.2.

The apparatus consists of two magnet pole pieces shaped so as to produce a strongly nonhomogeneous field. Particles travel in the x-direction through the region of a B-field pointing in the z-direction. But the field strength depends upon z. It is weaker at lower z and stronger at higher z.

Such a field causes a physical displacement of the entire dipole in the z-direction. The displacement arises, quite simply, because one end of the dipole finds itself in a more intense field region than does the other. So the force on this pole is greater than on the other. This pole is either attracted or repelled more than its companion. And thus the whole dipole is deflected up or down as it travels along.

Here is the quantitative description (see Problems 6.1 through 6.4). The energy associated with a magnetic dipole of moment, **M,** in a magnetic field $\mathbf{B} = \mathbf{e}_z B$ is

$$H = -\mathbf{M} \cdot \mathbf{B} = -M_z B \tag{6-3}$$

This energy is lowest when the magnetic moment is oriented in the **B**-direction. There is a torque causing a precession around this direction. The torque arises because the force on one pole is always opposite to the force on the other. If B has no position dependence, there is no net force on the magnetic dipole; there is only the torque. But if B has a position dependence, then there is a net force. It is in the z-direction and of magnitude

$$-\frac{\partial H}{\partial z} = F_z = M_z \frac{\partial B}{\partial z} \tag{6-4}$$

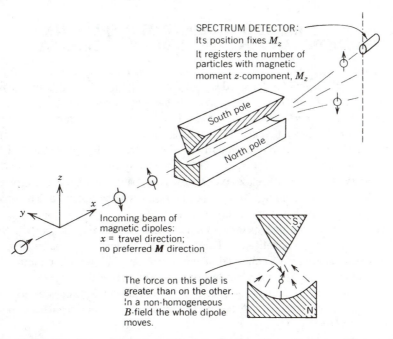

SPECTRUM DETECTOR:
Its position fixes M_z
It registers the number of particles with magnetic moment z-component, M_z

South pole

North pole

z
x
y

Incoming beam of magnetic dipoles:
x = travel direction;
no preferred **M** direction

The force on this pole is greater than on the other. In a non-homogeneous B-field the whole dipole moves.

Figure 6.2 The Stern–Gerlach apparatus displays the spectrum of magnetic moments, M_z.

From this equation we conclude that particles will be deflected up or down in direct proportion to M_z: their deflection in the field-direction signals their component of magnetic moment in that direction. The deflection comes from the acceleration that the force in Equation 6-4 produces: the total z-displacement depends on the time spent in the nonhomogeneous field region (see Problem 6.1).

This apparatus directly measures the z-component of magnetic moment; it measures M_z. A detector placed at the proper position records how many particles of the beam have the specific moment M_z. The Stern–Gerlach apparatus lays out the magnetic moment spectrum on a real space line. Each point on the line corresponds to a specific magnetic moment. And the position-moment ratio, the scale, is accessible; you calibrate the apparatus via a standard dipole.

This apparatus does for magnetic moments what a prism does for light. It displays the spectrum. A prism converts the various wavelength components of light into positions on a line. A Stern–Gerlach apparatus converts the z-component magnetic moments into positions on a line. They both yield a spectrum, an intensity as a function of position.

Take a moment to appreciate the incredible ingenuity of Stern and Gerlach. They invented an apparatus that visually displays the spectrum of the magnetic moments of atomic particles. They first used this apparatus in 1921 on a beam of neutral silver atoms. They found the result, then inexplicable, that the beam split into two, making a two-line spectrum.

6.2 BECAUSE ANGULAR MOMENTUM IS QUANTIZED, MAGNETIC MOMENTS HAVE DISCRETE SPECTRA

Suppose a beam of atom-sized sphere shells with their charged beads are passed through the Stern–Gerlach apparatus. What is the measurement result?

Classically we expect a *white spectrum*, all moments ranging from some $-M$ to a $+M$. That would mean that the beam consisted of particles with some magnetic moment, M. The orientation of each particle with respect to the z-direction is random, anything from completely *up* $(M_z = M)$ to completely *down* $(M_z = -M)$. A continuous range of M_z between these extremes would result. That's the classical expectation. It's shown in Figure 6-3. In fact, such a result is never observed. The spectrum of magnetic moments is never a continuous one.

What is observed are line spectra. Magnetic moments have a discrete spectrum.

The beam may pass undeflected. This is the case of zero magnetic moment. This result is expected both classically and quantum mechanically when the particles have no magnetic moment. The spectrum has one *line* and it is at the origin.

But then one may see a two-line spectrum. The beam splits. Half goes up and half goes down by exactly the same distance. The whole spectrum on the detection screen consists of two lines.

Then there is a three-line spectrum, one undeflected beam plus two oppositely and equally deflected beams; $+M$ and $-M$. Four-line spectra are possible, and so are five, and so on, but never a continuous distribution of moments. Line spectra are all

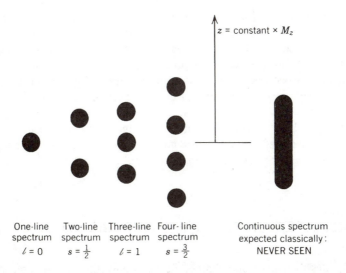

One-line spectrum   Two-line spectrum   Three-line spectrum   Four-line spectrum   Continuous spectrum expected classically: NEVER SEEN

$\ell = 0$ $s = \frac{1}{2}$ $\ell = 1$ $s = \frac{3}{2}$

Figure 6.3 Spectra observed in a Stern–Gerlach apparatus. Line z-position indicates M_z; line density (blackness) indicates how many particles have that M_z. Half the multiplicity-minus-one is the total angular momentum index.

that is ever seen, no matter what kind of atomic or molecular beam is investigated (see Figure 6-3).

This result is the experimental confirmation of our angular momentum calculation. To the measurement corresponds the operator $\hat{M}_z$. The measurement reveals that the operator has a discrete spectrum of eigenvalues. By virtue of Equation 6-1, to probe the operator $\hat{M}_z$ is to probe the operator $\hat{L}_z$. And indeed, this operator does have a discrete spectrum of eigenvalues.

MULTIPLICITY IN MAGNETIC MOMENT SPECTRA: THE NUMBER OF LINES REVEALS ℓ

The total angular momentum being zero means $\ell = 0$. If $\ell = 0$, there results the one-line spectrum $m = 0$.

If $\ell = 1$, we can find only $m = +1, 0,$ or -1: this three-line spectrum issues from the three eigenvalues of the $\hat{L}_z$ operator.

The number of lines in the spectrum is called the *multiplicity*. The multiplicity tells us the total angular momentum of the particles in the beam. Because there are $2\ell + 1$ states of m for every ℓ, the number of lines reveals the value of ℓ, one-half the multiplicity-minus-one.

But on this basis there should exist no two-line and four-line spectra. These multiplicities correspond to half-integer ℓ's. A two-line spectrum can come only from $\ell = 1/2$, a four-line one only from $\ell = 3/2$.

These are not allowed for orbital angular momentum. There can be no spatial wave function described by a half-integer angular momentum; such functions are not single valued. Hence, $\langle\theta,\phi|\ell=1/2,m\rangle$ is always zero. To measure the angular position of a circulating object with half-integer angular momentum is impossible.

Nevertheless a two-line spectrum is seen experimentally. Just such a spectrum is seen for a beam of free electrons. Free electrons are orbiting nothing; they are unbound, going around no center of attraction. Free electrons are like beads without any shells. There is no orbital momentum. The only angular momentum they could possibly have is that of spin: each individual electron could be spinning. That they always exhibit a magnetic dipole moment means that electrons possess an intrinsic *spin* angular momentum.

6.3 THE ELECTRON SPIN QUANTUM NUMBER IS 1/2

Besides their mass and charge, electrons have the intrinsic property of spin. Spin characterizes an electron just as surely as its mass and charge do. Spin is an *internal quantum number* for the electron.

It comes from an operator that follows the same rules as the angular momentum does but that has no representation in real physical space: you can't cast spin states into $\langle\theta,\phi|$ language. Although you can measure an electron's spin angular momentum, you cannot stop it to measure a body-position orientation. Because there is no space representation, there is no wave function to be made single valued; the spin angular momentum quantum number need not be an integer. It isn't an integer.

Like orbital angular momentum, this intrinsic spin angular momentum produces a magnetic moment. The *two-ness* of the magnetic moment spectrum means that the spin angular momentum quantum number is 1/2: multiplicity 2 implies angular momentum index 1/2. Had we seen a four-line magnetic moment spectrum, we would have deduced that the spin of the particle was 3/2.

The symbol s is used to index the spin angular momentum. By tradition, the symbol ℓ is not used for spin. Rather than saying $\ell = 1/2$, we say $s = 1/2$; ℓ is reserved for the characterization of orbital angular momenta.

Here is the mathematical portrayal of these ideas. Like the angular momentum vector **L**, from whose components the operators $\hat{L}_x$, $\hat{L}_y$, and $\hat{L}_z$ are built, there is a spin vector **S**, from whose components the operators $\hat{S}_x$, $\hat{S}_y$, and $\hat{S}_z$ are built. This group of operators obeys exactly the same key angular momentum operator relations that were explored in Chapter 5 in Equations 5-48 and 5-49. Hence

$$[\hat{S}_x,\hat{S}_y] = i\hbar\hat{S}_z \quad \text{(all cyclical permutations)} \tag{6-5}$$

and

$$[\hat{S}_x,\hat{S}^2] = [\hat{S}_y,\hat{S}^2] = [\hat{S}_z,\hat{S}^2] = 0 \tag{6-6}$$

Here are these ideas put formally:

$$\hat{S}^2 \, |s=1/2,m\rangle \;\; = \;\; \frac{1}{2}\left(\frac{1}{2}+1\right)\hbar^2 \, |s=1/2,m\rangle \tag{6-7}$$

$$\hat{S}_z \, |s=1/2,m\rangle \;\; = \;\; m\hbar|s=1/2,m\rangle, \qquad m \; = \; \pm 1/2 \tag{6-8}$$

ELECTRONS ARE FOUR-DIMENSIONAL PARTICLES

That the electron has an internal quantum number means that it has a fourth dimension to its state. A state like $|\psi\rangle$, which is representable only in 3-D space, $\langle x,y,z|\psi\rangle = \psi(x,y,z)$, is not enough. Such a state is insufficient to describe an electron fully. There are two possible electron states with the same wave function $\langle x,y,z|\psi\rangle$: $|\psi, +1/2\rangle$ and $|\psi, -1/2\rangle$.

"Where you find an electron" is decoupled from "whether it is spinning up or down." One can speak of the position state $|X,Y,Z\rangle = |\mathbf{R}\rangle$ with spin up; this is the state $|\mathbf{R}, +1/2\rangle$. The same position may be occupied by an electron with spin down: it would be in the state $|\mathbf{R}, -1/2\rangle$. The space of position is independent of the space of spin.

Thus a basis state for an electron is four dimensional. You must specify at least four measurement results to define its state fully.

The position-spin basis is $\langle x,y,z,m|$ or $\langle \mathbf{r},m|$. The state $|\psi, +1/2\rangle$, represented in this basis, looks like this

$$\langle \mathbf{r},m|\psi, +1/2\rangle \; = \; \langle \mathbf{r}|\psi\rangle \, \langle m|+1/2\rangle \tag{6-9}$$

It can be decomposed into a product of brackets because the state $|m=1/2\rangle$ has no representation in position space and the state $|\psi\rangle$ has no representation in spin space. Spin and position measurements are independent and compatible. Any position-related operator commutes with any spin operator.

$$0 \; = \; [\hat{z},\hat{S}_z] \; = \; [\hat{x},\hat{S}_z] \; = \; [\hat{x},\hat{S}^2] \; = \; \dots \tag{6-10}$$

We write $|\psi,m\rangle$ instead of $|\psi,s=1/2,m\rangle$ because all electrons are characterized by $s \; = \; 1/2$. The s never changes from this value. Thus there is no point to carrying it along in the kets and bras.

That we don't exhibit it in kets and bras portrays this physical concept: like its mass m and its charge $-e$, an electron spin, $s \; = \; 1/2$, is one of its intrinsic properties. Together, these quantities define what is meant by an electron.

6.4 THE SPIN OPERATOR IS THE HALF-UNIT ANGULAR MOMENTUM OPERATOR

Equations 6-5 and 6-6 give the essential properties of the spin operators $\hat{S}_x$, $\hat{S}_y$, and $\hat{S}_z$. Being angular-momentum-like, they follow all the rules for angular momentum. Just substitute $\hat{S}$'s for $\hat{L}$'s in Chapter 5.

That the electron has an intrinsic spin of 1/2 means that the index, s, for the eigenvalue of the operator $\hat{S}^2$, is fixed at 1/2. This value expresses the experimental fact that electron beams passed through a Stern–Gerlach apparatus always split into two.

Through Equations 5-52 and 5-53 you can define spin raising and lowering operators $\hat{S}_+$ and $\hat{S}_-$ in terms of $\hat{S}_x$ and $\hat{S}_y$. Their key features are easily deduced by examining the $\ell = 1/2$ case of (5-76). Here they are (see Problem 6.5):

$$\hat{S}_+ \, |-1/2\rangle \; = \; \hbar|+1/2\rangle \tag{6-11}$$

$$\hat{S}_+ \, |+1/2\rangle \; = \; 0 \; = \; \hat{S}_- \, |-1/2\rangle \tag{6-12}$$

$$\hat{S}_- \, |+1/2\rangle \; = \; \hbar|-1/2\rangle \tag{6-13}$$

THE g-FACTOR OF THE ELECTRON IS 2

The connection between magnetic moment and angular momentum given in Equation 6-1 and Figure 6-1 pertains only to orbital angular momentum. Because spin angular momentum has no expression in θ, ϕ language, the basis of the derivation of Equation 6-1 is removed. The proportionality of dipole moment to angular momentum given in this equation need not apply to spin.

Associating its spin angular momentum with its magnetic moment is how the experimental results on electrons is interpreted. Once we accept this association, the proportionality constant is directly measurable: it need not be that of Equation 6-1.

In the case of orbital angular momentum, the proportionality constant is cast in terms of Bohr magnetons (see Equations 6-1 and 6-2). For spin, the proportionality constant is in units of g Bohr magnetons. Spin angular momentum is viewed as a factor g more effective in producing magnetic moment than is orbital angular momentum. That the M_z measurement result is proportional to the eigenvalue, $m\hbar$, of $\hat{S}_z$, is expressed mathematically, in terms of this g-factor, by

$$M_z \; = \; -g \, \frac{e}{2\,m} \, S_z \; = \; \pm g \, \frac{e\hbar}{2\,m} \, \frac{1}{2} \tag{6-14}$$

The left-hand equality says that the z-component of magnetic moment is proportional to the z-component of spin angular momentum. It expresses the interpretation of all electron magnetic moment experiments.

The right-hand equality records the eigenvalues, $S_z = \pm\hbar/2$, of the $\hat{S}_z$ operator.

The equality of the extremes confirms the measurement finding of multiplicity 2. Equation 6-14 shows that there are only two spin dipole moment measurement results in the spectrum of M_z.

To discover the g-factor of the electron you measure its M_z with the Stern–Gerlach apparatus. Using some known magnetic moment, you calibrate the apparatus, you find the magnetic-moment-to-displacement ratio. Then the displacement of an electron in the same apparatus marks its magnetic moment z-component, M_z. The up-spin and the down-spin displacements are equal: both of amount $ge\hbar/4\,m$ or $g/2$ Bohr magnetons. Setting this equal to the measured value of M_z, you discover g. It is 2 for the spinning electron, the measured displacements are each one Bohr magneton. For the electron, spin angular momentum is twice as effective as is orbital angular momentum in producing a magnetic moment.

WHAT YOU MEASURE IS WHAT YOU KNOW

We explore the world by measuring things; the measurement results are our perceptions of the physical world around us. The idea, in quantum mechanics, is to base everything that we can call *known* on measurement results. The measurement results define what is known.

That's how the spin of the electron is to be perceived: the Stern–Gerlach measurement results define this property of the electron. The myriad other experiments that look at the magnetic moment of the electron confirm it. All electron magnetic moment measurement results expose its property of spin. They are consistent with this equation:

$$\hat{\mathbf{M}} = -g\,\frac{e}{2\,m}\,\hat{\mathbf{S}} \qquad\qquad (6\text{-}15)$$

where $g = 2$ and the $\hat{\mathbf{S}}$ operator has all the attributes of an angular momentum vector operator of index 1/2.

Though it's called *spin*, this property has no counterpart in classical physics. We visualize a physical extent to a classical rotating object, it has some spatial representation like $\langle\theta,\phi|$. But electrons have a half-integer intrinsic angular momentum. It follows that there can be no way to measure the shape of the electron, that you can never mark a position on it by values of θ and ϕ!

It seems that the quest to organize our perceptions tells us how limited our perceptions are. Through quantum mechanics, from looking we learn what we may not see. We can't *know* both the position and the momentum of a particle; we can't *know* the shape of an electron.

There's an alternative way to view these prohibitions: We can't expect to observe meaningless things. What we learn from quantum mechanics is not *what we may not see;* we learn *what questions are meaningless.*

Medieval people asked the sensible question, "How far is it to the end of the earth?" It turned out to be a meaningless question.

Quantum mechanics teaches that to demand the position of a momentum state particle is also meaningless. A fixed momentum means a wave train of fixed λ: such a wave train is indefinitely long; it has no position. You can't know the position of a momentum state particle because it hasn't any.

That the electron have a shape must be a meaningless notion.

THE ROTATED STERN–GERLACH APPARATUS MEASURES A TRANSFORMATION MATRIX

What measures a state can prepare a state. The same Stern–Gerlach apparatus that measures M_z can also function as a *magnetic moment selector*. Replace the detector in Figure 6-2 by a hollow tube and block off all particles not entering it. The beam emerging from the tube is a monomagnetic-moment beam: one of pure magnetic moment–z-component, M_z. In technical language this is called a *polarized beam,* all of its particles have the same M_z.

A polarized beam of dipoles is like a monochromatic beam of light. Pass white light through a prism; screen off all exiting wavelengths except those that pass through a narrow acceptance angle; and what comes through is a narrow portion of the whole spectrum, a monochromatic light beam.

We polarize a beam of electrons. All spin z-components of the electrons in this beam are equal: all have $+\hbar/2$, and none has $-\hbar/2$.

We pass this beam through another Stern–Gerlach apparatus, one that is rotated by 90 degrees so that its new field axis, the z'-axis, is in the y-direction. This apparatus is shown in Figure 6-4. It makes our measurements.

It measures the fraction of the incoming beam that has $S_{z'} = S_y = +\hbar/2$ and the fraction that has $S_{z'} = S_y = -\hbar/2$. Each electron is deflected either up or down in the z'-direction. Depending upon its deflection direction it is a spin-up′ or spin-down′ electron. The apparatus registers how many up′ (up-prime) and how many down′ spins there are in the beam. The probabilities of spin up′ or down′ are proportional to these numbers.

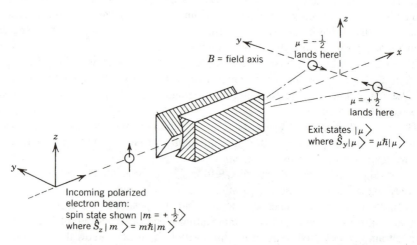

Figure 6.4 An incoming known-z-spin electron exits revealing its y-spin.

WHEN WORDS LOOK THE SAME, TAG THE LANGUAGES

Each electron in the incoming beam is in one of the states eigen to the $\hat{S}_z$ operator. We'll label these states m or M using the Latin alphabet. These label the states *entering* the apparatus of Figure 6-4.

Each electron in the outgoing beam is in one of the states eigen to the $\hat{S}_y$ operator. We'll label these states μ using the Greek alphabet. The Greek alphabet labels the states of the electrons *exiting* the apparatus in Figure 6-4, it indexes the spin components along the $z' = y$ axis. Thus

$$\hat{S}_z \, |m\rangle = m\hbar \, |m\rangle \tag{6-16}$$

and

$$\hat{S}_y \, |\mu\rangle = \mu\hbar \, |\mu\rangle \tag{6-17}$$

Being the z'-axis, the y-axis has equal status with the z-axis: the eigenvalues of $\hat{S}_y$ cannot be different from those for $\hat{S}_z$. They both have the same measurement spectrum, $+\hbar/2$ and $-\hbar/2$.

What is different are the states: the one corresponding to $m = 1/2$ is incompatible with the state $\mu = 1/2$. The former is labeled by the z-component of spin, and we know nothing about its y-component. The latter is labeled by the y-component of spin, and we know nothing about its z-component.

There are two distinct languages here, that generated by $\hat{S}_z$ and that generated by $\hat{S}_y$. The former is indexed by $m;$ the latter by μ. In these two languages the *words,* $\pm 1/2$, look the same but have very different meanings: the state labeled $m = 1/2$ is quite different from the state labeled $\mu = 1/2$ (see Figure 6-5).

Because the words look the same, we must exercise care in writing brackets. The meaning of $\langle 1/2|1/2\rangle$ is not self-evident; it could have several meanings. Its meaning is unique only if it is clear in which space each index is one-half. For example, the bracket $\langle \mu = 1/2|m = 1/2\rangle$ is not unity. The bracket $\langle m = 1/2|M = 1/2\rangle$ is unity; it's a self-space bracket. Both have 1/2 in the bra and in the ket, but the two brackets are not equal.

6.5 THE TRANSFORMATION MATRIX: AMPLITUDES FOR ALL POSSIBLE EXPERIMENTAL RESULTS

Now we can portray the experimental results mathematically. The apparatus measures $|\langle \mu|m\rangle|^2$. It measures the probability to find some spin $\mu\hbar$ in the y-direction if the incoming beam had spin $m\hbar$ in the z-direction. The incoming beam pictured in Figure 6-4 has $m = 1/2$.

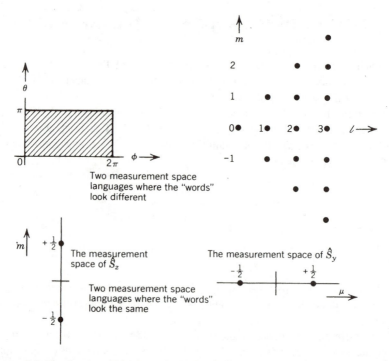

Figure 6.5 The basis *words* look the same, but they have different meanings.

Every transformation matrix element $\langle \mu | m \rangle$ is the amplitude for an experimental event measured by the rotated Stern–Gerlach apparatus. All possible experimental results are described by the transformation matrix. There is an input spectrum and an output spectrum: the transformation matrix exhibits what you see of the output spectrum as a function of what you inject at the input.

ONLY TWO QUESTIONS ARE ANSWERED BY QUANTUM MECHANICS

The entire business of practical quantum mechanics is devoted to obtaining transformation matrices! That's what we have been doing throughout these pages. For the harmonic oscillator we calculated the transformation matrix elements $\langle x | E \rangle$. A second transformation matrix for the same problem had elements $\langle p | E \rangle$. These are amplitudes for particular measurement events. The experimental outcome of measuring the energy of an oscillating mass prepared in a particular position state is governed by the amplitude $\langle E | x \rangle$.

Another example of a transformation matrix element is $\langle \theta, \phi | \ell, m \rangle$. It is the amplitude for this experimental measurement event: that θ and ϕ be found in a measurement of the angular position of a particle with orbital angular momentum quantum numbers ℓ and m.

The rotated Stern–Gerlach experiment follows the same pattern. The apparatus, acting on a prepared state, yields a spectrum of measurement results. The arrangement exemplifies the archetypical experimental event of quantum mechanics. All of our examples fit this pattern (see Figure 6-6).

It follows that there are only two questions that are answered through quantum mechanics. They are

1. What is the spectrum of possible results in an experiment?

2. What is the probability of finding each result in this spectrum?

These two questions consume all of the mathematics of quantum theory. The implication is that there are no other questions that one can ask of nature: these two are the only ones.

Using the rotated Stern–Gerlach experiment as a model, you can see how it works.

To every conceivable physical observable there corresponds a mathematical operator, like $\hat{S}_y$. An observable has a spectrum of possible measurement results. The

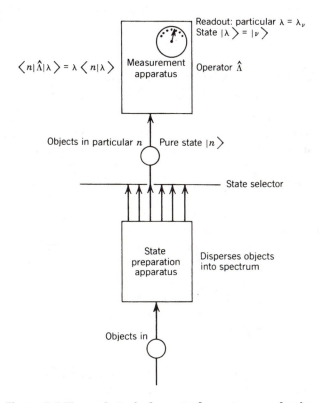

Figure 6.6 The archetypical event of quantum mechanics.

eigenvalues of its operator counterpart is this spectrum. Like $\mu\hbar$. Each eigenvalue marks a state. Like $|\mu\rangle$.

The measurement is performed on a prepared system, one whose state is known. State preparation requires a physical apparatus. It is a measurement apparatus converted to state-preparation use. Close off all channels but one. Letting this channel pass is *state preparation*. To the observable characterizing the state-preparation apparatus there corresponds an operator, like $\hat{S}_z$. An initial state is prepared that is an eigenstate of this operator. Like $|m\rangle$.

The questions answered by the measurement apparatus are (1) What is the spectrum produced? The μ? and (2) What is the probability of finding each μ having prepared *m?*

Where the particles fall yields the spectrum. The number of particles detected by the counter set to measure each μ is recorded. This number divided by the number entering the apparatus is just the probability, $|\langle\mu|m\rangle|^2$.

Theory produces the spectrum—the values of μ—and the set of amplitudes, $\langle m|\mu\rangle$. These consume all that the mathematics of quantum theory tells us. The eigenvalues are the spectrum, and from the amplitudes are deduced the probabilities of finding particular results in the spectrum.

To obtain the eigenvalues and the transformation matrix you must solve the eigenvalue problem. Thus, any conceivable finding in an experiment corresponds to a solution to an eigenvalue problem.

CAST THE EIGENVALUE PROBLEM IN THE LANGUAGE YOU WANT TO SPEAK

The eigenvalue problem associated with the rotated Stern–Gerlach apparatus is that of Equation 6-17. To solve this eigenvalue problem we cast it in the language that we know how to speak; the language, $\langle m|$, of the preparation state. The critical equation to be solved is

$$\langle m|\hat{S}_y|\mu\rangle = \mu\hbar \langle m|\mu\rangle \qquad (6\text{-}18)$$

It yields both the spectrum of eigenvalues, μ, and the transformation matrix elements, $\langle m|\mu\rangle$.

6.6 MATRICES SOLVE THE EIGENVALUE PROBLEM IN FINITE BASIS SPACES

What distinguishes this problem from the previous ones encountered is a finite denumerable state space. There are only a finite number of eigenstates: two of them, $\mu = +1/2$ and $\mu = -1/2$.

When there are a finite number of states, we may visually display every element of the matrix. Quantities such as $\langle x|p \rangle$ or $\langle x|\hat{H}|p \rangle$ are called matrix elements, but you can't draw the matrix. You can draw the matrix of elements connected with Equation 6-18.

Suppose we had a beam of $\ell = 1$ atomic magnetic dipoles, charged beads in a shell with purely orbital angular momentum. For these the Stern–Gerlach apparatus must show three lines: $m = -1$, 0, and $+1$. For the case $\ell = 1$, the entire space of the observable $\hat{L}_z$ consists of three eigenstates. For $\ell = 2$, the entire space comprises $2\ell + 1 = 5$ eigenstates.

The spin problem of Equation 6-18 is a special case of a large class of problems, those for which the number of eigenvalues is finite. When there are only a finite number of eigenvalues, the eigenvalue problem reduces to a matrix algebra procedure: to solve the eigenvalue problem is to diagonalize a matrix.

The connection between matrices and quantum mechanics was introduced into physics in 1925 by one of the founding fathers of quantum mechanics, W. Heisenberg.

TWO $N \times N$ MATRICES ARE EQUAL: A WAY OF WRITING N^2 EQUATIONS

Equation 6-18 is really a summary of four equations. Putting $m = 1/2$ and $\mu = 1/2$ is the first of them. Putting $m = -1/2$ and $\mu = 1/2$ is another. The other two cases are with $\mu = -1/2$.

All of these four equations can be written together as a single matrix equation. The term $\langle m = 1/2|\hat{S}_y|\mu = 1/2 \rangle$ is written as the 1,1-element of the matrix on the left. The term $\hbar/2 \langle m = 1/2|\mu = 1/2 \rangle$ is the 1,1-element of the matrix on the right. You assign each term to an element of the matrix.

Equality of two matrices means that every element of one is equal to every corresponding element of the other. Thus the equality of two $N \times N$ matrices corresponds to N^2 different equations. The four equations of (6-18) are conveyed by one matrix equation involving 2×2 matrices. Written symbolically it is

$$\left(\langle m|\hat{S}_y|\mu \rangle \right) = \left(\langle m|\mu \rangle \right) \begin{pmatrix} \mu\hbar & 0 \\ 0 & \searrow \end{pmatrix} \qquad (6\text{-}19)$$

The right-hand side of this equation is the matrix that is the product of the two shown. You perform a matrix multiplication to get the single matrix that is this product.

Matrix multiplication is a case of something much easier done than said, as illustrated by Figure 6-7. The 3×3 matrix (A) is multiplied by the 3×3 matrix (B) to form the 3×3 matrix (AB). The 1,2-element of the product matrix, called $(AB)_{12}$, is the sum of products shown and is manufactured from the elements of the first row of (A) and the second column in (B).

Carry out this process of multiplication for the pair of matrices on the right-hand

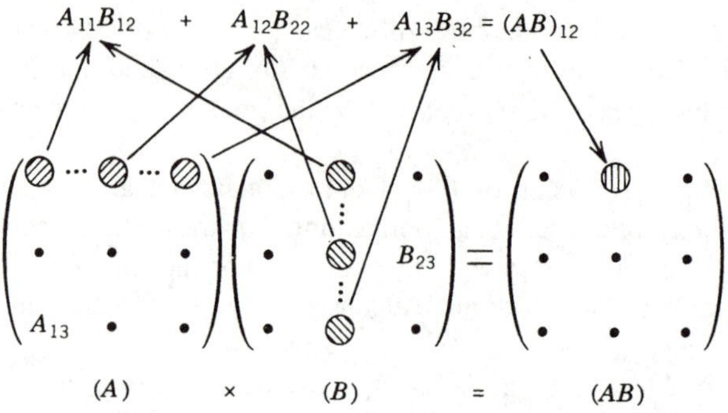

$$A_{11}B_{12} \quad + \quad A_{12}B_{22} \quad + \quad A_{13}B_{32} = (AB)_{12}$$

$$(A) \qquad \times \qquad (B) \qquad = \qquad (AB)$$

EXAMPLE:

$$\begin{pmatrix} \alpha & \beta \\ \\ \gamma & \delta \end{pmatrix} \begin{pmatrix} \dfrac{\hbar}{2} & 0 \\ \\ 0 & -\dfrac{\hbar}{2} \end{pmatrix} = \begin{pmatrix} \dfrac{\alpha\hbar}{2} & -\dfrac{\beta\hbar}{2} \\ \\ \dfrac{\gamma\hbar}{2} & -\dfrac{\delta\hbar}{2} \end{pmatrix}$$

Figure 6.7 The 3 × 3 matrix (A) multiplied by the 3 × 3 matrix (B) results in the 3 × 3 matrix (AB). The rule for getting the elements of the (AB) matrix is illustrated for the case of the one–two element of the product matrix (AB): the element $(AB)_{12}$.

side of (6-19). Then set every element of the product matrix equal to the corresponding element of the matrix on the left. You get precisely the set of four equations defined by (6-18). Thus (6-19) is merely the portrayal, in matrix notation, of 6-18.

6.7 THE KET-BRA SUM IS MATRIX MULTIPLICATION

Insert a ket-bra sum in the matrix element $\langle m|\hat{S}_y|\mu\rangle$ at the second vertical; between $\hat{S}_y$ and $|\mu\rangle$. Insert one that is in the space of $\hat{S}_z$ eigenvalues. To use the symbol m in the ket-bra sum insert would cause confusion because the matrix element itself already contains this symbol. We must use another symbol as the summation dummy variable, some other index for $\hat{S}_z$. We'll use M. Both m and M refer to the same basis.

$$\hat{S}_z|M\rangle = M\hbar|M\rangle \qquad \text{and} \qquad \hat{S}_z|m\rangle = m\hbar|m\rangle \tag{6-20}$$

The insertion of the M-space ket-bra sum into (6-18) produces

$$\sum_M \langle m|\hat{S}_y|M\rangle \langle M|\mu\rangle = \mu\hbar \langle m|\mu\rangle \tag{6-21}$$

But the sum on the left is precisely the prescription for a matrix multiplication. It is a statement of the rule for getting the m,μ element of the matrix produced by the product of two matrices: the matrix of $\hat{S}_y$ in S_z-index space multiplied by the transformation matrix, the one whose elements are $\langle M|\mu\rangle$. In matrix notation, what (6-21) says is this:

$$
\left(\begin{array}{c} \langle m|\hat{S}_y|M\rangle \end{array}\right)\left(\begin{array}{c} \langle M|\mu\rangle \end{array}\right) = \left(\begin{array}{c} \langle m|\mu\rangle \end{array}\right)\left(\begin{array}{cc} \mu\hbar & 0 \\ 0 & \diagdown \end{array}\right) \tag{6-22}
$$

The m, μ element of the product of the two matrices on the left of (6-22) is precisely what is displayed on the left of (6-21). Equation 6-22 is the matrix algebra statement of the eigenvalue problem in Equation 6-18.

There is a firm tradition for solving such matrix algebra problems. By writing our eigenvalue problem in matrix algebra form, we need merely resort to this mathematical tradition for solutions.

TO DIAGONALIZE A MATRIX IS
TO SOLVE THE EIGENVALUE PROBLEM

The $(\hat{S}_y)$ matrix on the left is a completely known quantity, it is one of the Pauli matrices of Problem 6.6.

The matrix that appears twice in (6-22) is the unknown: it appears to the immediate left and to the immediate right of the equals sign. This is the transformation matrix that we want to discover.

On the extreme right in (6-22) is a *diagonal matrix;* all off-diagonal terms are zero. The eigenvalues of the $\hat{S}_y$ operator are listed along the diagonal; the 1,1 element is $+\hbar/2$; the 2,2 element is $-\hbar/2$. This diagonal matrix represents the $\hat{S}_y$ operator, just as does the matrix at the extreme left: it is the matrix of the $\hat{S}_y$ operator in the space of its own eigenstates. Its elements are $\langle\mu'|\hat{S}_y|\mu\rangle$!

The matrix of an operator portrayed in its own space is always diagonal. This is the matrix counterpart of the dictum: *In its home space the hat comes off.* Thus

$$
\langle\mu'|\hat{S}_y|\mu\rangle = \mu\hbar\,\delta_{\mu,\mu'} \tag{6-23}
$$

Portrayed in a foreign space, the matrix of an operator is not generally diagonal: $\langle m|\hat{S}_y|M\rangle \neq$ const. $\times\,\delta_{mM}$.

The matrix algebra form of the eigenvalue problem can be viewed in this way: given the matrix of an operator in a foreign space, to find the transformation matrix that will diagonalize it, to find the transformation matrix that will bring it into its home space. That is the content of the eigenvalue problem; to produce the lexicon that effects the translation from a familiar language into the new language of the experimental observations.

The Solution For s = 1/2

A clearly understood rule in constructing matrix arrays must precede any calculations. You may not put an element in any old place in the matrix. The ordering of the elements must be carefully preserved.

The tradition in physics is to order the elements for angular momentum from the highest to the lowest. That's the way it's shown in Figure 6-8. This is just opposite to the tradition in mathematics, where the first or lowest element goes in the first position, rising from there with the position number. The eigenstates of energy are ordered by physicists in this second way, from low to high.

The origin of the elements in the $(\hat{S}_y)$ matrix shown in Figure 6-8 is the subject of Problem 6.6. All eigenvalue problems begin with some known matrix of numbers, like this one, the matrix to be diagonalized.

The figure makes evident the intimate connection between Dirac bracket notation and the elements of a matrix. The state index of the ket labels the column, and the state index of the bra labels the row in which the matrix element is to appear.

Given any matrix, the diagonalization procedure can generate the eigenvalues as well as the transformation matrix. But in our present example we already know the eigenvalue spectrum. We need only to obtain the transformation matrix, the four elements $\langle m|\mu \rangle$.

The calculation proceeds one column at a time. The first column is characterized by $\mu = 1/2$. There are two equations for this value of μ, one from each row of this first column.

Both of these equations tell us the same thing, that

$$\langle m = -1/2|\mu = 1/2 \rangle = i\langle m = 1/2|\mu = 1/2 \rangle \qquad (6\text{-}24)$$

This is a statement relating the various amplitudes of the $\mu = 1/2$ state among themselves. It tells us the amplitude for $m = -1/2$ *relative* to that for $m = +1/2$.

To get absolute amounts we need to implement the certainty condition, that

$$\langle \mu = 1/2|\mu = 1/2 \rangle = 1 \qquad (6\text{-}25)$$

Inserting a ket-bra sum over m at the vertical, it is easy, using (6-24), to discover that

$$|\langle m = -1/2|\mu = 1/2 \rangle|^2 = 1/2 = |\langle m = 1/2|\mu = 1/2 \rangle|^2 \qquad (6\text{-}26)$$

Now consider the second value of μ: the case $\mu = -1/2$. Again there are two equations for it. Each comes from a row in the second column of the product matrix on each side of the equation in Figure 6-8. Both of these equations also make a single statement, the same statement twice, that

$$\langle m = -1/2|\mu = -1/2 \rangle = -i\langle m = 1/2|\mu = -1/2 \rangle \qquad (6\text{-}27)$$

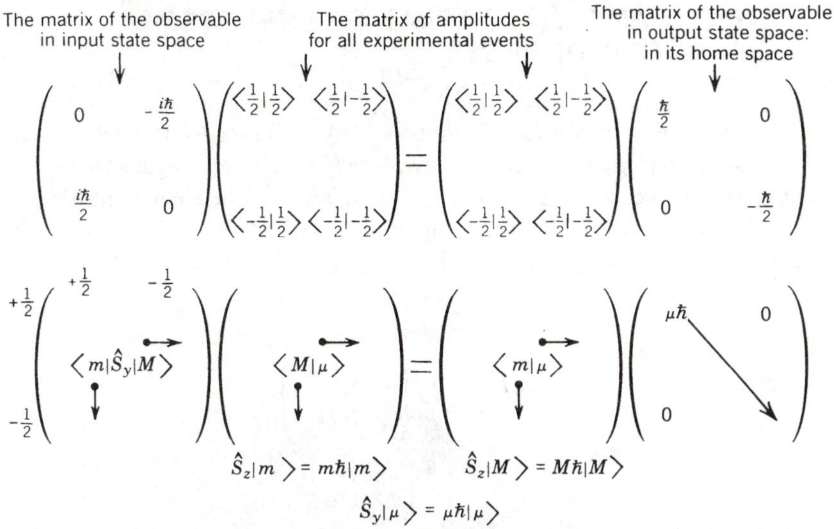

The matrix of the observable
in input state space

The matrix of amplitudes
for all experimental events

The matrix of the observable
in output state space:
in its home space

$$\begin{pmatrix} 0 & -\frac{i\hbar}{2} \\ \frac{i\hbar}{2} & 0 \end{pmatrix} \begin{pmatrix} \langle\frac{1}{2}|\frac{1}{2}\rangle & \langle\frac{1}{2}|-\frac{1}{2}\rangle \\ \langle-\frac{1}{2}|\frac{1}{2}\rangle & \langle-\frac{1}{2}|-\frac{1}{2}\rangle \end{pmatrix} = \begin{pmatrix} \langle\frac{1}{2}|\frac{1}{2}\rangle & \langle\frac{1}{2}|-\frac{1}{2}\rangle \\ \langle-\frac{1}{2}|\frac{1}{2}\rangle & \langle-\frac{1}{2}|-\frac{1}{2}\rangle \end{pmatrix} \begin{pmatrix} \frac{\hbar}{2} & 0 \\ 0 & -\frac{\hbar}{2} \end{pmatrix}$$

$$\begin{array}{c} +\frac{1}{2} \\ \\ -\frac{1}{2} \end{array}\begin{pmatrix} \overset{+\frac{1}{2} \quad -\frac{1}{2}}{\langle m|\hat{S}_y|M\rangle} \end{pmatrix} \begin{pmatrix} \langle M|\mu\rangle \end{pmatrix} = \begin{pmatrix} \langle m|\mu\rangle \end{pmatrix} \begin{pmatrix} \mu\hbar & 0 \\ 0 & \end{pmatrix}$$

$$\hat{S}_z|m\rangle = m\hbar|m\rangle \qquad \hat{S}_z|M\rangle = M\hbar|M\rangle$$
$$\hat{S}_y|\mu\rangle = \mu\hbar|\mu\rangle$$

Figure 6.8 **The spin 1/2 rotated Stern-Gerlach experiment eigenvalue problem. The lower equation establishes the meaning of the symbols in the upper matrix equation.**

This tells us how the m-space amplitudes of the $\mu = -1/2$ state are related to each other, not what the amplitudes themselves are, only their relative strengths.

To get absolute amounts we need to implement this certainty statement:

$$\langle\mu = -1/2|\mu = -1/2\rangle = 1 \tag{6-28}$$

With the help of a ket-bra sum over m inserted at the vertical, we discover

$$|\langle m = -1/2|\mu = -1/2\rangle|^2 = 1/2 = |\langle m = 1/2|\mu = -1/2\rangle|^2 \tag{6-29}$$

Combining equations 6-24, 6-26, 6-27, and 6-29 permits us to construct the grand result as the display of the $\langle m|\mu\rangle$ matrix (see Problem 6.8):

$$\begin{pmatrix} \langle m|\mu\rangle \end{pmatrix} = \begin{pmatrix} \frac{1}{\sqrt{2}} & \frac{1}{\sqrt{2}} \\ \frac{i}{\sqrt{2}} & \frac{-i}{\sqrt{2}} \end{pmatrix} \tag{6-30}$$

The absolute square of each element in this matrix is the probability for the outcome of an experiment: every experimental event possible with the apparatus of Figure 6-4 is covered by the elements of this matrix.

ALMOST ANY $N \times N$ ARRAY OF NUMBERS CAN BE DIAGONALIZED

Make up an $N \times N$ array of numbers. Any such array constitutes a matrix, it probably can be diagonalized. Almost any $N \times N$ matrix can diagonalized.

Here's what is meant. Call the matrix of numbers Λ. The element in the nth row and the mth column call Λ_{nm}. That Λ can be diagonalized means that a matrix U can be found that solves this special problem.

$$\Lambda U = U \begin{pmatrix} \lambda & & 0 \\ & \searrow & \\ 0 & & \end{pmatrix} \tag{6-31}$$

Like Λ, the matrix U is also an N-square array of numbers. Given any matrix of numbers, Λ, a set $\lambda_1, \lambda_2, \ldots \lambda_N$ can be found. What is listed along the diagonal in the matrix on the right are these λ's. All other elements are zero. And in most cases, every element of a U matrix that produces this diagonal matrix can also be found.

Notice that the form of this equation is just like that of Figure 6-8, the latter is a special case of the general problem in Equation 6-31.

The set of λ's is unique. Only one such set can be found that satisfies (6-31). Hence for any matrix Λ (a square array of $N \times N = N^2$ numbers), there is always a group of N numbers that characterize the matrix Λ. These are the N λ's. These numbers are the eigenvalues of Λ. When a matrix is constructed from an operator (say $\hat{\Lambda}$, like $\hat{S}_y$), these are the eigenvalues of that operator.

The procedure for finding the λ's and the U matrix is quite a straightforward one.

6.8 DIAGONALIZING A MATRIX IS A MECHANICAL PROCEDURE

Conceptually break up the U matrix into a set of single-column matrices. Each column is called a *vector;* in Equation 6-32, each vector is in m-space. Dirac notation is particularly useful in describing this decomposition, the right-hand matrix shows it.

$$U = \begin{pmatrix} U_{11} & U_{12} & U_{m3} & \cdots \\ U_{21} & U_{22} & \downarrow & \cdots \\ U_{31} & \vdots & & \\ \vdots & & & \end{pmatrix} = \begin{pmatrix} \langle m|\lambda_1 \rangle & \langle m|\lambda_2 \rangle & \cdots \\ \downarrow & \downarrow & \cdots \\ & & \cdots \end{pmatrix} \tag{6-32}$$

The first vector is the column of elements $U_{11}, U_{21}, U_{31}, \ldots$ These all have the family name of 1. And m ranges over each member of the family, every element in the first column is connected to the first of the λ's, to λ_1.

The second vector $U_{12}, U_{22}, U_{32}, \ldots U_{N2}$ has the family name 2 or λ_2, and again m ranges over all members of the family. This conceptual decomposition into vectors is the content of Equation 6-32.

Equation 6-31 can be viewed as a condensed notation for N different column matrix equations, one for each of the column vectors. Valid for each of the λ's, λ_1, λ_2 and so on, Equation 6-31 is entirely equivalent to

$$
\begin{pmatrix} \langle n|\hat{\Lambda}|m\rangle \end{pmatrix}
\begin{pmatrix} \langle m|\lambda\rangle \\ \vdots \end{pmatrix}
= \lambda
\begin{pmatrix} \langle n|\lambda\rangle \\ \vdots \end{pmatrix}
\tag{6-33}
$$

The index m is a dummy index, it is summed over. Instead of the traditional Λ_{nm} is written $\langle n|\hat{\Lambda}|m\rangle$, but they're the same. Equation 6-31 is just Equation 6-33 done N times, once for each different λ. This equation and (6-31) are equivalent.

Here's how we get the values of λ. We recognize that Equation 6-33 is a condensed way of writing N homogeneous equations for N unknowns. The N unknowns are the N elements of the column vector. Pretending, for the moment, that λ is not an unknown, these N equations are

$$
\sum_{m=1}^{N} [\langle n|\hat{\Lambda}|m\rangle - \lambda\delta_{nm}] \langle m|\lambda\rangle = 0
\tag{6-34}
$$

The first equation is when $n = 1$, the second when $n = 2$, the third when $n = 3$, and the Nth when $n = N$.

A set of nonhomogeneous linear equations is a familiar sight in physics. They often come into physical problems looking like a set of simultaneous equations for the unknowns $x, y, z, \ldots$

$$
\begin{aligned}
a_1 x + a_2 y + a_3 z + \ldots &= A \\
b_1 x + b_2 y + b_3 z + \ldots &= B \\
c_1 x + c_2 y + c_3 z + \ldots &= C \\
. + . + . + \ldots &= . \\
. + . + . + \ldots &= .
\end{aligned}
\tag{6-35}
$$

The solutions of this simultaneous set of equations is a ratio of determinants. For example, the value of x is given by

$$x = \frac{\begin{vmatrix} A & a_2 & a_3 & \ldots \\ B & b_2 & b_3 & \ldots \\ C & c_2 & c_3 & \ldots \\ \ldots & \ldots & \ldots & \end{vmatrix}}{\begin{vmatrix} a_1 & a_2 & a_3 & \ldots \\ b_1 & b_2 & b_3 & \ldots \\ c_1 & c_2 & c_3 & \ldots \\ \ldots & \ldots & \ldots & \end{vmatrix}} \tag{6-36}$$

The solution for y has A, B, C and so forth in the second column of the numerator determinant instead of in the first column.

In (6-34), the $\langle m = 1 | \lambda \rangle$ is x; the $\langle m = 2 | \lambda \rangle$ is y; and so on. These are the *unknowns*. And the system is a homogeneous one: $A = B = C = \ldots = 0$. Thus, when you apply the principle shown in (6-36) to solve (6-34), you find that the determinant in the numerator is zero. Hence the unknown $\langle m = 1 | \lambda \rangle = x$ can be finite only if the determinant in the denominator is also zero. That this determinant be zero is exactly what picks out the values of λ. The values of λ are determined by the equation:

$$\begin{vmatrix} \Lambda_{11} - \lambda & \Lambda_{12} & \Lambda_{13} & \ldots \\ \Lambda_{21} & \Lambda_{22} - \lambda & \Lambda_{23} & \ldots \\ \Lambda_{31} & \cdot\cdot & \cdot\cdot\cdot\cdot\cdot & \\ \cdot\cdot & \cdot\cdot & \cdot\cdot\cdot\cdot\cdot & \\ \cdot\cdot & \cdot\cdot & \cdot\cdot & \end{vmatrix} = 0 \tag{6-37}$$

An $N \times N$ determinant like this condenses into an Nth order equation in λ.

$$\lambda^N + \alpha\lambda^{N-1} + \beta\lambda^{N-2} + \ldots + \gamma = 0 \tag{6-38}$$

Such an equation has N roots, λ_1, λ_2 . . . and λ_N. They are the eigenvalues of the matrix Λ.

For the notational convenience of writing these ideas in more tractable form, it is useful to be acquainted with the unit matrix I, also called the *identity matrix*. It is that special matrix in which all off-diagonal terms are zero and every diagonal element is the number one. Thus

$$I \equiv \begin{pmatrix} 1 & & 0 \\ & 1 & \\ 0 & & \searrow \end{pmatrix} \tag{6-39}$$

Using this unit matrix, Equations 6-33 or 6-34 can be rewritten as

$$\left(\begin{matrix} \\ \Lambda - \lambda I \\ \\ \end{matrix} \right) \left(\begin{matrix} \langle m|\lambda\rangle \\ \downarrow \\ \end{matrix} \right) = 0 \qquad (6\text{-}40)$$

The difference between two matrices is itself a matrix. The symbol $(\Lambda - \lambda I)$ means the difference matrix obtained by subtracting λI from Λ. This matrix is just the array of numbers inside the determinant of Equation 6-37. That equation rewritten is

$$det|\Lambda - \lambda I| = 0 \qquad (6\text{-}41)$$

Now we come to the problem of getting the U matrix. This matrix is simply *the column eigenvectors all in a row*. To get any one of these column eigenvectors, we go back to (6-34) or (6-40). We insert the value of λ (say λ_3) for which we want the vector ($\langle m|\lambda_3\rangle$) and solve.

For the given λ, Equation 6-34 is a set of linear equations in N unknowns. But because the set of N equations is homogeneous in the unknowns, we cannot, in fact, determine any of these unknowns absolutely. We can find them only relatively. We can find each one of them only in terms of one of them.

For example, in (6-35) if $A = B = C = \ldots = 0$, we could never determine x as a particular number, y as another, and so on. All we could do is to determine y in terms of x and then z in terms of x, and so on. To determine x, y, and z absolutely requires some other condition, beside (6-35).

It is the same with the set in (6-34). We can expect to find $\langle 2|\lambda\rangle$ only in terms of $\langle 1|\lambda\rangle$ and then to get $\langle 3|\lambda\rangle$ in terms of $\langle 1|\lambda\rangle$ and on through all the indices until we have finally $\langle N|\lambda\rangle$ in terms of $\langle 1|\lambda\rangle$.

To do this we rewrite (6-34) to look like a set of nonhomogeneous equations. We put $\langle 1|\lambda\rangle$ as the driving term on the right-hand side. We pretend that we know $\langle 1|\lambda\rangle$. Here is the rewritten equation set:

$$\sum_{m=2}^{N} [\langle n|\hat{\Lambda}|m\rangle - \lambda\delta_{nm}] \langle m|\lambda\rangle = [\lambda\delta_{n1} - \langle n|\hat{\Lambda}|1\rangle] \langle 1|\lambda\rangle \qquad (6\text{-}42)$$

The term $[\lambda - \langle 1|\hat{\Lambda}|1\rangle]\langle 1|\lambda\rangle$ is like the A of (6-35), and the term $-\langle 2|\hat{\Lambda}|1\rangle \langle 1|\lambda\rangle$ is like the B of (6-35), and so forth.

We are treating $\langle 1|\lambda\rangle$ as if it were known, like A and B and C. Therefore (6-42) describes N equations in only $N - 1$ unknowns. Hence, of the N equations available, one is redundant. We don't need it. The last equation gives no information not already present in the other $N - 1$ equations we choose to solve.

We extract the elements, $\langle m|\lambda\rangle$, of the vector in terms of its first element $\langle 1|\lambda\rangle$ by using just the formulas exemplified by (6-36). Every element is a quotient of two determinants. But now the determinants are not $N \times N$ ones; they are $(N-1) \times (N-1)$ determinants.

They are the ones that arise in solving

$$
\begin{pmatrix}
\Lambda_{12} & \Lambda_{13} & \cdots & \Lambda_{1N} \\
\Lambda_{22}-\lambda & \Lambda_{23} & \cdots & \Lambda_{2N} \\
\cdot\cdot & & \cdots & \cdot\cdot \\
\cdot\cdot & & \cdots & \cdot\cdot \\
\Lambda_{N-1,2} & & \cdots & \Lambda_{N-1,N}
\end{pmatrix}
\begin{pmatrix}
\langle 2|\lambda\rangle \\
\langle 3|\lambda\rangle \\
\cdot\cdot \\
\cdot\cdot \\
\langle N|\lambda\rangle
\end{pmatrix}
= \langle 1|\lambda\rangle
\begin{pmatrix}
\lambda-\Lambda_{11} \\
-\Lambda_{21} \\
\cdot\cdot \\
\cdot\cdot \\
-\Lambda_{N-1,1}
\end{pmatrix}
\tag{6-43}
$$

Compare this driven set $(A \neq 0,\ B \neq 0)$ with (6-35). By the isomorphism, from the solutions characterized by (6-36), you can solve (6-43).

It's evident that each $\langle m|\lambda\rangle$ will be found to be directly proportional to $\langle 1|\lambda\rangle$. As long as $\langle 1|\lambda\rangle$ is not zero, this scheme can work. If $\langle 1|\lambda\rangle$ is zero, it may also be made to work; we get everything in terms of, say, $\langle 2|\lambda\rangle$ or $\langle 3|\lambda\rangle$.

When two or more of the λ's are equal, the system is degenerate. Degeneracy means that the apparatus, to which the Λ martix corresponds, is insufficiently selective, the states $|1\rangle$ and $|2\rangle$ are not distinguished by it if $\lambda_1 = \lambda_2$. Here we are concerned only with the nondegenerate case, in which all the states of the spectrum are distinguishable from one another. No two λ's are equal.

This concludes the mechanics of the diagonalization procedure. It doesn't conclude the story. All the λ's, but not all the elements of the U matrix, are deduced by this process.

The eigenvalue equation, (6-32), determines all of the N^2 matrix elements of U but N of them. In the scheme outlined, all the elements of U come to us in terms of the first row of U elements: they come relative to $\langle 1|\lambda\rangle$.

For the application to physics relative values of the U matrix elements are not enough; we need absolute values. The scheme is incomplete. For the application to physics a principle is lacking.

The principle we have not yet used is that of the certainty condition, that the self-space bracket yield certainty. The matrix algebra counterpart to this principle is connected to the concept of an inverse matrix.

A MATRIX AND ITS INVERSE: THE PRODUCT IS THE UNIT MATRIX

An $N \times N$ matrix may have an inverse, it is the multiplier matrix necessary to produce the unit matrix, I. The matrix inverse to U is written U^{-1}, and the definition of U^{-1} is

$$
U^{-1}U = I = UU^{-1}
\tag{6-44}
$$

Given a matrix, U, that has one, you can always calculate its inverse. From U you can get every element of U^{-1} by virtue of Equation 6-44. This equation amounts to N sets of driven coupled equations. The unknown variables are the elements of U^{-1}.

Each set is exactly solvable. Each element of U^{-1} is a ratio of determinants manufactured from the U matrix itself. Each element is obtained in just the same way that (6-36) is obtained from (6-35) (see Problem 6.12).

The matrix elements of U are $U_{m\lambda}$. In Dirac notation they are called $\langle m|\lambda\rangle$ (see Equation 6-32). The matrix elements of the U^{-1} matrix may also be written in Dirac Notation:

$$(U^{-1})_{\lambda m} = \langle\lambda|m\rangle \tag{6-45}$$

Suppose the index for λ is v: the vth value that λ has is λ_v. Because its index is its name, we may tag the states by v: $|\lambda\rangle=|v\rangle$. In this way the bracket position in the matrix is made evident. Thus $U_{m\lambda} = \langle m|\lambda\rangle = \langle m|v\rangle$ is the element in the mth row and the vth column of the U matrix. The symbols $(U^{-1})_{\lambda m} = \langle\lambda|m\rangle = \langle v|m\rangle$ are equivalent: each means that element in the vth row and the mth column of the U^{-1} matrix.

Using v and μ to tag two of the λ's, we can rewrite the matrix Equation 6-44 in Dirac notation.

$$\sum_m \langle v|m\rangle \langle m|\mu\rangle = \delta_{v\mu} \tag{6-46}$$

The left-hand term is the v,μ element of the product matrix $U^{-1}U$: it is $(U^{-1}U)_{v\mu}$. The right-hand term is the v,μ element of the unit matrix: it is $(I)_{v\mu}$. Equation 6-46 is just the left-hand equality of (6-44) rewritten.

But we can view (6-46) in another way. The left-hand term is the self-space bracket $\langle v|\mu\rangle$ with a ket-bra sum inserted at the vertical. Thus, the equation may be read as a statement about the self-space bracket; it says that $\langle v|\mu\rangle$ is a Kronecker-δ function of v and μ.

Comparing these two views of (6-46), we see that the existence of an inverse matrix, Equation 6-44, is what guarantees the certainty condition, Equation 6-46.

6.9 FOR PHYSICAL STATES
THE TRANSFORMATION MATRIX IS UNITARY

Each element of the matrix U is associated with a Dirac bracket. It's displayed that way in Equation 6-32. Such a bracket connects two sets of basis states. For physical systems these states list the measurement spectra of observables: the brackets connect physical states.

Physical states are ones for which the reverse amplitude is related to the forward amplitude: one is the complex conjugate of the other. This amplitude reversal principle is what weds the mathematical diagonalization scheme to physics. The Dirac bracket statement of it is

$$\langle\lambda|m\rangle = \langle m|\lambda\rangle^* \tag{6-47}$$

Its counterpart in matrix algebra is

$$U^{-1} = U\dagger \tag{6-48}$$

By the bracket, $\langle \lambda | m \rangle$, is meant an element of the inverse transformation matrix U^{-1}: $(U^{-1})_{\lambda m} = \langle \lambda | m \rangle$. The quantity $\langle m | \lambda \rangle^*$ is also a matrix element, it is the νth row and m^{th} column element of the hermitian conjugate matrix, $U\dagger$: $(U\dagger)_{\lambda m} = \langle m | \lambda \rangle^* = \langle m | \lambda_\nu \rangle^*$. Hence (6-47) is the statement about the U matrix shown in (6-48); the two statements are equivalent.

A matrix that has the special property shown in (6-48) is called a *unitary matrix*. Hence all matrices that transform between physical basis states are *unitary transformation matrices*. Physical state Dirac brackets are unitary transformation matrix elements.

Imposing the physical state condition of (6-48) in addition to the eigenvalue problem statement, (6-31), every element of the U matrix is calculable. Except for the nonphysical phase factors, nothing indeterminate remains.

The combination of (6-44) and (6-48) gives us the certainty statement. It assures the connection between amplitudes and probabilities of events. For the rotated Stern–Gerlach case of spin 1/2 particles, you see its use in Equations 6-25 and 6-28.

The three equations, (6-31), (6-44), and (6-48), constitute the matrix algebra portrayal of quantum mechanics. Here they are summarized together with the connection to Dirac notation:

Dirac Notation	*Matrix Algebra*	
$\langle \lambda | m \rangle = \langle m | \lambda \rangle^*$	$U^{-1} = U\dagger$	(6-49)
$\sum_m \langle \lambda_\mu | m \rangle \langle m | \lambda_\nu \rangle = \delta_{\mu\nu}$	$U^{-1}U = I$	(6-50)
$\langle m | \hat{\Lambda} | \lambda \rangle = \lambda \langle m | \lambda \rangle$	$\Lambda U = U \begin{pmatrix} \lambda & & 0 \\ & \searrow & \\ 0 & & \end{pmatrix}$	(6-51)

PROBLEMS

6.1 The original Stern–Gerlach experiment used silver atoms. Each one had a magnetic moment whose B-field component was one Bohr magneton—58 microelectron-volts per tesla. The kinetic energy of beam atoms was 1/5 electron-volt, and they traveled a distance of 3 cm through the region of magnetic field with a gradient of 200 kilogauss/cm. Calculate the beam deflection at 30 cm beyond the magnet.

Answer: 2.7 mm. It's a difficult experiment.

6.2 How do we know that we are measuring the M_z with which the particle entered the Stern–Gerlach apparatus? How is it that M_z remains constant? What about the torque tending to align **M** with **B?** Doesn't it change M_z?

Answer: Indeed there is such a torque. But torque produces a change in the vector angular momentum, and in this system the magnetic moment vector is proportional to the angular momentum vector. One is never present without the other. The torque **M** x **B** contains the angular momentum. Hence the dynamics of the angular momentum vector, **L**, is governed by

$$\frac{d\mathbf{L}}{dt} = -\frac{e\mathbf{B}}{2m} \times \mathbf{L} \qquad\qquad (A)$$

This equation says that the angular momentum vector precesses around the **B**-field direction so that L_z and M_z remain constant.

6.3 Show from Equation A of Problem 6.2 that the component of **L** in the **B**-direction does not change, thus proving that $dM_z/dt = 0$. Show also that L^2 remains constant: that $dL^2/dt = 0$. Deduce the frequency of precession.

Answer: Because **B** is in the z direction, $(\mathbf{B} \times \mathbf{L})_z = 0$, which implies by virtue of Equation A that $dL_z/dt = 0$; $dL^2/dt = 2\,\mathbf{L} \cdot d\mathbf{L}/dt = 2\mathbf{L} \cdot \mathbf{\Omega} \times \mathbf{L} = 0$ where $\mathbf{\Omega} = -e\mathbf{B}/2m$

6.4 Even if the beam consists of charged particles, the Stern–Gerlach apparatus can analyze their magnetic moment. For charged particles there is a lateral deflection in the y-direction due to the motion of charge through a B-field. The magnetic moment readout is in the B-field direction—the z direction. Thus the two effects are distinguishable being in perpendicular directions. The y-deflection can be compensated, if need be, by a crossed electric field to cancel it.

What are the direction and magnitude of the crossed electric field necessary to compensate for the transverse deflection due to its charge of a monoenergetic electron beam in the Stern–Gerlach apparatus?

6.5 Derive the equations that show the effects of the $\hat{S}_-$ and $\hat{S}_+$ operators on eigenstates of $\hat{S}_z$: Equations 6-11, 6-12, and 6-13.

6.6 Calculate the matrix elements of the operator $\hat{S}_y$ between all eigenstates of the operator $\hat{S}_z$, and put them in a 2 × 2 matrix array. Choose the traditional placement convention.

$$
\begin{array}{c}
\phantom{+\frac{1}{2}} \quad +\frac{1}{2} \quad\ -\frac{1}{2} \\
\begin{array}{c} +\frac{1}{2} \\[1.5em] -\frac{1}{2} \end{array}
\left(
\begin{array}{cc}
\langle m|\hat{S}_y|M\rangle &
\end{array}
\right)
\end{array}
$$

This matrix multiplied by $2/\hbar$ is one of the three Pauli spin matrices. Find and exhibit all three Pauli spin matrices: $\dfrac{2}{\hbar}(\hat{S}_x)$, $\dfrac{2}{\hbar}(\hat{S}_y)$, and $\dfrac{2}{\hbar}(\hat{S}_z)$.

Key to the solution: Calculate the matrix elements of $1/2i\,(\hat{S}_+ - \hat{S}_-)$ instead of $\hat{S}_y$.

6.7 Show that the single Equation 6-24 is obtained twice in solving the matrix algebra problem of Figure 6-8.

Solution:

$$\begin{pmatrix} 0 & -i\hbar/2 \\ i\hbar/2 & 0 \end{pmatrix}\begin{pmatrix} \langle m = \tfrac{1}{2}|\mu = \tfrac{1}{2}\rangle \\ \langle m = -\tfrac{1}{2}|\mu = \tfrac{1}{2}\rangle \end{pmatrix} = \frac{\hbar}{2}\begin{pmatrix} \langle m = \tfrac{1}{2}|\mu = \tfrac{1}{2}\rangle \\ \langle m = -\tfrac{1}{2}|\mu = \tfrac{1}{2}\rangle \end{pmatrix}$$

These equations both are (6-24).

6.8 Derive (6-27), (6-29), (6-24), and (6-26), and from them produce the transformation matrix of (6-30).

Solution: Notice that (6-24) combined with (6-26) means that for any real α at all: $\langle m = \tfrac{1}{2}|\mu = \tfrac{1}{2}\rangle = (1/\sqrt{2})\exp i\alpha$ and $\langle m = -\tfrac{1}{2}|\mu = \tfrac{1}{2}\rangle = (i/\sqrt{2})\exp i\alpha$.

6.9 What is the probability of an exit $+y$-direction spin if an $s = 1/2$ particle enters the rotated Stern–Gerlach apparatus with spin down in the z-direction?
     Answer: You want $\langle \mu = \tfrac{1}{2}|m = -\tfrac{1}{2}\rangle$. It equals $\langle m = -\tfrac{1}{2}|\mu = \tfrac{1}{2}\rangle^* = -i/\sqrt{2}$.

6.10 Consider the matrix

$$\begin{pmatrix} e^{i\alpha}/\sqrt{2} & e^{i\beta}/\sqrt{2} \\ ie^{i\alpha}/\sqrt{2} & -ie^{i\beta}/\sqrt{2} \end{pmatrix}$$

Where α and β are any arbitrary real numbers. Show that this matrix serves as the one whose elements are $\langle m|\mu\rangle$ just as well as that given in Equation 6-30; that is, show that it diagonalizes the $\hat{S}_y$ matrix of Figure 6-8.

6.11 Diagonalize each of the three Pauli spin matrices constructed in Problem 6.6. Your results should be compatible with the known eigenvalues of the three operators involved: $+\hbar/2$ and $-\hbar/2$.

6.12 Get the inverse of the matrix

$$\begin{pmatrix} 2 & 5i \\ 6 & 3 \end{pmatrix}$$

Is this matrix a unitary one?

Answer:

$$\frac{1}{1-5i}\begin{pmatrix} 1/2 & -5i/6 \\ -1 & 1/3 \end{pmatrix}$$

Not unitary.

6.13 Here is the matrix (A)

$$(A) = \begin{pmatrix} 1 & 2 \\ 1 & 2 \end{pmatrix}$$

Its eigenvalues are real. Find them. Find the transformation matrix U that diagonalizes A. Find the inverse of U: U^{-1}. And find the hermitian conjugate of U: $U\dagger$. Is this U a unitary transformation matrix? Is A hermitian?

Answer:

$$\lambda = 0 \text{ and } 3$$

$$U = \begin{pmatrix} -2/\sqrt{5} & 1/\sqrt{2} \\ 1/\sqrt{5} & 1/\sqrt{2} \end{pmatrix}$$

$$U^{-1} = \begin{pmatrix} -\sqrt{5}/3 & \sqrt{5}/3 \\ \sqrt{2}/3 & \sqrt{8}/3 \end{pmatrix}$$

$$U\dagger = \begin{pmatrix} -2/\sqrt{5} & 1/\sqrt{5} \\ 1/\sqrt{2} & 1/\sqrt{2} \end{pmatrix}$$

This U is not unitary; A is not hermitian.

6.14 For a matrix M that has an inverse, M^{-1}, show that the inverse of its hermitian conjugate is the hermitian conjugate of its inverse.

Hint: $(M\dagger)^{-1} = (M^{-1})\dagger$ because $I\dagger = I$.

6.15 Consider an $N \times N$ matrix Λ. Its eigenvalues λ and a transformation matrix V may be found that ensures

$$\Lambda V = V \begin{pmatrix} \lambda & & 0 \\ & \searrow & \\ 0 & & \end{pmatrix}$$ (B)

Now consider the hermitian conjugate matrix $\Lambda^\dagger$. It too has eigenvalues and a transformation matrix.

Show that the eigenvalues of $\Lambda^\dagger$ are λ^* and that the transformation matrix that diagonalizes $\Lambda^\dagger$ is $(V^\dagger)^{-1}$: the inverse of the hermitian conjugate of the V that diagonalizes Λ.

Solution: Take the dagger of (B) and sandwich the resulting equation between $(V^\dagger)^{-1}$ and $(V^\dagger)^{-1}$.

From this result deduce that physical states are generated only by hermitian operators. To do it you must notice what constitutes a physical state: that $V^{-1} = V^\dagger$ and $\lambda = \lambda^*$.

6.16 Suppose the atomic magnetic dipoles entering the rotated Stern–Gerlach apparatus of Figure 6-4 are characterized by a pure orbital angular momentum $\ell = 1$ instead of a spin. The incoming beam has been selected to have z-component of angular momentum m: m may be $-1, 0$, or $+1$. Set up the eigenvalue problems to describe this experiment, and deduce from the solutions what spectrum emerges from the apparatus and the probability of finding each state in the spectrum.

6.17 An electron has four degrees of freedom. The one that enters the rotated Stern–Gerlach apparatus has a four-variable state $|P,m\rangle$. The momentum $P = (P_x = \sqrt{2\,m\,E},\ P_y = P_z = 0)$ characterizes the motion part of the state of an electron in a monoenergetic beam moving in the $+x$ direction. The exiting electron is in the state $|P,\mu\rangle$ because the effect of the deflection on P is negligible.

In the superposition rule, the ket-bra sum must include all states of the system. For traveling free electrons, the proper statement of the rule is

$$\sum_{\mathbf{p},m} |\mathbf{p},m\rangle\,\langle \mathbf{p},m| = 1$$

Show that the contracted sum in Equation 6-21, $\Sigma |m\rangle\,\langle m| = 1$ with no $\mathbf{p}$ in it, is, in fact, appropriate to our case. You thus prove that the momentum P, being fixed, may be suppressed in writing the states.

Solution: Note that

$$\langle p,m|P,\mu\rangle = \langle p|P\rangle\,\langle m|\mu\rangle$$

and

$$\langle P,m|\hat{S}_y|p,M\rangle = \langle P|p\rangle\,\langle m|\hat{S}_y|M\rangle$$

so

$$\sum_{p,M}\langle P,m|\hat{S}_y|p,M\rangle\,\langle p,M|,P\mu\rangle = \mu\hbar\langle P,m|P,\mu\rangle$$

reduces to

$$\langle P|P\rangle \sum_{M} \langle m|\hat{S}_y|M\rangle\,\langle M|\mu\rangle = \langle P|P\rangle\,\mu\hbar\langle m|\mu\rangle$$

Chapter 7

Time

THERE ARE NONINSTANTANEOUSLY SUBSEQUENT MEASUREMENTS

The events that we have discussed called for *instantaneously subsequent* measurements. The second measurement follows *immediately* upon the first.

A measurement shows that the atom is at position x with rotational quantum numbers ℓ and m. There is a probability to find that the atom has linear momentum p while rotating with orbital angular momentum characterized by ℓ and μ. The amplitude for this event is $\langle \mu\ell p | m\ell x \rangle$.

You must make the μ, ℓ, p measurement upon an atom whose condition is known to be m, ℓ, x just preceding the proposed measurement. Thus the Dirac bracket describes a pair of measurements that are performed *almost simultaneously*.

A STATE EVOLVES IN TIME

Suppose some time elapses between the two measurements. One measurement is made at time $t = 0$. The second measurement is made later, at time t, *not instantaneously subsequent*.

The state measured at $t = 0$ may evolve. By the time t rolls around, the original state may no longer be the state of the system. At the time t, just previous to the second measurement, the state of the system may be quite different from that measured at $t = 0$.

Suppose we represent the whole set of initial state quantum numbers by $|\alpha\rangle$. The state $|m\ell x\rangle$, just mentioned, is an example of $|\alpha\rangle$. The state measured at the time $t = 0$ is $|\alpha\rangle$. After a time t has elapsed, the system evolves into some other state; call it $|\beta\rangle$.

177

THERE'S A FORWARD-EVOLUTION-IN-TIME OPERATOR

That a state evolves in time means that it is a function of time. One can construct a table in which for each moment of time there corresponds a specific recordable physical state of the system into which the initial state evolves.

The time evolution of a state is a matter of physical law. The rule by which $|\beta\rangle$ evolves from $|\alpha\rangle$ after a time t is governed by the following law of quantum mechanics:

$$|\beta\rangle = \exp(-it\hat{H}/\hbar)|\alpha\rangle \qquad (7\text{-}1)$$

The operator $\hat{H}$ is the hamiltonian for the system. The exponential operator in this equation is the forward-displacement-in-time-by-t operator. It produces the new state into which a given state will evolve after time t if the system is left undisturbed.

This profound and far-reaching law is the Dirac notation expression of two principles at the foundation of physics. One is from classical physics, and the other from quantum mechanics.

THE CONSERVATION OF ENERGY MEANS AN ENERGY EIGENSTATE IS STATIONARY

The classical principle is that the energy of a closed system is an invariant of the motion, that the energy doesn't change with time. An isolated system, not interacting with the universe, is a closed one. For such a system, the hamiltonian has no explicit dependence on time. It is equal to the energy, and that energy is conserved; it is time independent. That is the classical principle.

In quantum mechanics the same principle is expressed this way: *energy eigenstates are stationary*.

A stationary state is one that persists in time. For such a state, measurements made later yield the same state as that found earlier, no matter what time has elapsed between measurements.

The quantum statement of the conservation of energy is that energy eigenstates are such stationary states.

Any state of the system consists of many labels in the ket, at least one for each degree of freedom. Let's call one such state $|E \ldots \rangle$. The dots show that there may be other quantum numbers.

Suppose that at time $t = 0$ the state of the system is $|E \ldots \rangle$. This is an eigenstate of energy with eigenvalue E. The conservation of energy assures us that after time t, the state of the system must be $\exp i\phi(t) |E \ldots \rangle$. The earlier and later states can differ only through a phase factor. That guarantees (probability $= 1$) that the state found later is the same as that found earlier; that the state is stationary.

That both the earlier and the later state are eigenstates of the hamiltonian with the same eigenvalue, E, expresses the conservation of energy.

The entire time dependence of an energy eigenstate resides in the phase, ϕ. It is only through this that it can depend on time: nothing else changes. Thus $\phi = \phi(t)$; the phase is a function of time. Of course, at $t = 0$, $\phi = 0$.

The quantum statement of the classical conservation of energy idea is that $|E \ldots \rangle$ evolves into $\exp i\phi(t) |E \ldots \rangle$. This is the first of the two principles upon which rests the time evolution operator of Equation 7-1.

Quantum Principle 2, completed:

7.1 PARTICLE ENERGY EXPRESSES WAVE FREQUENCY

The second of the two principles embodied in Equation 7-1 is this: to the energy, E, of the system corresponds a frequency, $\nu = \omega/2\pi$. The relationship between the two is given by

$$E = \hbar\omega \qquad (7\text{-}2)$$

This is a purely quantum mechanical notion. It has no counterpart in classical mechanics. It completes de Broglie's statement of the duality between particles and waves. It associates particle energy with wave frequency. Earlier, in Equation 2-6, particle momentum was associated with wave vector. It is these two equations that bring the signature of quantum mechanics, Planck's constant, h, into the theory. The entire structure rests on these two statements of de Broglie, that particle momentum means wave vector and that particle energy means wave frequency.

That atoms exhibit characteristic spectra derives from the relationship between energy and frequency. An example is the ubiquitous neon light.

The atoms of the gas of neon in the lamp are excited by an electrical discharge. They radiate off their excitation energy as visible light in a characteristic red glow. This light, rather than being white, exhibits a line spectrum of color. Only certain frequencies emerge and not others. These frequencies, multiplied by Planck's constant, correspond to energy-level differences in the neon atom. The energy, E, lost by a relaxing atom, is carried off in radiation as $h\nu$. Atomic spectra were long known but unexplained before quantum mechanics.

THE PHOTOELECTRIC EFFECT
DEMONSTRATES WAVE-PARTICLE DUALITY

The key experiment demonstrating the wavelength-momentum connection is the diffraction of particles, the Davisson–Germer experiment (see Section 2.4). A critical experiment revealing the frequency-energy relation is the photoelectric effect.

It is interpreted as a clear demonstration of Equation 7-2. Albert Einstein was awarded the Nobel Prize in 1921 for this interpretation.

The experiment is diagramed in Figure 7-1. Monochromatic electromagnetic waves of a prepared frequency, ν, are incident upon a metal surface. Electrons are found to be ejected from the metal. Electrons may be knocked out of a metal surface by a light beam.

These electrons come out with a range of kinetic energies extending up to some measurable maximum amount, K. The key experimental fact lies in this observation: the kinetic energy, K, is found to be directly correlated to the incident radiation frequency, ν. Figure 7-1 shows the correlation determined by experiment.

Below a certain threshold of frequency, ϕ/h, no electrons are ejected. Above this

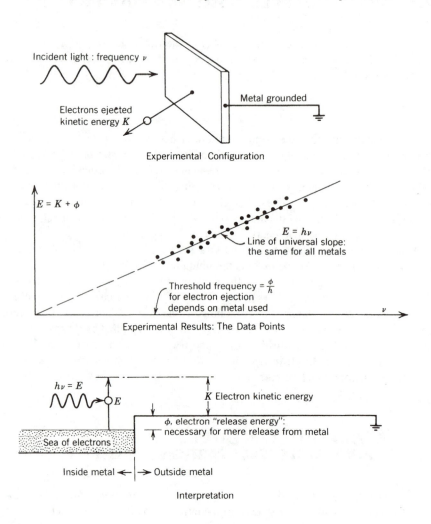

Figure 7.1 **The photoelectric effect.**

frequency, electrons are ejected, and their kinetic energy maximum, K, is directly proportional to v. The threshold depends upon the nature of the metal. The proportionality constant does not; it is the *same for all metals*. It is universal. And it is found to be just Planck's constant.

Here is the interpretation of these facts:

1. A metal consists of a sea of electrons bathing the fixed ion cores of the atoms. The fixed position ion cores express the solid structural integrity of the metal; the electron sea is fluid. An electrical current is a flow in this electron sea. Metals are good conductors because of it.

2. To expel one of these electrons from the metal, a minimum amount of energy, ϕ, is necessary. This amount is required just to raise it from sea level to ground level where the electron can sit outside the metal (see Figure 7-1).

3. The ejected electrons coming from the surface of the sea are the ones with kinetic energy, K. These have the maximum portion possible of their absorbed energy in kinetic form. Of the energy E that they have absorbed, an amount $K = E - \phi$ is the kinetic energy.

4. That the experimental points define the line shown on the graph is the confirmation of the de Broglie statement that $E = hv$. It shows that to each E there corresponds a frequency $v = E/h$. It demonstrates that frequency and energy are equivalent. Frequency is the wave characterization of a state, and energy is its particle characterization.

5. The photoelectric effect has a further significance: it demonstrates the particle nature of light. The ejected electrons absorb energy only in units of hv. This is interpreted as the energy of a light particle, the amount delivered up to the electron that absorbs it.

Particles of light are called *photons*. Each photon has energy hv, all of which it loses to the electron when it is absorbed.

7.2 THE TIME EVOLUTION OPERATOR: ENERGY EIGENSTATES ARE ITS HOME SPACE

Energy conservation and the energy-frequency relation comprise the two physical principles upon which (7-1) is founded. To derive that equation from these principles requires only to consummate the logic mathematically. The argument parallels that for deriving the pure momentum state wave function, Equation 2-8.

With every energy there is to be associated a unique single frequency $\omega = E/\hbar$. Eigenstates of energy develop in time at this fixed frequency.

The only continuous functions of time that have fixed frequency, ω, are linear combinations of $\sin \omega t$ and $\cos \omega t$. Only pure waves have a specific frequency. All other functions contain many frequencies. Thus the function of time $\exp i\phi(t)$ must be a linear combination of $\sin \omega t$ and $\cos \omega t$ where $\omega = E/\hbar$.

The only meaningful linear combination is the one where $\phi = -\omega t$, the function

$\exp(-i\omega t)$. Because ω is positive, for a free particle this choice ensures that the $+t$ direction is forward in time (see Problem 2.6). The only other possibility, $\exp +i\omega t$, is unacceptable: there is at least one case in which it doesn't work, that of free particles (see Problems 2.6, 7.1, and 7.2).

The ideas are displayed mathematically via the superposition theorem. We resolve the initial state $|\alpha\rangle$ into some complete set that embraces the eigenstates of energy.

$$|\alpha\rangle = \sum_{E \ldots} |E \ldots\rangle \langle E \ldots |\alpha\rangle \qquad (7\text{-}3)$$

After a time t, each energy state, $|E \ldots\rangle$, evolves into $\exp(-itE/\hbar)|E \ldots\rangle$. Hence, the later state $|\beta\rangle$, into which $|\alpha\rangle$ evolves, must be given by

$$|\beta\rangle = \sum_{E \ldots} \exp\left(\frac{-itE}{\hbar}\right) |E \ldots\rangle \langle E \ldots |\alpha\rangle \qquad (7\text{-}4)$$

This is the result we wanted to prove. This equation embodies the meaning of (7-1). The two are identical. The exponential operator in (7-1) means just what is shown in (7-4).

The meaning of an operator is founded upon its home space effect, where you remove the hat. Thus, the time evolution operator is defined by its effect on energy eigenstates. The proof that (7-1) and (7-4) are equivalent derives merely from this observation:

$$\sum_{E \ldots} \exp\left(\frac{-itE}{\hbar}\right) |E \ldots\rangle \langle E \ldots |\alpha\rangle$$

$$= \exp\left(\frac{-it\hat{H}}{\hbar}\right) \sum_{E \ldots} |E \ldots\rangle \langle E \ldots |\alpha\rangle \qquad (7\text{-}5)$$

Now we'll apply (7-1) to a physical example.

7.3   A NEUTRON IN A MAGNETIC FIELD: ITS STATE EVOLVES IN TIME

A neutron has a spin. For the neutron, $s = 1/2$, as for the electron. Even though the neutron has no net electrical charge, it does exhibit a magnetic moment. That its magnetic dipole moment shows a multiplicity of 2 leads us to assign it a spin angular momentum index $s = 1/2$. The operator corresponding to the magnetic moment of the neutron is $\hat{\mathbf{M}}$ where

$$\hat{\mathbf{M}} = -3.83 \frac{e}{2\mathcal{M}} \hat{\mathbf{S}} \qquad (7\text{-}6)$$

In this equation the mass, $\mathscr{M}$, is that of a proton, and the vector operator, $\hat{\mathbf{S}}$, is that of spin angular momentum for index $s = 1/2$. This just parallels the electron case discussed in Section 6.3. The g-factor of the neutron is -3.83; it exhibits a magnetic moment z-component of 1.91 nuclear magnetons (see Equations 6-14 and 6-15). The nuclear magneton is the unit in which the magnetic moments of nuclear particles are tabulated. It is to a Bohr magneton (Equation 6-2) what the ratio of the electron mass is to the proton mass.

$$1 \text{ nuclear magneton} \equiv e\hbar/2\,\mathscr{M} = 3.15 \text{ micro eV/Mega gauss} \qquad (7\text{-}7)$$

$$\approx 1/1840 \text{ Bohr magnetons}$$

A beam of neutrons passes through a uniform magnetic field. The experimental configuration is shown in Figure 7-2. A beam of incoming neutrons is prepared with known y-direction spin component. Each neutron travels in the $+x$-direction with momentum $\hbar K$. It enters the region of uniform magnetic field at time $t = 0$. The effect of the magnetic field on the spin, the interaction, begins at $t = 0$. From this time until the neutron leaves the field, the spin orientation part of the hamiltonian is

$$\hat{H} = -\hat{\mathbf{M}} \cdot \mathbf{B} = (3.83eB/2\,\mathscr{M})\,\hat{S}_z = \Omega\,\hat{S}_z \qquad (7\text{-}8)$$

The left-hand equalities express the energy of the neutron magnetic moment when it is in a magnetic field B pointing in the $+z$-direction. It is the application of (6-3) to our spin one-half neutron case.

The right-hand equality defines the symbol Ω. It represents a collection of parameters

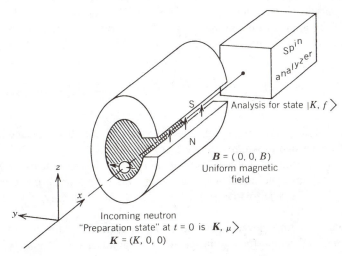

Figure 7.2 **A neutron spends time t in a magnetic field.**

that together carry the dimensions of angular frequency (see Problems 6.2 and 6.3). Thus

$$3.83eB/2 \, \mathscr{M} \equiv \Omega = (2\pi \times 2.9 \, KHz/\text{gauss})B \qquad (7\text{-}9)$$

The neutron enters the field at time $t = 0$ in a spin state $|\mu = 1/2\rangle$. This is the initial state of the system, the $|\alpha\rangle$ of Equation 7-1. It's an eigenstate of the operator $\hat{S}_y$ with eigenvalue $\mu = 1/2$.

$$\hat{S}_y \, |\mu\rangle = \mu\hbar \, |\mu\rangle \qquad (7\text{-}10)$$

As in Chapter 6, we index the eigenstates of $\hat{S}_y$ by μ.

The neutron state evolves in time. After a time t spent in the field, the spin state of the neutron is given by a new quantum number label, β. It is in the state $|\beta\rangle$ where

$$|\beta\rangle = \exp\left(\frac{-it\Omega\hat{S}_z}{\hbar}\right) |\mu = 1/2\rangle \qquad (7\text{-}11)$$

The state $|\mu = 1/2\rangle$ is not an eigenstate of $\hat{S}_z$. Label the eigenstates of $\hat{S}_z$ by m. What defines $|m\rangle$ is

$$\hat{S}_z \, |m\rangle = m\hbar \, |m\rangle \qquad (7\text{-}12)$$

Through Equation 7-8, m also labels the eigenstates of $\hat{H}$. To exhibit the state $|\beta\rangle$ we write the initial state, $|\alpha\rangle = |\mu = 1/2\rangle$, in terms of the hamiltonian's eigenstates $|m\rangle$. That's the prescription for carrying out the calculation, the one laid down in Equation 7-4.

It is easily accomplished using the ket-bra sum superposition theorem. You need only the unitary transformation elements, $\langle m|\mu = 1/2\rangle$, but these have already been calculated. They are displayed in the matrix of Equation 6-30. Here is what you get when you carry out this calculation (see Problem 7.5):

$$|\beta\rangle = |m = 1/2\rangle \frac{1}{\sqrt{2}} \exp\left(\frac{-it\Omega}{2}\right) + |m = -1/2\rangle \frac{i}{\sqrt{2}} \exp\left(\frac{it\Omega}{2}\right) \qquad (7\text{-}13)$$

Both $|\alpha\rangle$ and $|\beta\rangle$ are examples of nonstationary states. Proof: the later state is not the same as the earlier one to within a phase factor.

At any time a state is a linear combination of all the eigenstates, $|m\rangle$, of the hamiltonian operator. The initial state, $|\mu = 1/2\rangle$, was such a linear combination, so is the later state, $|\beta\rangle$. The relative phases associated with each energy eigenstate are what change with time.

The formula for time development, in (7-1), appears forbiddingly abstract. It is simply a notational recipe for a quite practical program of calculation. With it one easily displays the later state that evolves from an earlier one. The calculation of (7-1) can procede via (7-4), and (7-13) is an example of it.

WE DAWDLE BETWEEN MEASUREMENTS
THE STATES EVOLVE

Knowing how states evolve in time, we are equipped to allow for measurements that are not instantaneously subsequent. Suppose the system is in a state $|\alpha\rangle = |i\rangle$ initially. What is the amplitude to find it in state $|f\rangle$ after a time t has elapsed?

To answer this question we must know the state of the system at the time t just previous to the $|f\rangle$ measurement. The $|f\rangle$ measurement is made instantaneously subsequent to the state of the system into which $|i\rangle$ evolves at time t. But it is just this state, $|\beta\rangle$, that (7-1) gives us. Thus the amplitude to find $|f\rangle$ at the time t, if we knew the state to be $|i\rangle$ at the time $t = 0$, is

$$\langle f|\beta\rangle = \langle f|\exp(-it\hat{H}/\hbar)|i\rangle \qquad (7\text{-}14)$$

Here's the idea as applied to our neutron spinning in a magnetic field. The amplitude to find a spin z-component of $1/2$ at a time t after the neutron entered the magnetic field with spin y-component $1/2$ is

$$\langle m=1/2|\exp(it\hat{H}/\hbar)|\mu=1/2\rangle = \frac{1}{\sqrt{2}}\exp\left(\frac{-it\Omega}{2}\right) \qquad (7\text{-}15)$$

This result is a direct application of Equation 7-14: apply the bra $\langle f| = \langle m=1/2|$ to the ket $|\beta\rangle$ given in Equation 7-13.

This equation tells us that the probability to find spin up, $+z$-direction, in a magnetic field remains constant in time: it is the absolute value squared of (7-15). This spin up state is an eigenstate of energy. That the probability of finding such a state does not vary with time means that the state is a stationary one. Thus Equation 7-15 exemplifies the principle that energy eigenstates are stationary (see Problem 7.3).

FIRST WRITE THE ANSWER, THEN SOLVE THE PROBLEM

Suppose we don't measure one of the energy eigenstates. Instead we ask for the amplitude to find a spin y-component of $1/2$ at the time t if the neutron entered the field at time $t = 0$ with exactly this same spin.

We can write down the answer immediately: $\langle\mu=1/2|\exp -i\Omega t\hat{S}_z/\hbar|\mu=1/2\rangle$. Dirac notation allows us to write down the form of the answer before solving the problem.

The problem is reduced to calculating a matrix element. This calculation is the substance of Problem 7.6, and the result is

$$\langle \mu = 1/2 | \exp(-it\hat{H}/\hbar) | \mu = 1/2 \rangle = \cos \frac{\Omega t}{2} \tag{7-16}$$

The amplitude oscillates in time at frequency $\Omega/2$. The probability of finding $|\mu = 1/2\rangle$ varies sinusoidally in time at the frequency Ω (see Problem 7.6). The state $|\mu = 1/2\rangle$ is not stationary, whether or not you find it changes with time.

This time variation issues from the relative phase differences among energy eigenstates in the superposition. The relative phase differences among terms in a superposition expansion have profound physical effects: the overall phase of the system state represented by this expansion has no physical effect at all.

7.4 THE SCHROEDINGER EQUATION: DIFFERENTIAL EVOLUTION IN TIME

A milestone in the early development of quantum mechanics was a famous formula known as the Schroedinger equation. It preceded Dirac's formulation of quantum mechanics. Out of it grew many of the clues leading to the conceptual structure portrayed in this text. In 1933 the Nobel Prize in Physics was awarded jointly to E. Schroedinger and P. A. M. Dirac for their contributions to quantum theory.

The Schroedinger equation is, in fact, a special case of the time evolution theorem of (7-1). It is the differential form of this theorem when applied to a wave function, to a state expressed in position language.

Here is the Schroedinger equation as it appears in traditional notation:

$$\hat{H}\psi = i\hbar \frac{\partial \psi}{\partial t} \tag{7-17}$$

The Dirac notation for $\psi = \psi(x,t)$ is $\langle x|\psi\rangle$ and for $\hat{H}\psi$ is $\langle x|\hat{H}|\psi\rangle$.

To derive this from (7-1) you need merely note that in the case of small time increment, Δt, the exponential operator is well represented by

$$\exp\left(\frac{-i\Delta t\hat{H}}{\hbar}\right) = 1 - \frac{i\Delta t\hat{H}}{\hbar} \tag{7-18}$$

The smaller the time increment, the more exact is this relation.

The proof is easy. Just make your calculations in a basis that embraces the eigenstates of $\hat{H}$, the basis $|E \ldots \rangle$. Then you can remove the hat on the operator, and you need confront only algebraic quantities. Make the algebraic expansion for the exponential function of small argument. The argument, $\Delta t E/\hbar$, must be small. Hence the proof applies to all states $|\psi\rangle$ for which $\langle E \ldots |\psi\rangle$ becomes negligibly small for large E.

Recasting the algebraic result back into operator formalism completes the proof. This program is the substance of Problem 7.7.

Use Equation 7-18 in this way. Call the state of the system at time t by the symbol $|\psi\rangle$. The state of the system at time $t + \Delta t$ is called $|\psi + \Delta\psi\rangle$. Then, by virtue of (7-18), (7-1) tells us that the difference between these two states, when cast in the language of position, is

$$\langle x|\psi + \Delta\psi\rangle - \langle x|\psi\rangle = \frac{-i\Delta t}{\hbar}\langle x|\hat{H}|\psi\rangle \tag{7-19}$$

Divide both sides of this equation by Δt and take the limit $\Delta t \to dt$. The result and (7-17) are equivalent. The meaning of (7-19) is just what is expressed in (7-17); (7-19) is its Dirac notation form. The Schroedinger equation, (7-17), is the differential method of expressing the time evolution of a state.

REPEATED MEASUREMENTS OF AN OBSERVABLE YIELD ITS AVERAGE VALUE

Let the operator $\hat{A}$ represent a measurement observable.

In our spinning neutron example, one may measure the B-field-direction spin component, the component of spin in the z-direction. The operator corresponding to this measurement is $\hat{S}_z$: $\hat{S}_z$ is an example of what $\hat{A}$ might be.

In any one measurement of an observable we can find only a point in its spectrum of eigenvalues; only one of the allowed values, A, that the operator $\hat{A}$ can produce. In the neutron example one may find either $+\hbar/2$ or $-\hbar/2$ as measurement results; these would be the allowed values of A if the operator $\hat{A}$ were $\hat{S}_z$.

The probability of finding any one particular value of A depends, of course, on the state of the system just previous to the A measurement. If we make our A measurement at $t = 0$, this state is $|\alpha\rangle$. In our spinning neutron example, this prepared, or known, state of the system is $|\mu = 1/2\rangle$. The probability (density) of finding the particular value A in an instantaneously subsequent measurement is $|\langle A|\alpha\rangle|^2$.

Now suppose that we repeat the experiment a large number of times. Each time we make a measurement we find some value in the spectrum of A's. We can deduce the average value of A from the experimental data. Experimentally, the average value is the sum over A of each value of A multiplied by the relative number of times it turns up in experiments. This relative number of times is simply the probability for the event: $|\langle A|\alpha\rangle|^2$.

Thus, by virtue of its meaning, the average value of A, called $\bar{A}$ or $\langle\hat{A}\rangle$ or $\langle A\rangle$, is

$$\langle A\rangle \equiv \sum_A A|\langle A|\alpha\rangle|^2 = \langle\alpha|\hat{A}|\alpha\rangle \tag{7-20}$$

The left-hand equality is the definition of *average value*. The right-hand equality is an alternative rendering of this result, a more elegant rendering (see Problem 7.8). What should be marked is the equality of the extremes; the average value of an operator is a diagonal matrix element.

THE OBSERVABLE'S AVERAGE EVOLVES IN TIME

If we don't make our measurements instantaneously subsequent to state preparation, the average value that we obtain might be different. Because the state evolves with time, so does the average value of measurements corresponding to $\hat{A}$.

The prepared state, $|\alpha\rangle$, evolves into $|\beta\rangle$ at time t. So if we make our repeated measurements of $\hat{A}$ at time t rather than at time $t = 0$, the average value of the A's that we find will be $\langle\beta|\hat{A}|\beta\rangle$.

When expressed in terms of the prepared state $|\alpha\rangle$, the time evolution of the average value becomes apparent.

$$\langle\beta|\hat{A}|\beta\rangle = \langle\alpha|\exp(it\hat{H}/\hbar)\,\hat{A}\,\exp(-it\hat{H}/\hbar)|\alpha\rangle \tag{7-21}$$

The proof of this equation is explored in Problems 7.11 and 7.12. It requires only the observation that for the time-evolution operator, hermitian conjugation is equivalent to time reversal, to going backward in time.

$$\left[\exp\left(\frac{-it\hat{H}}{\hbar}\right)\right]\dagger = \exp\left(\frac{it\hat{H}}{\hbar}\right) \tag{7-22}$$

Here is the content of Equation 7-21 expressed in differential form:

$$\frac{d}{dt}\langle\beta|\hat{A}|\beta\rangle = \langle\beta|\frac{\partial\hat{A}}{\partial t} + \frac{1}{i\hbar}[\hat{A}, \hat{H}]|\beta\rangle \tag{7-23}$$

That this equation follows from (7-21) is deduced in Problem 7.14. Here we want to focus on the significance of this very important result.

It describes how the average value of an operator changes with time. The time rate of change of this average is the average of a new operator bracketed between the same states. The new operator is the sum of the pair on the right-hand side of (7-23).

The first of the pair is the derivative with respect to time of whatever time dependence is explicitly exhibited by the operator $\hat{A}$. Such a time dependence corresponds to varying your measurement apparatus with time during the experiment! As an example, revolving your spin analyzer with time yields an explicit $\hat{A}$ time dependence. If the rotation period is T, then it measures $\hat{A} = -\hat{S}_x \sin 2\pi t/T + \hat{S}_y \cos 2\pi t/T$. We treat only the case in which you make measurements that stay fixed in time: $\partial\hat{A}/\partial t = 0$.

The second term of the pair is the operator produced by the commutation of $\hat{A}$ with

$\hat{H}$, the commutator of $\hat{A}$ with $\hat{H}$. The average value of a fixed measurement observable, $\hat{A}$, changes with time if $\hat{A}$ doesn't commute with $\hat{H}$.

This equation is the key to two important physical ideas. The first concerns the *classical limit* of quantum mechanics, and the second concerns *invariant observables*.

7.5 THE CLASSICAL LIMIT OF QUANTUM MECHANICS: WHAT HAPPENS ON THE AVERAGE

Suppose you apply the machinery of quantum mechanics in the classical domain. You must get the classical answer. The machinery of quantum mechanics applied in the classical domain is Equation 7-23.

By the classical domain I mean where the quantum of action, $\hbar$, does not enter. Classical measurements are those that don't see quantum effects. The experiments are too rough; they see only average results. That's why we expect (7-23) to yield the classical results for observables. That equation is the quantum theory expression for *any* of the equations of motion of classical physics (see Problem 7.15).

This notion is called the Correspondence Principle: it ensures that newtonian mechanics appears as the classical limit of quantum mechanics.

The spinning neutron in the magnetic field of Figure 7-2 illustrates the idea.

Although neutron spin itself is a nonclassical quantity, its operator follows the rules of angular momentum. Angular momentum is a classical quantity. Classical relations regarding angular momentum must also be obeyed by spin. Any relation involving angular momentum that is true in classical physics must be true on the average in quantum physics, even for spin.

The hamiltonian for our neutron in the magnetic field is that of Equation 7-8. Take $\hat{A} = \hat{S}_z$, $\hat{S}_x$ and $\hat{S}_y$ in succession, and evaluate the commutator $[\hat{A}, \hat{H}]$ in each case. Then execute Equation 7-23. The result is a set of three equations, one for the time derivative of each component of the vector operator $\hat{\mathbf{S}} = (\hat{S}_x, \hat{S}_y, \hat{S}_z)$. The three equations governing the average of $\hat{\mathbf{S}}$ can be summarized in a single vector equation if we attribute a vector nature to the frequency Ω. Define the vector $\mathbf{\Omega}$ as having magnitude Ω and pointing in the *B*-field direction. Then here's what (7-23) says about the average of the neutron spin:

$$\frac{d}{dt} \langle \hat{\mathbf{S}} \rangle = \mathbf{\Omega} \times \langle \hat{\mathbf{S}} \rangle \tag{7-24}$$

This is the classical expectation for the angular momentum of a spinning dipole in a magnetic field. The classical result is given as Equation A in Problem 6.2. It says that the angular momentum vector precesses around the field direction with precession frequency Ω. So, too, is it with the spin angular momentum—see Problems 6.3 and 7.16. The quantum average is the classical result.

WHAT HOLDS STEADY AMIDST TURMOIL: THE INVARIANTS OF THE SYSTEM

There is a class of operators for which the average of the measurement results never changes; the average result is time independent. Such operators represent the invariants of the system. An observable whose average value stays constant in time is an invariant of the motion.

Invariants are the foundation stones upon which physical understanding is built. To perceive what it is that holds steady amidst apparent turmoil is to perceive the structure of the process.

The energy of a closed system is just such an invariant. Its invariance is enshrined in the phrase "conservation of energy". The invariance of energy has already been incorporated into the theory, in Equation 7-4.

7.6 INVARIANTS COMMUTE WITH THE HAMILTONIAN

Equation 7-23 tells us what critical properties invariants must have. Given that $\partial \hat{A}/\partial t = 0$, all $\hat{A}$ for which $[\hat{A},\hat{H}] = 0$ have stationary time averages; for them $d\langle \hat{A} \rangle/dt = 0$. They are invariants.

Our spinning neutron in a magnetic field provides an example. The case $\hat{A} = \hat{S}^2$ is one in which the $\hat{A}$ operator is not explicitly time dependent and commutes with the hamiltonian governing the system; it commutes with the $\hat{H}$ of Equation 7-8. Thus the total angular momentum of spin is an invariant of the system: $S^2 = s(s + 1)\hbar^2 = 3\hbar^2/4$ is preserved in a magnetic field. The field cannot change the neutron's $s = 1/2$ spin. Precisely this same result is reached classically in Problems 6.2 and 6.3.

For contrast, consider the operator $\hat{S}_y$. Although like $\hat{S}^2$, it is not explicitly time dependent, unlike $\hat{S}^2$, it does not commute with $\hat{H}$. Hence $S_y = \mu\hbar$ is not invariant. Indeed it isn't: an entering $\mu = 1/2$ neutron can be found at the exit with $\mu = -1/2$.

FROM INVARIANT OPERATORS COME STATIONARY MEASUREMENT RESULTS

A measurement result corresponding to an operator that is an invariant of the system is stationary: it persists in time. You get the same result later as you got earlier.

Suppose at t = 0 the state is $|\alpha\rangle = |A. . . \rangle$, an eigenstate of the operator $\hat{A}$. Then so is the state $|\beta\rangle$ to which it evolves an eigenstate of $\hat{A}$ if $\hat{A}$ is an invariant; if it commutes with $\hat{H}$. Equation 7-21 shows it. Since $\hat{A}$ commutes with $\hat{H}$ so does it with any function of $\hat{H}$. Hence the bracketed operator on the right reduces to $\hat{A}$. Thus a later $\hat{A}$-measurement yields the same eigenvalue as the earlier one.

The spinning neutron is a case in point. Both the operators $\hat{S}_z$ and $\hat{S}^2$ commute with

$\hat{H}$. Thus $\hat{S}_z$ and $\hat{S}^2$ measurement results are stationary; an entering neutron characterized by m and s exits with *the same* m and s.

IN AN ENERGY EIGENSTATE, ALL THE LABELS MARK INVARIANTS

Each of the many labels of a multidimensional state indexes an operator. And all of these operators must be compatible with one another. No state exists that isn't simultaneously eigen to all of its own labels' operators. Thus all of these operators commute with one another.

It follows that in an energy eigenstate, all the labels correspond to invariants of the system. In the state $|E \ldots \rangle$, the dots must label only other invariants. The operators, to which they correspond, being part of the ket, must commute with $\hat{H}$. But if they commute with $\hat{H}$, they are invariants. Thus all of the indices in an energy eigenstate label invariants.

THE STATE PREPARED IS EIGEN TO SOME OPERATOR: THE STATE EVOLVED, TO SOME OTHER OPERATOR

A system is prepared in a state $|\alpha\rangle$; this state is an eigenstate of some operator. It must be so because the label α is a measurement observable. The state-preparation apparatus derives from some measurement observable that has α as a point in its spectrum of measurement results. To this measurement observable corresponds the physical operator, say $\hat{\alpha}$, of which $|\alpha\rangle$ is an eigenstate.

The new state $|\beta\rangle$, into which $|\alpha\rangle$ evolves after time t, is also an eigenstate of some operator: some entirely different operator. It is not a matter of a new value for the old quantum number. In fact, the value remains unchanged: What this value represents, changes. It represents a new observable.

The value of the quantum label itself doesn't change at all; it is the language in which it is a quantum label that changes. The operator to which $|\beta\rangle$ is eigen may be quite different from the operator to which $|\alpha\rangle$ is eigen. There is a physical operator, $\hat{\beta}$, to which the time-evolved state $|\beta\rangle$ is eigen. This operator represents a different observable than does $\hat{\alpha}$.

THE LANGUAGE OF THE QUANTUM LABEL CHANGES WITH TIME

Our spinning neutron in the magnetic field of Figure 7-2 provides an object example of the idea. The prepared state of the neutron is $|\mu = 1/2\rangle$, an eigenstate of the $\hat{S}_y$ operator. In this example $\alpha = \mu = 1/2$, and $\hat{\alpha}$ is $\hat{S}_y$.

The state of the neutron after time t is $|\beta\rangle$. It is an eigenstate of an entirely different operator, the operator $\hat{S}_\eta$.

This is a spin operator, just like $\hat{S}_x$ and $\hat{S}_y$, but it refers neither to the x- nor the y-direction but to the η-direction shown in Figure 7-3. This direction is displaced by an angle ϕ from the original y-axis direction; it amounts to a rotated y-axis. The η-axis is the y-axis rotated about z by an angle ϕ.

The $\hat{S}_\eta$ operator is the spin angular momentum operator for this axis, and $|\beta\rangle$ is an eigenstate of this operator if $\phi = \Omega t$. To verify this assertion it's useful to label the eigenvalues of $\hat{S}_\eta$. We'll index them by η. Thus

$$\hat{S}_\eta|\eta\rangle = \eta\hbar|\eta\rangle \tag{7-25}$$

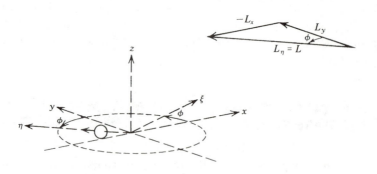

The classical view: The measured value is $L = L_\eta$, so $L_\xi = 0$ and

$$L = L_\eta = -L_x \sin \phi + L_y \cos \phi$$
$$0 = L_\xi = L_x \cos \phi + L_y \sin \phi$$

The quantum view: The measurement corresponding to $\hat{L}_\eta$ yields $L_{\bar{\eta}} = \eta\hbar$. The average of $\hat{L}_\xi$ measurements on a system in state $|\eta\rangle$ is zero.

$$\eta\hbar = \langle\eta|\hat{L}_\eta|\eta\rangle = -\langle\eta|\hat{L}_x|\eta\rangle \sin \phi + \langle\eta|\hat{L}_y|\eta\rangle \cos \phi$$
$$0 = \langle\eta|\hat{L}_\xi|\eta\rangle = \langle\eta|\hat{L}_x|\eta\rangle \cos \phi + \langle\eta|\hat{L}_y|\eta\rangle \sin \phi$$

Hence the operators transform as vectors do:

$$\hat{L}_\eta = -\hat{L}_x \sin \phi + \hat{L}_y \cos \phi$$

$$\hat{L}_\xi = \hat{L}_x \cos\phi + \hat{L}_y \sin \phi$$

Figure 7-3 To angular momentum corresponds a vector operator: Its components transform as do those of a vector.

Only the usual two values, $\pm 1/2$, span the spectrum of η because the neutron always carries spin $s = 1/2$.

The state $|\beta\rangle$, which evolves from $|\mu = 1/2\rangle$ after a time t, is the $\eta = +1/2$ state of the operator $\hat{S}_\eta$ where the angle ϕ is just Ωt. That's the assertion. Put mathematically it is

$$|\langle \eta = 1/2, (\phi = \Omega t)|\beta\rangle|^2 = 1 \qquad (7\text{-}26)$$

Here's the physical statement. You orient your spin measurement apparatus, the spin analyzer of Figure 7-2, to measure spin components in the η-direction, of neutrons that have spent a time t in the magnetic field. You will find all the spins pointed in the $+\eta$ direction. That $\eta = 1/2$ is a certainty; neutrons will never be found with spin $\eta = -1/2$.

Thus neutrons that were in the state $|\mu = 1/2\rangle$ find themselves, after a time t, in the state $|\eta = 1/2\rangle$. The state signature has changed from μ to η. There exists an experiment that defines the new state, as there existed one that defined the original state. The time-evolved state is eigen to some operator, albeit one different from that characterizing the initial state.

Now we will explore the proof that the assertion is true. We want to validate (7-26), to do this we need to appreciate why and how the operator $\hat{S}_\eta$ is related to $\hat{S}_x$ and $\hat{S}_y$. The rest of the proof requires no new insights, just a careful use of things already explored.

7.7 ANGULAR MOMENTUM VECTOR OPERATORS ROTATE AS THEIR VECTOR COUNTERPARTS DO

The angular momentum operator for the η-direction is related to those for the x- and y-directions by

$$\hat{S}_\eta = -\sin \phi \, \hat{S}_x + \cos \phi \, \hat{S}_y \qquad (7\text{-}27)$$

Angular momentum operators transform just as their classical vector component counterparts do. You merely put hats on the classical observables.

Here's the argument in support of (7-27). Suppose a classical rotator had an angular momentum of magnitude $L = L_\eta$ pointing in the η-direction, as illustrated in Figure 7-3. Then classical measurements of its x- and y-components must yield the L_x and L_y determined from the plane geometry of the figure.

But classical measurements correspond to averages quantum mechanically. And our spin operators are angular momentum operators. Thus we must expect that a spin angular momentum known to be $S_\eta = \eta\hbar$ in the η-direction will exhibit an average value of its x-component, $\langle \eta|\hat{S}_x|\eta\rangle$, and of its y-component, $\langle \eta|\hat{S}_y|\eta\rangle$, commensurate with the classical result.

$$\langle\eta|\hat{S}_\eta|\eta\rangle = \eta\hbar = -\langle\eta|\hat{S}_x|\eta\rangle \sin\phi + \langle\eta|\hat{S}_y|\eta\rangle \cos\phi \qquad (7\text{-}28)$$

This is an application of the Correspondence Principle: the classical result is stated in the language of quantum mechanics. It is on this basis that we wrote Equation 7-27. This mathematical statement about the operator $\hat{S}_\eta$ expresses what observable it is; it measures the spin component in the η-direction.

The rest of the proof of (7-26) requires only that we obtain the unitary transformation matrix elements, $\langle m|\eta\rangle$, between states of $\hat{S}_z$ and $\hat{S}_\eta$. With these it is a straightforward matter to calculate $\langle\eta|\beta\rangle$ directly; you insert an m-space ket-bra sum and use (7-13).

To get these matrix elements we need to solve an eigenvalue problem, (7-25) cast into $\langle m|$ language. The procedure is that outlined around Figure 6-8. What is called there μ corresponds here to η. We must diagonalize the m-space matrix of $\hat{S}_\eta$, the matrix whose elements are $\langle m|\hat{S}_\eta|M\rangle$.

We first construct this matrix in m-space. It was to do just this that (7-27) was needed. Bracket that equation between $\langle m|$ and $|M\rangle$. Because we know $\hat{S}_x$ and $\hat{S}_y$ in m-space (see Problem 6.6 on the Pauli matrices), we can easily find the $\hat{S}_\eta$ matrix:

$$\left(\underset{\downarrow}{\langle m|\hat{S}_\eta|M\rangle}\right) = \begin{pmatrix} 0 & \dfrac{-i\hbar e^{-i\phi}}{2} \\ \dfrac{i\hbar e^{i\phi}}{2} & 0 \end{pmatrix} \qquad (7\text{-}29)$$

In Figure 6-8, the matrix $\hat{S}_y$ is displayed: it is the special $\phi = 0$ case of the $\hat{S}_\eta$ matrix in (7-29). The solution for the problem in the figure is the $\langle m|\mu\rangle$ matrix given in Equation 6-30. We need $\langle m|\eta\rangle$, the unitary transformation matrix that diagonalizes the $\hat{S}_\eta$ matrix in (7-29). It is

$$\left(\underset{\downarrow}{\langle m|\eta\rangle}\right) = \begin{pmatrix} \dfrac{1}{\sqrt{2}} & \dfrac{1}{\sqrt{2}} \\ \dfrac{ie^{i\phi}}{\sqrt{2}} & \dfrac{-ie^{i\phi}}{\sqrt{2}} \end{pmatrix} \qquad (7\text{-}30)$$

With this result it is easy to calculate $\langle\eta|\beta\rangle$. Split it with an m-space ket-bra sum, and use (7-13) (see Problem 7.21). You will get

$$\langle\eta=1/2|\beta\rangle = \exp\left(\frac{-i\phi}{2}\right) \cos\frac{1}{2}(\Omega t - \phi) \qquad (7\text{-}31)$$

and

$$\langle \eta = -1/2 | \beta \rangle \; = \; -i\exp\left(\frac{-i\phi}{2}\right) \, \sin \frac{1}{2}(\Omega t - \phi) \qquad (7\text{-}32)$$

When $\phi = \Omega t$, the state $|\beta\rangle$ is the state $|\eta = 1/2\rangle$; they differ only through a phase factor. Making an experimental measurement of the spin component in the η-direction, $\phi = \Omega t$, at time t, must yield a value of $\hbar/2$ with certainty. Equation 7-26 is proved true.

IF YOU CHOOSE THE RIGHT MEASUREMENT, THERE IS NO UNCERTAINTY

A central feature of quantum mechanics is *uncertainty*. The result in any measurement is uncertain. For a particular result there is only a probability. This is the signature of uncertainty. If the result were certain, there would be no probability associated with it.

That a result will certainly happen means with *probability one;* that it will certainly not happen means with *probability zero*. Any probability between these characterizes uncertainty.

In quantum mechanics what we calculate are amplitudes for events, and these yield probabilities, generally between zero and one. Only the probable outcome of a measurement is predicted by quantum mechanics; uncertainty about the outcome is inherent in its structure.

THERE'S ALWAYS A LANGUAGE OF CERTAINTY

Uncertainty does characterize measurement results in almost all cases, but there is a class of measurements whose results are always certain. For every closed system there is a language of certainty. If your measurements are in this language, all the probabilities for events will be either zero or one. Events either must happen or cannot happen.

A system is always in some state. In the self-space of this state, all events are certain. The language of certainty is the one in which the system state resides. The state of the system is one of the words in this language. In this space the amplitudes for events are self-space brackets; the only probabilities they yield are zero or one: certainties. Measurement results in the eigenspace of the system's state are predictable with certainty.

Thus if you choose the right measurements, there is no uncertainty about the result. Pigheaded insistence upon measuring other observables causes the uncertainty. It is only when you insist on looking at observables foreign to the state that uncertainty arises (see Figure 7-4).

Regardless of what measurement you perform on it, the system is always in some state, with certainty. The right measurement discovers this state: it finds it with certainty. Other measurements yield probabilities; the probability distribution of the state in the measurement space you insist on observing.

The truth of these assertions hangs entirely on this supposition: that the ket $|\beta\rangle$ into which $|\alpha\rangle$ evolves is itself an eigenstate of some physical operator. Were it so, then to this operator corresponds the certainty measurement.

A STATE IS ALWAYS EIGEN TO SOME OPERATOR

There is, indeed, such a physical operator. Call it $\hat{\beta}$. This operator can be displayed. In terms of the time elapsed and the original operator $\hat{\alpha}$, from which $|\alpha\rangle$ came, here it is:

IN HOME SPACE THERE'S NO UNCERTAINTY

The home space of the state	$|a,b,g\rangle$
has labels	$a,\ b,\ g$
that index the physical observables	$\hat{A},\ \hat{B},\ \hat{G}$
each of which is compatible with any other.	$[\hat{A},\ \hat{B}] = 0$ $[\hat{A},\ \hat{G}] = 0$ $[\hat{B},\ \hat{G}] = 0$
A foreign space label indexes the measurement space of an incompatible observable.	λ $\hat{\Lambda}|\lambda\rangle = \lambda|\lambda\rangle$ $[\hat{\Lambda},\ \hat{A}] \neq 0$
The foreign space measurement carries uncertainties.	$0 < |\langle\lambda...|a,b,g\rangle|^2 < 1$
Home space measurements carry none.	$\langle A,b,g|a,b,g\rangle = 0$ or 1 self-space bracket

---

Figure 7.4 Poem

$$\hat{\beta} = \exp\left(\frac{-it\hat{H}}{\hbar}\right) \hat{\alpha} \exp\left(\frac{it\hat{H}}{\hbar}\right) \quad (7\text{-}33)$$

The state $|\beta\rangle$ is an eigenstate of this operator; it has as eigenvalue just the same α as labeled the original state (see Problem 7.22). The operator $\hat{\beta}$ has real eigenvalues; it is hermitian (see Problem 7.23). There can exist a measurement corresponding to $\hat{\beta}$, and this measurement yields no uncertainty.

Of course, a good experimenter in the laboratory is not governed by obstinancy; he or she measures whatever possible. Uncertainty is not entirely due to pigheadedness; it's also due to impotence. An experimenter may not be able to make the measurement corresponding to $\hat{\beta}$.

THOUGH IT BE EVER CHANGING, THERE'S NO SPACE LIKE HOME

Our spinning neutron in the uniform magnetic field provides an object example. A neutron enters the field region in the state $|\mu = 1/2\rangle$; that's the α state. After having spent time t in the field, the neutron finds itself in the analyzer (see Figure 7-2).

If the experimenter orients the spin analyzer to measure $\hat{S}_y$, he or she will encounter uncertainty. The neutron is found with either $\mu = +1/2$ or $\mu = -1/2$.

If the experimenter reorients the spin analyzer so as to measure $\hat{S}_\eta$, he or she will encounter certainty. No neutron measured can be in the $\eta = -1/2$ state. Every neutron that enters the analyzer must be in the $\eta = +1/2$ state, with certainty (see Problem 7.24).

The home space language of the state comes from $\hat{S}_\eta$ experiments; $\hat{S}_y$ experiments generate a foreign language. If you insist, in the wrong language, on knowing something, you will get an imperfect answer. Some things are not comprehensible in one language that are easily comprehended in another: to get an answer with certainty you must speak the language in which such an answer can be delivered.

COUNTING CYCLES MEASURES FREQUENCY: TO WITHIN ONE COUNT IS HIGH ACCURACY

In 1944 the Nobel Prize in Physics went to I. I. Rabi for his ingenious use of the time dependence in quantum effects: he invented the magnetic resonance technique of measurement.

Rabi appreciated that measurements of frequency can be made with greater accuracy than can any other type of measurement. To measure a frequency, you count, using an instrument called a counter. It counts the number of cycles in some fixed time governed by the experiment; the time may be a second, or an hour, or a day. The longer you can count, the higher the accuracy, and the higher the frequency you can count, the higher the accuracy will be, also. That's because both increase the total number of counts. This number divided by the total time of counting is the frequency. The accuracy of the result is the number you have counted to within ± 1/2 count.

Rabi realized that a good way to measure magnetic moments is through the characteristic frequency Ω. This frequency divided by the magnetic field strength, yields the magnetic moment: compare Equations 7-8 or 7-9 with 7-6.

It is easier to measure the frequency Ω than to measure the very small beam deflection distances produced in the Stern–Gerlach apparatus (see Problem 6.1). Rabi's insight was validated. As a practical laboratory tool, the magnetic resonance technique of examining magnetic moments quickly superseded the Stern–Gerlach technique of measurement.

Rabi introduced a second magnetic field into the apparatus, a time-dependent oscillatory field. Besides the uniform field of the pole pieces, there is added the auxilliary field shown in Figure 7-5 which varies with time. It oscillates sinusoidally at whatever frequency, $\omega/2\pi$, the experimenter chooses. He varies the frequency until some strong effect is observed. This occurs at *resonance,* when $\omega = \Omega$. Thus, Ω is that frequency ω, at which a resonance is observed. That's how you measure Ω. You scan a frequency spectrum until you encounter the signal.

The practical art of carrying out resonance experiments has been a monument to ingenuity. It is rarely ω that is scanned; rather, one sweeps through a range of Ω by slowly varying the static B-field. The original experiments of I. I. Rabi involved the focusing and defocusing of beams of atoms or molecules. Later F. Bloch and E. M. Purcell developed the nuclear magnetic resonance technique illustrated in Figure 7-5.

The specimen container amounts to a box of protons whose magnetic moments are to be measured. In fact these protons are the nuclei of the hydrogen atoms comprising the water in the container.

Like electrons and neutrons, protons are also spin 1/2 particles. Protons exhibit

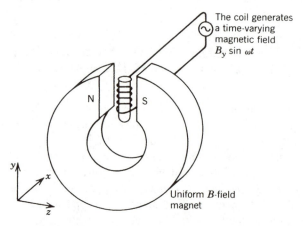

The coil generates a time-varying magnetic field $B_y \sin \omega t$

Uniform B-field magnet

Figure 7.5 The magnetic resonance apparatus.

magnetic moment components of ± 2.79 nuclear magnetons: they have a g-factor of 5.58. For the proton

$$\hat{\mathbf{M}} = +5.58 \frac{e}{2 \mathcal{M}} \hat{\mathbf{S}} \tag{7-34}$$

so in a given field B its characteristic frequency, Ω, is not the same as that of the neutron.

A coil surrounds the sample. Being electrically driven at the frequency $\omega/2\pi$, by an outside power supply, this coil produces the time varying magnetic field that adds to the static one in which the protons sit.

At the resonance condition, energy is exchanged between the protons and the oscillating magnetic field. This is perceived by the coil power supply as impedence loading. The coil's inductance develops a resistive component. The energy absorbed by the protons is registered as extra power that the power supply must deliver. It happens only at resonance.

7.8 TWO SYSTEMS INTERACT: THEY EXCHANGE ENERGY

Consider one of the protons. The applied oscillating magnetic field amounts to a radiation bath in which it is immersed. The system being investigated has two distinct parts, the proton and the radiation field.

The hamiltonian for the proton alone is

$$\hat{H}_P = -\left(\frac{5.58eB}{2 \mathcal{M}}\right)\hat{S}_z = -\Omega\hat{S}_z \tag{7-35}$$

This is equation 7-8 with 3.83 replaced by -5.58. The B-field in the equation remains, as before, a strong nonvarying one produced by the static field pole pieces shown in Figure 7-5. The proton in this static B-field without the coil field is one part of the system.

The coil's field is the other. The varying field of the coil is the radiation bath. There is a hamiltonian for the radiation field alone; we haven't studied it, but there *is* a hamiltonian for an isolated radiation field. Call it $\hat{H}_R$.

THE ENERGY IN A RADIATION FIELD
CAN CHANGE ONLY IN QUANTIZED UNITS

The essential property that we need to appreciate here about a radiation field is that monochromatic electromagnetic radiation at frequency $\omega/2\pi$ comes quantized in energy units of $\hbar\omega$.

This arises as the content of Item 5 in Einstein's interpretation of the photoelectric effect. The energy that each electron receives is hv: no electron absorbs energy $0.6\ hv$ or $5.7\ hv$ from the radiation field, only hv. This means that the radiation field can give up energy only in these units. We accept it as a property of nature that the energy of electromagnetic radiation can change only by multiples of hv.

The radiation field in which our proton is bathed is a monochromatic one; only the experimenter's chosen frequency $v = \omega/2\pi$ is present. The fact that we need to appreciate is that its energy, E_R, can change only in units of hv: $\Delta E_R = hv$.

BETWEEN TWO SYSTEMS
THERE IS AN INTERACTION HAMILTONIAN

The two parts of the system, proton and radiation field, interact with each other. One feels the presence of the other; energy can be exchanged between them.

The interaction effect can be cast into an interaction hamiltonian, $\hat{H}_I$, an extra term that adds to the two interaction-free isolated-systems part. The interaction-free hamiltonian is $\hat{H}_0 = \hat{H}_P + \hat{H}_R$. The interaction hamiltonian is defined by writing the total energy of the compound system as the sum of two parts in three terms.

$$\hat{H} = \hat{H}_0 + \hat{H}_I = \hat{H}_P + \hat{H}_R + \hat{H}_I \tag{7-36}$$

The interaction energy may be small compared with the energy of the component isolated parts. Here's the meaning of *small* in physical terms. Imagine the interaction *turned off*, that $\hat{H}_I = 0$. You do this by separating the systems; being isolated from one another, they have no interaction. Measure the energy of this compound non-interacting system. Its average is $\langle \hat{H}_0 \rangle$.

Now *turn the interaction back on*, and measure the energy of the system. Its average is $\langle \hat{H} \rangle$.

The difference divided by the original of these two energies must be small compared with unity. This condition, put mathematically, is

$$\langle \hat{H}_I \rangle = \langle \hat{H} \rangle - \langle \hat{H}_0 \rangle \ll \langle \hat{H}_0 \rangle \tag{7-37}$$

The right-hand inequality describes the thought. The inequality of the extremes embodies the meaning of *small*, that averages over $\hat{H}_I$ must be small compared with averages over $\hat{H}_0$. Interacting systems with this property are the kinds we envision in this analysis, and the proton in a radiation field is one.

OUR MEASUREMENT OBSERVABLES:
THE SYSTEM AS ISOLATED PARTS

We make measurements on the system proton-in-a-radiation-field. What we measure are proton properties and radiation field properties, we look at changes in the two bare noninteracting systems. We have no apparatus to measure the system-as-a-whole. We think in terms of proton and radiation, not in terms of the single thing *protodiation*.

Mathematically, the thought is that we don't examine the system in its self-space basis. The basis states by which our measurements describe the system are those eigen to $\hat{H}_0$: we don't measure the eigenstates of $\hat{H}$. No one even knows how to measure the eigenstates of $\hat{H}$. It's not pigheadedness but impotence that fixes the observables.

For our proton in the radiation field, we examine spin z-component states of the proton together with the energy state of the radiation field: we examine the states $|m,E_R\rangle$ where

$$-\Omega\hat{S}_z|m,E_R\rangle = \hat{H}_P \, |m,E_R\rangle = -m\hbar\Omega \, |m,E_R\rangle \tag{7-38}$$

and

$$\hat{H}_R \, |m,E_R\rangle = E_R \, |m,E_R\rangle \tag{7-39}$$

The states $|m,E_R\rangle$ are eigenstates of $\hat{H}_0$. They are not eigenstates of $\hat{H}$; we never measure those. We measure eigenstates of $\hat{H}_0$.

OUR APPARATUS MEASURES NONSTATIONARY STATES: TRANSITIONS APPEAR BETWEEN THEM

To those who insist upon these measurements, there appears to be a time variation in the system; the states evolve in time. The time variation, or the transition rate between states, is an artifact of our viewpoint. If we fail to measure invariants of the whole system, things will appear to change with time. So it is with the proton-plus-radiation system. Looking at the separate parts of a coupled system, we are examining nonstationary states: there are transitions between these states.

Here is the idea applied to our protodiation example. Call the initial state of the radiation-proton system $|i\rangle$. Specifically we take the ket $|i\rangle$ to be $|m=1/2,E_R\rangle$. This is the state: z-component proton spin up and energy E_R in the radiation field. In this state the total energy of the system is

$$E_i = E_R - \frac{\hbar\Omega}{2} \tag{7-40}$$

After a time t we examine the system for the state $|f\rangle$. We are interested in the amplitude to find a spin flip accompanied by a decrease of energy in the radiation field—by $\hbar\omega$. We are looking for the final state $|f\rangle = |m=-1/2,E_R-\hbar\omega\rangle$. In this state the total energy of the system is

$$E_f = E_R - \hbar\omega + \frac{\hbar\Omega}{2} \tag{7-41}$$

That the radiation field loses energy means that we're considering *radiative absorption;* the lost radiation field energy is absorbed by the proton.

Because of the interaction, embodied in $\hat{H}_I$, there exist transitions between these states. The amplitude to find the final state $|f\rangle$ at the time t if we knew the state to be $|i\rangle$ at $t = 0$ is $\langle f|\exp(-it\hat{H}/\hbar)|i\rangle$. The probability for the event is the square of the absolute value of this quantity. The probability per unit time that the event take place multiplied by the number of protons that can take part is the transition rate; it is the number of transitions from i to f per unit time. This rate is experimentally accessible. It is just the number of protons per unit time that suffer a spin flip.

It is from the bracket $\langle f|\exp(-it\hat{H}/\hbar)|i\rangle$ that the rate of transition between the states is calculated. Since nothing in the hamiltonian operator is time dependent, the time dependence of the matrix element can be extracted. Momentarily putting aside how this is accomplished, here is the result of the calculation:

$$i\hbar \, \langle f| \exp(-it\hat{H}/\hbar) \, |i\rangle = \langle f|\hat{H}_I|i\rangle \, \exp\left(\frac{-it\overline{E}}{\hbar}\right) \frac{2}{\Delta} \sin \frac{t\Delta}{2} \qquad (7\text{-}42)$$

Notice that the transition amplitude is proportional to the interaction, $\hat{H}_I$. If there were no interaction, the transition rate between the states $|i\rangle$ and $|f\rangle$ must be zero, as, indeed, (7-42) implies.

In this equation the symbol $\overline{E}$ represents the average of the initial and final state energies, the average of E_i and E_f. The Δ is the frequency counterpart of the energy difference $E_f - E_i$. Performing the subtraction yields $\hbar$ multiplied by the difference frequency, $\Omega - \omega$. Thus

$$\overline{E} \equiv \frac{1}{2}(E_f + E_i) \qquad \text{and} \qquad \hbar\Delta \equiv E_f - E_i = \hbar(\Omega - \omega) \qquad (7\text{-}43)$$

7.9 THE EXPONENTIAL OF AN OPERATOR-SUM: TIME MANUFACTURES ORDER

To prove the amplitude for transition result shown in (7-42) requires the following five key notions:

1. You must appreciate that the product law of exponentials is not valid for operators unless they commute: for two operators $\hat{A}$ and $\hat{b}$ that do not commute, $\exp(\hat{A} + \hat{b})$ has no evident meaning because $\exp \hat{A} \exp \hat{b} \neq \exp \hat{b} \exp \hat{A}$ (see Problem 7.27).

However, if the operators have a common incremental factor, a smallness parameter, like the Δt of Equation 7-18, then that equation is applicable, and during the small time interval, the noncommutability of the pair need not disturb us. Thus

$$\exp\left[\frac{-i\Delta t(\hat{H}_0 + \hat{H}_I)}{\hbar}\right] = 1 - \frac{i\Delta t(\hat{H}_0 + \hat{H}_I)}{\hbar} \qquad (7\text{-}44)$$

This expression for the exponential operator has the same meaning for either order of $\hat{H}_0$ and $\hat{H}_I$. For small enough Δt the exponential of two operators becomes a simple sum.

LATER OPERATORS TO THE LEFT

2. To make this result useful we write $it\hat{H}/\hbar$ in such a way as to divide the whole time t into N tiny increments $\Delta t_n = t/N$. Temporarily calling $i\hat{H}_0/\hbar \equiv \hat{A}$ and $i\hat{H}_I/\hbar \equiv \hat{b}$ for convenience, this means

$$
t(\hat{A} + \hat{b}) = \Delta t_N(\hat{A} + \hat{b}) + \Delta t_{N-1}(\hat{A} + \hat{b}) + \ldots + \Delta t_2(\hat{A} + \hat{b}) + \Delta t_1(\hat{A} + \hat{b}) \quad (7\text{-}45)
$$

Remember the idea behind an operator product, that the later one is to the left. The operator encountered first is always to the right. Ordering speaks of time. Any operator product is time ordered: the right-hand one operates *first in time;* the left one *later in time*.

TO ORDER OPERATORS IS TO ARRANGE THEM IN TIME

Now we can utilize (7-44) for each of the increments $\Delta t_n(\hat{A} + \hat{b})$. The result is the N-term product

$$
\exp[-t(\hat{A} + \hat{b})] = [1 - \Delta t_N(\hat{A} + \hat{b})] \ldots [1 - \Delta t_2(\hat{A} + \hat{b})][1 - \Delta t_1(\hat{A} + \hat{b})] \quad (7\text{-}46)
$$

For each small interval, Δt_n, the linear approximation is valid, but the operator for the next such time interval must keep its place of order. It operates later and is therefore to the left. The operator for the later time interval is at the left of the earlier because it operates later.

3. Now we reconstitute the product in (7-46). Because $\hat{b}$ is small, we do it in orders of $\hat{b}$. The zeroth-order term has no $\hat{b}$ in it. The first order term is a gathering of everything that contains a $\hat{b}$ to the first power. The collection of all terms containing $\hat{b}^2$ is the second order (see Problem 7.28).

The zeroth-order term, written down and reconstituted, is this:

$$
(1 - \Delta t_N\hat{A}) \ldots (1 - \Delta t_2\hat{A})(1 - \Delta t_1\hat{A}) = \exp(-\Delta t_N\hat{A} \ldots -\Delta t_2\hat{A} - \Delta t_1\hat{A})
$$
$$
= \exp(-t\hat{A}) \quad (7\text{-}47)
$$

The sum of all the parts of (7-46) that multiply a first power in $\hat{b}$ is the first-order term. Here it is

$$-\sum_{n=1}^{N}(1-\Delta t_N \hat{A}) \ldots (1-\Delta t_{n+1}\hat{A}) \, \Delta t_n \hat{b} \, (1-\Delta t_{n-1}\hat{A}) \ldots (1-\Delta t_1 \hat{A}) \quad (7\text{-}48)$$

The sum sweeps the $\Delta t_n \hat{b}$ term through all the N different increments into which t has been divided.

You can see, easily enough, how it works by examining a simple case. The $\hat{\varepsilon}$ sweeps through the three possible positions in the first-order term in the expansion of the triple product $(\hat{F} + \hat{\varepsilon})^3$.

$$(\hat{F} + \hat{\varepsilon})(\hat{F} + \hat{\varepsilon})(\hat{F} + \hat{\varepsilon}) = \hat{F}^3 + \hat{\varepsilon}\hat{F}^2 + \hat{F}\hat{\varepsilon}\hat{F} + \hat{F}^2\hat{\varepsilon} + \ldots \quad (7\text{-}49)$$
$$\text{zeroth} \quad | \quad \text{first order} \quad | \quad \text{higher}$$

To reconstitute the sum in (7-48) into an integral we let $N \to \infty$ taking $t_n = \tau$ so $\Delta t_n \to d\tau$. In writing the integral we must be sure to keep the ordering demanded by the successive time increments. The sum in (7-48) becomes this integral:

$$-\int_{0}^{t} e^{-(t-\tau)\hat{A}} \, d\tau \hat{b} \, e^{-\tau \hat{A}} \quad (7\text{-}50)$$

This is the first-order term in $\hat{b}$.

In this calculation we keep terms only out to the first order. It is valid, therefore, only if the interaction energy is small, when the effect of $\hat{b}$ is small.

4. Combining the results in the preceding three steps, the amplitude for the transition can be written

$$i\hbar \, \langle f| \exp(-it\hat{H}/\hbar) \, |i\rangle = \int_{0}^{t} d\tau \langle f| \exp[-i(t-\tau)\hat{H}_0/\hbar] \, \hat{H}_I \exp(-i\tau\hat{H}_0/\hbar) \, |i\rangle \quad (7\text{-}51)$$

This is the answer to first order; there is no zeroth-order term. The zeroth-order term is $\exp(-it\hat{H}_0/\hbar)$ bracketed by the two states $\langle f|$ and $|i\rangle$, and it vanishes.

$$\langle f| \exp(-it\hat{H}_0/\hbar) \, |i\rangle = \exp(-itE_i/\hbar) \, \langle f|i\rangle = 0 \quad (7\text{-}52)$$

The reason for this is that the amplitude $\langle f|i\rangle$ is a self-space bracket; it's made from two states in the same basis. These two states refer to two entirely different points in the spectrum of basis states. One point is *proton spin up, radiation energy E_R*, and the other point is *proton spin down, radiation energy $E_R - \hbar\omega$*.

We are asking for the amplitude for a transition to a state *different* from the original one. That's why the zeroth-order term disappears. It would be nonzero only if $|f\rangle = |i\rangle$, and we are seeking transitions between states where $|f\rangle \neq |i\rangle$.

5. The bracket in the integral of (7-51) can be evaluated because the states $|i\rangle$ and $|f\rangle$ both are eigenstates of $\hat{H}_0$. These states are points in its home space. In its home space the operator wears no hat, so

$$\exp(-i\tau\hat{H}_0/\hbar)\,|i\rangle = |i\rangle\,\exp(-i\tau E_i/\hbar) \tag{7-53}$$

and

$$\langle f|\exp[-i(t-\tau)\hat{H}_0/\hbar] = \exp[-i(t-\tau)E_f/\hbar]\,\langle f| \tag{7-54}$$

Applying these two results to (7-51) it requires no more than elementary integration to produce (7-42). This completes the derivation of that equation.

7.10 FERMI'S GOLDEN RULE IS THE TRANSITION PROBABILITY PER UNIT TIME

Equation 7-42 gives us the amplitude for the transition to take place. Its absolute square is the probability of the event; that the state $|f\rangle$ could be found after a time t if the state were $|i\rangle$ at $t = 0$. This probability is

$$|\langle f|\exp(-it\hat{H}/\hbar)\,|i\rangle|^2 = |\langle f|\hat{H}_I|i\rangle|^2 \left(\frac{2}{\hbar\Delta}\right)^2 \sin^2\frac{t\Delta}{2} \tag{7-55}$$

The proton of Figure 7-5 finds itself in an added time-varying magnetic field applied at frequency ω. According to (7-55), the probability that a spin flip occurs oscillates back and forth in time at the difference frequency $\Delta = \Omega - \omega$. If the magnetic field frequency is far from Ω ($\Delta = $ large), the probability will oscillate in time but always will remain small; the process is unlikely. As ω gets closer to Ω, the probability becomes larger and oscillates less frequently. The system, proton plus electromagnetic radiation, is in resonance when $\omega = \Omega$, when $\Delta = 0$.

In this condition there is a net rate of transitions; the probability per unit time that a transition takes place approaches a limit. After a long time period ($t \to \infty$), if you measure the number of transitions that take place in Δt, you will find a result proportional to Δt. The quantity of physical significance that we want is this limiting transition rate.

This transition rate is the product of two things: the number present that could undergo transitions multiplied by the probability per unit time for one transition. The former is an experimental design parameter; it is the latter that we want to calculate. We calculate the transition rate per unit one possible. This is simply the probability per unit time that the transition takes place. It is this limiting transition probability rate that we seek.

Call it R. Here's what R means:

$$R = \lim_{t \to \infty}\left[\frac{d}{dt}|\langle f|\exp(-it\hat{H}/\hbar)\,|i\rangle|^2\right] \tag{7-56}$$

This probability per unit time provides the critical link between theory and experiment. The rate, R, can be deduced from both experiment and theory. Because of the key role that R plays in the interpretation of experiments, the equation for R has acquired a name, Fermi's Golden Rule.

To get R, you perform the calculation defined by Equation 7-56.

Get the time derivative of (7-55) and take the prescribed limit. This limit is the subject of part e in Problem 2.13. It is a Dirac delta-function of the argument Δ. The meaning of $\hbar\Delta$ is the energy difference between state i and state f (see Equation 7-43). In terms of this energy difference, the result takes the form of the traditional equation labeled as Fermi's Golden Rule. It is

$$R = \frac{2\pi}{\hbar} |\langle f|\hat{H}_I|i\rangle|^2 \, \delta(E_f - E_i) \qquad (7\text{-}57)$$

The rate diverges at $E_f = E_i$. This is the ideal resonance formula; nothing happens unless you match the imposed frequency, ω, to the natural one of the system, Ω. At the match the system is *driven at resonance*.

Another way to view (7-57) is that it specifies a selection rule on energy. The transition does not occur unless energy is conserved. Though the initial and final states differ, $|i\rangle \neq |f\rangle$, their energies must be the same for transitions to occur; the δ-function in energy selects only such states to interact. Interactions between states of differing energy are forbidden. For a process to take place, there must be an exchange of energy between the parts of the system: what is lost by one part is gained by the other.

In applying the Golden Rule to particular problems, there always arises some spectral window that must be folded into the Golden Rule function of (7-57). Precise monochromaticity is never realized; a window of frequencies partake. Invariably an integral over dE_f or over $d\omega$ is required when the Golden Rule is applied to practical cases. Thus no δ-function appears in the specific application of R to laboratory results. Laboratory resonances have finite widths.

PROBLEMS

7.1 Show that the "translation-forward-in-position-by-length-b" operator, $\hat{T}$ (see Equation 3-3), and the "take-the-derivative-with-respect-to-x" operator, $\hat{D}$ (see Equation 3-4), are related by

$$\hat{T} = e^{-b\hat{D}}$$

Solution principles:

a. The wave vector (momentum) states $|k\rangle$ are eigen to both $\hat{D}$ and $\hat{T}$ with eigenvalues ik and $\exp(-ikb)$.

b. A function of an operator is defined by what it does in its eigenspace: Remove The Hat theorem. Thus $e^{-b\hat{D}} |k\rangle = |k\rangle e^{-ikb}$.

c. Any state $|\psi\rangle$ is a superposition of wave vector states: $|\psi\rangle = \Sigma |k\rangle\langle k|\psi\rangle$.

7.2 A state displaced in position yields the same momentum measurement results as does the undisplaced state (see Figure P-7-2). Prove this. It results from the association of momentum with wave vector—the de Broglie relation.

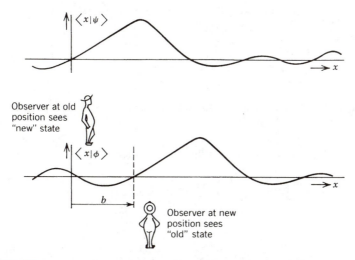

Figure P7.2 All momentum measurement results are the same for two observers. Changing your physical location doesn't change your momentum measurement result.

Solution principle: Call the original state $|\psi\rangle$. Its position representation is $\langle x|\psi\rangle = \psi(x)$. Call the displaced state $|\phi\rangle$. Its position representation is $\langle x|\phi\rangle = \phi(x) = \psi(x-b)$. The proof consists in showing that $|\langle p|\phi\rangle|^2 = |\langle p|\psi\rangle|^2$ for all p and any $|\psi\rangle$. But this is true because, using Problem 7.1,

$$\langle p|\phi\rangle = \langle p|e^{-b\hat{D}}|\psi\rangle = \langle p|\psi\rangle \exp(-ibp/\hbar)$$

7.3 Prove that you get the same result no matter when you make an energy measurement on a closed system. Using the meanings attached to α and β in the text, the translation of this thought into mathematics is this: show that

$$|\langle E \ldots |\alpha\rangle|^2 = |\langle E \ldots |\beta\rangle|^2$$

7.4 What is the characteristic frequency, $\Omega/2\pi$, for a neutron in a magnetic field of 1 kilogauss? In what region of the electromagnetic spectrum is this: radio, microwave, infrared, visible, x-ray?

7.5 Show that (7-13) does indeed follow from (7-11).

Solution: Use

$$|\mu = 1/2\rangle = \frac{1}{\sqrt{2}}|m = 1/2\rangle + \frac{i}{\sqrt{2}}|m = -1/2\rangle$$

7.6 A neutron spinning initially in the $+y$-direction, ($\mu = 1/2$), spends time t in a z-direction magnetic field of strength B. What is the probability of finding the intial spin ($\mu = 1/2$) at time t?

Hint:

$$\langle \mu | \beta \rangle = \sum_m \langle \mu | m \rangle \langle m | \beta \rangle.$$

Using the $|\beta\rangle$ given in (7-13) the answer is $\frac{1}{2}(1 + \cos \Omega t)$

7.7 Prove that

$$\exp\left(\frac{-i\Delta t \hat{H}}{\hbar}\right) = 1 - \frac{i\Delta t \hat{H}}{\hbar}$$

when applied to any state $|\psi\rangle$ for which all brackets $\langle E \ldots | \psi \rangle$ vanish above some $E < E_{max} = \hbar / \Delta t$.

Answer: What we must prove is that for any such $|\psi\rangle$,

$$\exp\left(\frac{-i\Delta t \hat{H}}{\hbar}\right) |\psi\rangle = (1 - i \, \Delta t \, \hat{H} / \hbar) \, |\psi\rangle$$

Thus

$$\exp\left(\frac{-i\Delta t \hat{H}}{\hbar}\right) \sum_{E \ldots} |E \ldots\rangle \langle E \ldots | \psi\rangle$$

$$= \sum_{E \ldots} \exp\left(\frac{-i\Delta t E}{\hbar}\right) |E \ldots\rangle \langle E \ldots | \psi\rangle$$

$$= \sum_{E \ldots} (1 - i\Delta t E / \hbar) |E \ldots\rangle \langle E \ldots | \psi\rangle$$

$$= (1 - i\Delta t \hat{H} / \hbar) \sum_{E \ldots} |E \ldots\rangle \langle E \ldots | \psi\rangle$$

The equality of the extremes does it. That in the sum all nonzero terms have E less than $E_{max} = \hbar / \Delta t$ guarantees that the expansion is valid.

7.8 Prove the equality of the extremes in Equation 7-20, by actually calculating the matrix element $\langle \alpha | \hat{A} | \alpha \rangle$ in the basis that embraces the eigenstates of the operator $\hat{A}$; use as basis states $|A \ldots\rangle$ where

$$\hat{A} |A \ldots\rangle = A |A \ldots\rangle$$

Notice that the symbol $|\langle A | \alpha \rangle|^2$ in (7-20) must mean in this notation

$$|\langle A | \alpha \rangle|^2 \equiv \sum_{\ldots} |\langle A \ldots | \alpha \rangle|^2$$

7.9 The state of the neutron in Figure 7-2 at time $t = 0$ is not completely specified by $|\mu = 1/2\rangle$; the state is really $|K, \mu = 1/2\rangle$. The neutron has momentum $k_x = K$, $k_y = 0$, $k_z = 0$.

The hamiltonian of Equation 7-8 is an incomplete description of the neutron passing through the field region. It should include the kinetic energy of the neutron. The hamiltonian should be

$$\hat{H} = \frac{\hbar^2}{2 \mathcal{M}} \hat{k}^2 + \Omega \, \hat{S}_z$$

Show that after a time t the state of the system, $|\beta\rangle$, is not really that of (7-13) but should be written

$$|\beta\rangle = |K, m = 1/2\rangle \frac{1}{\sqrt{2}} \exp\left[-it\left(\frac{\Omega}{2} + \frac{\hbar K^2}{2 \mathcal{M}}\right)\right]$$

$$+ |K, m = -1/2\rangle \frac{i}{\sqrt{2}} \exp\left[it\left(\frac{\Omega}{2} - \frac{\hbar K^2}{2 \mathcal{M}}\right)\right]$$

Key:

$$|K, \mu\rangle = \sum_{k,m} |k, m\rangle \langle k, m | K, \mu\rangle = \sum_m \sum_k |k, m\rangle \langle k | K\rangle \langle m | \mu\rangle$$

$$= \sum_m |K, m\rangle \langle m | \mu\rangle$$

7.10 Show that the probability of finding neutrons with momentum K and spin y-component $\mu = 1/2$ at time t is the same as deduced in Problem 7.6; the physical conclusions about the spin state probabilities are not affected by the inclusion of the neutron linear momentum in the problem.

$$dK |\langle K, \mu = 1/2 | \exp(-it\hat{H}/\hbar) | K, \mu = 1/2\rangle|^2 dK = |\langle \mu = 1/2 | \beta\rangle|^2$$

7.11 Equation 7-22 is a special case of this more general statement about the hermitian conjugate of an operator function:

$$\text{if } \hat{f} = f(\hat{q}) \text{ then } \hat{f}^\dagger = f^*(\hat{q}^\dagger)$$

where $\hat{q}$ generates a complete set of states $|q\rangle$. It's true for $\hat{q} = \hat{p}$ or $\hat{x}$ or $\hat{H}$ but not for $\hat{a}$ or $\hat{L}_+$ or $\hat{S}_-$.

Prove this and that (7-22) is, indeed, an application of the idea.

Answer:
a. Note that if $\hat{q}|q\rangle = q|q\rangle$ then $\hat{q}^\dagger|q\rangle = q^*|q\rangle$. The eigenspace of $\hat{q}^\dagger$ is q, and its eigenvalues are q^*. For a complete space q this is easily proved. It follows from

$$\langle q | \hat{q}^\dagger | Q\rangle = \langle Q | \hat{q} | q\rangle^* = q^* \langle q | Q\rangle = Q^* \langle q | Q\rangle$$

where both q, Q are in q − space

b. Now taking note that $[f(q)]^* = f^*(q^*) \neq f^*(q)$ we find

$$\langle q | \hat{f}^\dagger | Q\rangle = \langle Q | \hat{f} | q\rangle^* = (f(q))^* \langle q | Q\rangle = f^*(q^*)\langle q | Q\rangle = \langle q | f^*(\hat{q}^\dagger) | Q\rangle$$

The equality of the extremes is good for all q and Q; hence, it's good for *any* pair of states, thus

$$\hat{f}^\dagger = f^*(\hat{q}^\dagger)$$

7.12 Establish the truth of (7-21).

Solution key: Together with (7-22) use

$$\langle\beta|\hat{A}\ \exp(-it\hat{H}/\hbar)|\alpha\rangle = \langle\alpha|[\hat{A}\ \exp(-it\hat{H}/\hbar)]^\dagger\ \exp(-it\hat{H}/\hbar)|\alpha\rangle^*$$

Note the implication that

$$\langle\beta| = \langle\alpha|\exp(it\hat{H}/\hbar)$$

7.13 Suppose a matrix element of $\hat{A}$ between two *different* states $|\alpha\rangle$ and $|a\rangle$ at time $t = 0$ is $\langle a|\hat{A}|\alpha\rangle$. The two states into which $|a\rangle$ and $|\alpha\rangle$ evolve at time t are called $|b\rangle$ and $|\beta\rangle$. Show that the matrix element of $\hat{A}$ evolves according to

$$\langle b|\hat{A}|\beta\rangle = \langle a|\exp(it\hat{H}/\hbar)\ \hat{A}\ \exp(-it\hat{H}/\hbar)|\alpha\rangle$$

Thus every matrix element of $\hat{A}$ evolves just as the average value of A does: through the time-evolution operator $\exp(it\hat{H}/\hbar)\ \hat{A}\ \exp(-it\hat{H}/\hbar)$.

7.14 Prove that (7-23) follows from (7-21).

Elements of the solution:
a. The meaning of the time derivative: we call $\exp(it\hat{H}/\hbar)\ \hat{A}\ \exp(-it\hat{H}/\hbar) = G(t,\hat{A})$ the operator whose average is taken on the right of (7-21). Then the meaning of the time derivative of the average of this operator is

$$d\langle\beta|\hat{A}|\beta\rangle/dt = \lim_{\Delta t\to 0} \frac{1}{\Delta t}\ \langle\alpha|G(t+\Delta t,\hat{A}+\Delta t\partial\hat{A}/\partial t) - G(t,\hat{A})|\alpha\rangle$$

b. In taking limits, the approximation of (7-18) becomes exact. Use it to carry out the prescription in this equation.
c. In doing so remember that the order of the operators $\hat{A}$ and $\hat{H}$ must be preserved. But any function of the operator $\hat{H}$ commutes with $\hat{H}$ itself: $[\hat{H}, f(\hat{H})] = 0$.
d. Note that: if $|\beta\rangle = \exp(-it\hat{H}/\hbar)\ |\alpha\rangle$ then $\langle\alpha|\exp(it\hat{H}/\hbar) = \langle\beta|$

7.15 Show that for a 1-D particle in a potential $V(x)$ for which the hamiltonian is

$$\hat{H} = \frac{\hat{p}^2}{2m} + V(\hat{x})$$

that

$$\frac{d}{dt}\langle\hat{x}\rangle = \frac{1}{m}\langle\hat{p}\rangle$$

and

$$\frac{d}{dt}\langle\hat{p}\rangle = \langle\hat{F}\rangle$$

where $\hat{F}$ is defined by its instruction in x-space

$$\langle x|\hat{F}|\psi\rangle = -\frac{dV(x)}{dx} \langle x|\psi\rangle$$

Solution: You do it by taking first $\hat{A} = \hat{x}$ and then $\hat{A} = \hat{p}$ in (7-23). Note that $\langle x|[\hat{p},\hat{V}] = -i\hbar(dV/dx)\langle x| = i\hbar\langle x|\hat{F}$ and that $[\hat{x},\hat{p}^2] = 2i\hbar\hat{p}$.

7.16 Show, for the neutrons in the magnetic field, the truth of the three equations represented by (7-24).

$$\frac{d}{dt}\langle\hat{S}_z\rangle = 0$$

$$\frac{d}{dt}\langle\hat{S}_x\rangle = -\Omega\langle\hat{S}_y\rangle$$

$$\frac{d}{dt}\langle\hat{S}_y\rangle = \Omega\langle\hat{S}_x\rangle$$

Solution: Follow the directions in the paragraph just preceding Equation 7-24.

7.17 Calculate the average value of the x-direction spin component for a neutron spin state $|\eta\rangle$; that is, for a neutron known to have a η-direction spin of $\eta\hbar$, calculate the expected average of findings for its x-direction spin component.

Solution: The answer to this question is $\langle\eta|\hat{S}_x|\eta\rangle$; it is this matrix element that is to be calculated. The simultaneous solution of the two operator-average equations in Figure 7-3 gives the formula $\langle\eta|\hat{L}_x|\eta\rangle = -\eta\hbar\sin\phi$. Verbally indicate how your answer makes physical sense and thus confirm the connection shown in Figure 7-3 between rotated angular momentum operators.

7.18 Using (7-27), show that the operator $\hat{S}_\eta$ in terms of $\hat{S}_+$ and $\hat{S}_-$ is

$$2i\hat{S}_\eta = e^{-i\phi}\hat{S}_+ - e^{i\phi}\hat{S}_-$$

7.19 Construct the $\hat{S}_\eta$ matrix, (elements $\langle m|\hat{S}_\eta|M\rangle$), and show that it is indeed the one of (7-29). Use the previous problem as an aid.

7.20 Diagonalize the $\hat{S}_\eta$ matrix of Equation 7-29, and show that its eigenvalues are, indeed, $\eta = +1/2$ and $\eta = -1/2$ and that a unitary transformation matrix that diagonalizes $\hat{S}_\eta$ is, indeed, that of Equation 7-30.

7.21 Deduce the result shown in (7-31) from (7-13) and (7-30).

7.22 Prove that $\hat{\beta}|\beta\rangle = \alpha|\beta\rangle$ where α labeled the original state $|\alpha\rangle$ from which $|\beta\rangle$ evolved. You thus will show that the value of the state label doesn't change with time even though its meaning does!

Solution: Apply (7-33) to (7-1).

7.23 Show that the operator $\hat{\beta}$ of Equation 7-33 is hermitian.
  The results in Problem 7.11 plus the relations in Equations 4-43 and 4-44 make it easy.

7.24 Apply the general Equation 7-33 to the case of the spinning neutron spending a time t in the uniform magnetic field pictured in Figure 7-2. Noting that $|\alpha\rangle$ is $|\mu = 1/2\rangle$ and $\hat{H} = \Omega\hat{S}_z$, show that (7-33) implies that

$$\hat{S}_\eta = \exp\left(\frac{-it\Omega\hat{S}_z}{\hbar}\right)\hat{S}_y \exp\left(\frac{it\Omega\hat{S}_z}{\hbar}\right)$$

Apply this to our spinning neutron. Show that when the operator given here for $\hat{S}_\eta$ operates on $|\eta = 1/2\rangle$, you do, indeed, get $|\eta = 1/2\rangle\,\hbar/2$. You may do it by calculating the matrix elements of the triple product operator bracketed between $\langle m|$ and $|M\rangle$ to see the same matrix as 7-29.

7.25 Demonstrate that the triple-product operator given in Problem 7.24 for $\hat{S}_\eta$ is exactly equivalent to that given in (7-27) with $\phi = \Omega t$; that is, show that

$$\exp\left(\frac{-it\Omega\hat{S}_z}{\hbar}\right)\hat{S}_y \exp\left(\frac{it\Omega\hat{S}_z}{\hbar}\right) = -\hat{S}_x \sin\Omega t + \hat{S}_y \cos\Omega t$$

Solution: Construct the 2×2 matrix of each operator in the covenient basis of eigenstates of $\hat{S}_z$. If the matrices are the same, so must be the operators. The matrix of the right-hand side is constructed in the text and exhibited as Equation 7-29. To construct the matrix of the left-hand side operator, merely consult Figure 6-8 and note that

$$\langle m|\exp(-it\Omega\hat{S}_z/\hbar)\,\hat{S}_y\,\exp(it\Omega\hat{S}_z/\hbar)|M\rangle = \exp[-it\Omega(m-M)]\,\langle m|\hat{S}_y|M\rangle$$

7.26 Prove that the states shown in (7-38) and (7-39) to be eigenstates of $\hat{H}_P$ and $\hat{H}_R$ separately, are also eigenstates of $\hat{H}_0$. Find the eigenvalues for $\hat{H}_0$ operating on these states. What is the physical interpretation of these eigenvalues?

7.27 Show that if $[\hat{A}, \hat{b}] \neq 0$ then

$$e^{\hat{A}}\,e^{\hat{b}} \neq e^{\hat{b}}\,e^{\hat{A}}$$

but that if $\hat{A}$ and $\hat{b}$ do commute, then $\exp(\hat{A} + \hat{b})$ has meaning.

Hint: If $[\hat{A}, \hat{b}] \neq 0$, then $e^{\hat{A}}|b \ldots\rangle = c|S\rangle$ where $e^{\hat{b}}|S\rangle \neq e^{\hat{b}}|S\rangle$.

7.28 Show that the second-order term in $\hat{b}$—when the product in (7-46) is reconstituted—is

$$\int_0^t \exp[-(t - t_2)\hat{A}]\,dt_2\,\hat{b} \int_0^{t_2} \exp[-(t_2 - t_1)\hat{A}]\,dt_1\hat{b}\,\exp(-t_1\hat{A})$$

Key: You must extract the second-order term sum from (7-46) and then reconstitute it into an integral.

Notice that in the product $(\hat{a}_2 + \hat{b}_2)(\hat{a}_1 + \hat{b}_1)$, the order of the operations must be preserved; it's equal to $\hat{a}_2\hat{a}_1 + \hat{a}_2\hat{b}_1 + \hat{b}_2\hat{a}_1 + \hat{b}_2\hat{b}_1$. There is no term like $\hat{a}_1\hat{b}_2$.

In collapsing the second-order sum of products, take your cue from the terms expressing (7-49) to the second order. *Two* $\hat{\varepsilon}$'s sweep through the three positions. The second-order term is $\hat{F}\,\hat{\varepsilon}\,\hat{\varepsilon} + \hat{\varepsilon}\,\hat{F}\,\hat{\varepsilon} + \hat{\varepsilon}\,\hat{\varepsilon}\,\hat{F}$.

Chapter 8

The Simplest Atom:
Two Particles Bound Together

RADIATION SPECTRA REVEAL
THE HAMILTONIAN'S EIGENVALUES

Each species of atom has a characteristic spectrum.

Consider a gas of atoms. An electric discharge takes place within the glass container holding them. It causes light to be emitted by the atoms. The radiated light exhibits a spectrum of frequencies; not all frequencies are present, only quite specific ones. These frequencies characterize the species of atom. That the particular species is, indeed, present in the gas may be deduced from its emission spectrum. The spectrum is the atom's signature, a fact known many years before the birth of quantum mechanics, although there had been no explanation for it.

The explanation requires quantum mechanics. Each atom has only a discrete set of internal energy states available to it. Through an electric discharge or vigorous heating, the atoms acquire the internal energy necessary to excite them into higher-level energy states. They radiate off this energy when they return to a lower-energy state. The radiation frequencies correspond to the differences between energy levels of the atoms. That there exist discrete energy levels, and not a continuous spectrum of them, is the reason for a discrete spectrum of emission frequencies: the concept requires quantum mechanics.

An energy measurement corresponds to the hamiltonian operator, $\hat{H}$. The atomic-energy levels are eigenvalues of $\hat{H}$. If we calculate the eigenvalues of $\hat{H}$, we can easily subtract larger ones from smaller ones to see whether the radiative emission spectrum is matched. Effectively, the signature radiation spectrum of an atom measures its hamiltonian (see Figure 8-1).

The atom constitutes an isolated system; its stationary states persist in time. Its energy eigenstates are stationary states of the atom. That is the reason for the central role played by the hamiltonian operator, $\hat{H}$, and all operators that commute with $\hat{H}$:

213

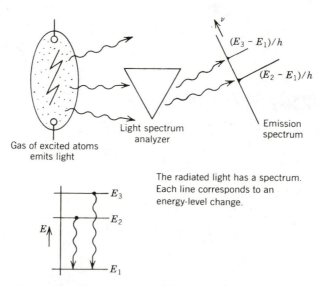

Figure 8.1 The Radiative emission experiment is a hamiltonian operator measurement.

their eigenstates are stationary. These are the states of the atom that characterize it when it is isolated and undisturbed by external influence; these are the ones that *don't* change with time. The emission spectrum shows us these eigenstates; the energy of transitions between them is emitted as electromagnetic radiation.

8.1 A CENTRAL FIELD POTENTIAL MAY BIND TWO PARTICLES

We will study a model atom, the simplest ideal model atom imaginable is a particle in the attractive central field of another particle. The hydrogen atom is the archetypical example: the electron circles the proton because of the electrostatic attractive force field between them. The pair, moving as a bound unit, constitutes the atom.

A central field is one for which the potential energy of interaction depends only on the magnitude of the distance between the particles:

$$V(\mathbf{r}_1, \mathbf{r}_2) = V(|\mathbf{r}_2 - \mathbf{r}_1|) \tag{8-1}$$

The electrostatic potential is of this class.

The classical hamiltonian for such a pair is well known. It consists of three energies: the kinetic energy of the first particle plus the kinetic energy of the second plus the potential energy of interaction between them. Hence the hamiltonian operator for this two-particle system is

$$\hat{H} = \frac{\hat{p}_1^2}{2m_1} + \frac{\hat{p}_2^2}{2m_2} + V(|\hat{\mathbf{r}}_2 - \hat{\mathbf{r}}_1|) \qquad (8\text{-}2)$$

8.2 DECOUPLING THE HAMILTONIAN REPLACES OLD PARTICLES BY NEW ONES

The hamiltonian of (8-2) is a six-dimensional one: a position-space state is $|x_1,y_1,z_1,x_2,y_2,z_2\rangle = |\mathbf{r}_1,\mathbf{r}_2\rangle$. The representation of $\hat{H}$ in position space is easily written down; we can exhibit $\langle\mathbf{r}_1,\mathbf{r}_2|\hat{H}$. You need only realize that

$$\langle\mathbf{r}_1,\mathbf{r}_2|\hat{p}_{1x}^2 = -\hbar^2\frac{\partial^2}{\partial x_1^2}\langle\mathbf{r}_1,\mathbf{r}_2| \qquad (8\text{-}3)$$

and that

$$\langle\mathbf{r}_1,\mathbf{r}_2|V(\hat{\mathbf{r}}_1,\hat{\mathbf{r}}_2) = V(\mathbf{r}_1,\mathbf{r}_2)\langle\mathbf{r}_1,\mathbf{r}_2| \qquad (8\text{-}4)$$

In this space the hamiltonian of (8-2) is a *coupled* one; the potential energy term couples the 3-D space of particle 1 to the 3-D space of particle 2.

The idea of coupling is easily grasped. You first consider what is meant by *uncoupled*. A pair of coordinates in the hamiltonian is said to be *uncoupled* if the eigenvalue problem can be separated into two distinct parts; one for each coordinate alone. Coupling in the hamiltonian refers to those coordinates which appear in such a way as to prevent this separation.

If there were no interaction potential, $V = 0$, then the hamiltonian would be a decoupled one: it would be a sum of six hamiltonian parts, each in an independent space and separately solvable (see Problem 8.1). It is the potential term in Equation 8-2 that contains the coupling term.

The only way an eigenvalue problem is ever solved exactly is when the N-coordinate problem can be reformulated as N one-coordinate problems. For the hamiltonian operator, this process of reduction is called *decoupling the hamiltonian*. No exact solution has ever circumvented this process.

To solve the eigenvalue problem

$$\hat{H}|E\ldots\rangle = E|E\ldots\rangle \qquad (8\text{-}5)$$

we must construct six independent single-variable problems from this single six-variable problem. We must decouple the hamiltonian.

There is a physical counterpart to this mathematical process. The two initial 3-D particles are replaced by two new 3-D particles. Instead of the particles of mass m_1 and m_2, two new ones arise: a freely traveling massive particle of mass $\mathcal{M}$ plus a light particle of mass m where

$$\mathcal{M} = m_1 + m_2 \qquad (8\text{-}6)$$

and

$$m = \left(\frac{1}{m_1} + \frac{1}{m_2} \right)^{-1} \tag{8-7}$$

The massive one sees no potential. The light one moves within a potential reaching out from a fixed point in space at the origin of coordinates.

This decoupling is effected mathematically by a coordinate transformation. You replace the original particle coordinates by new ones: those of the center-of-mass plus those of relative position. Together with these, the total momentum and the relative velocity part of the transformation are contained in the following four statements:

$$(m_1 + m_2)\hat{\mathbf{R}} = \mathcal{M}\,\hat{\mathbf{R}} = m_1\hat{\mathbf{r}}_1 + m_2\hat{\mathbf{r}}_2 \tag{8-8}$$

$$\hat{\mathbf{P}} = \hat{\mathbf{p}}_1 + \hat{\mathbf{p}}_2 \tag{8-9}$$

$$\hat{\mathbf{r}} = \hat{\mathbf{r}}_2 - \hat{\mathbf{r}}_1 \tag{8-10}$$

$$\left(\frac{1}{m_1} + \frac{1}{m_2} \right)\hat{\mathbf{p}} = \frac{\hat{\mathbf{p}}}{m} = \frac{\hat{\mathbf{p}}_2}{m_2} - \frac{\hat{\mathbf{p}}_1}{m_1} \tag{8-11}$$

These equations define four new vector operators, $\hat{\mathbf{R}}$, $\hat{\mathbf{P}}$, $\hat{\mathbf{r}}$, and $\hat{\mathbf{p}}$ in terms of the four old vector operators, $\hat{\mathbf{r}}_1$, $\hat{\mathbf{p}}_1$, $\hat{\mathbf{r}}_2$, and $\hat{\mathbf{p}}_2$. The definitions precisely parallel those for the classical case. In classical mechanics the vector $\mathbf{R}$ is the center-of-mass coordinate; the vector $\mathbf{P}$ is the total momentum; the relative coordinate vector is $\mathbf{r}$; and the momentum vector $\mathbf{p}$ embodies a statement about the relative velocity $\mathbf{p}/m$ (see Figure 8-2).

The transformation equations are just as effective in quantum mechanics as they are in classical mechanics. They accomplish the same task too; they decouple the hamiltonian.

In terms of the new coordinates, the kinetic energy remains a sum of two separate parts; the decoupling in the kinetic energy is preserved by the transformation. The potential energy is decoupled in that it becomes a function of only one of the new coordinates and not of the other.

$$V(|\hat{\mathbf{r}}_2 - \hat{\mathbf{r}}_1|) = V(|\hat{\mathbf{r}}|) = V(\hat{r}) \tag{8-12}$$

The potential energy is independent of $\hat{\mathbf{R}}$.

Using the transformation Equations 8-8, 8-9, 8-10, and 8-11, the hamiltonian of Equation 8-2 becomes

$$\hat{H} = \frac{\hat{P}^2}{2\,\mathcal{M}} + \frac{\hat{p}^2}{2\,m} + V(\hat{r}) \tag{8-13}$$

To prove that this is, indeed, the same hamiltonian as that of (8-2) is a simple matter. Substitute the equations for $\hat{\mathbf{P}}$, $\hat{\mathbf{r}}$, and $\hat{\mathbf{p}}$ of (8-9), (8-10), and (8-11) into this equation. You retrieve (8-2) (see Problem 8.2).

$$\mathcal{M} = m_1 + m_2 \qquad \frac{1}{m} = \frac{1}{m_1} + \frac{1}{m_2}$$

$$\mathcal{M} \, \mathbf{R} = m_1 \mathbf{r}_1 + m_2 \mathbf{r}_2$$

$$\mathbf{r} = \mathbf{r}_2 - \mathbf{r}_1$$

$$\mathbf{P} = \mathbf{p}_1 + \mathbf{p}_2$$

$$\mathbf{p}/m = \mathbf{p}_2/m_2 - \mathbf{p}_1/m_1$$

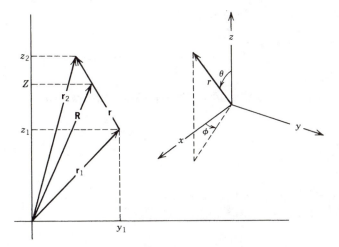

Figure 8.2 **The classical two-particle transformation of coordinates. The center-of-mass and relative coordinates are linear combinations of the individual particle coordinates.**

The new coordinate operators $\hat{\mathbf{r}}$ and $\hat{\mathbf{R}}$ are independent of each other, just as were the old coordinate operators $\hat{\mathbf{r}}_1$ and $\hat{\mathbf{r}}_2$. The new momenta $\hat{\mathbf{p}}$ and $\hat{\mathbf{P}}$ are also an independent pair, as were the old $\hat{\mathbf{p}}_1$ and $\hat{\mathbf{p}}_2$. (see Problem 8.2) The new coordinates give us two distinct new particles, these are an uncoupled pair.

That they do not interact is clear from the form of the hamiltonian. Equation 8-13 is the sum of two parts: the first term is one part, and the last two terms are the other. Everything in one part commutes with everything in the other: the two parts don't see each other. Each part refers to a distinct particle because the pair $\hat{\mathbf{R}},\hat{\mathbf{P}}$ is independent of the pair $\hat{\mathbf{r}},\hat{\mathbf{p}}$. All of the six operators $\hat{\mathbf{r}},\hat{\mathbf{p}}$ are compatible with any of the six comprising $\hat{\mathbf{R}}$ and $\hat{\mathbf{P}}$. Simultaneous eigenstates of $\hat{\mathbf{r}}$ and $\hat{\mathbf{R}}$ exist. The hamiltonian of (8-13) may be cast in the basis $\langle \mathbf{r},\mathbf{R}|$.

The hamiltonian of (8-2) represents two particles coupled by an interaction between them. That of (8-13) represents the same system viewed in another way: two new particles with no interaction between them. The two new particles are uncoupled. One particle moves along with constant momentum $\mathbf{P}$. It has mass $\mathcal{M}$. The other is a mass m moving in a potential $V(r)$ reaching out from a point in space at $r = 0$.

If we agree to take, as state labels, the energies of each of these new particles, then the total energy eigenstate, the solution to (8-5) for the hamiltonian of (8-13), will be

$$|E \ldots \rangle = |E(\mathcal{M})\ldots,E(m)\ldots\rangle \tag{8-14}$$

This state solves two separate eigenvalue problems simultaneously: one for $E(\mathcal{M})$

$$\frac{\hat{P}^2}{2\mathcal{M}}|E \ldots\rangle = E(\mathcal{M})|E \ldots\rangle \tag{8-15}$$

and one for $E(m)$.

$$\left[\frac{\hat{p}^2}{2m} + V(\hat{r}) \right] |E \ldots\rangle = E(m)|E \ldots\rangle \tag{8-16}$$

Because the state $|E \ldots\rangle$ also solves Equation 8-5, it follows that

$$E = E(\mathcal{M}) + E(m) \tag{8-17}$$

See also Problem 8.3.

THE PARTICLE-IN-A-BOX HAS NO MOMENTUM

We fix our attention on the *whole-atom* part of the problem, the part governed by Equation 8-15.

The hamiltonian describes a particle of mass $\mathcal{M}$ free to move anywhere in the three dimensions X, Y, and Z. For the practical purpose of a laboratory experiment, *anywhere* really means *within the experimental chamber where the measurements are carried out*.

The volume of the chamber is L^3. It is, of course, macroscopic in size.

The mass $\mathcal{M}$ is a *particle-in-a-box*. The box is the experimental chamber. The walls of the box constitute an infinite potential barrier; the particle cannot get out of this square well.

This idea can be meaningful only if the size of the box drops out of all atomic-scale physical deductions. In fact, this is the case. The physical results persist in the limit that L^3 approaches infinity. The procedure is justified by its results.

Because the physical results must be independent of the shape as well as the size of the box, we choose a cubical box of side L.

With these ideas in mind we can solve the eigenvalue problem of (8-15) in X-, Y-, Z-coordinate space.

The wave function for a particle in a box must drop to zero at every boundary of the box because the wave function is identically zero outside any boundary. It must reach this value continuously from within the boundary. Thus the solutions to (8-15) are the following (see Problem 8.4):

$$\sqrt{8/L^3} \; \sin \; \pi\nu_1\frac{X}{L} \; \sin \; \pi\nu_2\frac{Y}{L} \; \sin \; \pi\nu_3\frac{Z}{L}$$

$$\text{where } 0 \leqslant X,Y,Z \leqslant L \qquad (8\text{-}18)$$

$$\langle XYZ|\nu_1\nu_2\nu_3\rangle =$$

$$0 \quad \text{where } X,Y,Z \text{ outside 0-to-}L$$

The ν_1, ν_2, and ν_3 are three positive integers, indexing the energy $E(\mathscr{M})$.

$$E(\mathscr{M}) = \frac{h^2}{8 \, \mathscr{M}L^2} \, (\nu_1^2 + \nu_2^2 + \nu_3^2) \qquad (8\text{-}19)$$

None of the solutions (8-18) are eigenstates of momentum. Standing wave states have no definite momentum (see Problem 8.6). Furthermore, the physical confinement box forbids a particle to have a fixed momentum: momentum state wave functions cannot be constructed because none have zero amplitude anywhere. (see Problem 3.15) Thus the system is not useful for thinking of a current or a beam of particles. Standing waves always have zero current.

It is more convenient to think in terms of traveling waves, then we can treat a current or beam of particles moving through the experimental chamber. In physics we use a computational trick. It allows traveling waves for calculations but preserves the box idea; we exchange the particle-in-a-box for the particle-through-a-box concept.

8.3 THE PARTICLE-THROUGH-A-BOX HAS PERIODIC BOUNDARY CONDITIONS

Instead of the material box of volume L^3, we imagine the X-, Y- and Z-coordinates to somehow bend back on themselves so that each makes a circular ring of length L. We invent a fictional mathematical box to replace the physical one. It is the three-dimensional counterpart of the 1-D bead-on-a-ring; each dimension is a ring.

The domain of X is only between 0 and L, but in this ring point-of-view, the two positions X and $X + L$ are at the same place. Similarly, the coordinate Y is constrained to be between 0 and L, but Y and $Y + L$ are at the same point on the Y-ring. Single valuedness requires that the wave functions for such rings be periodic in the ring length, which is why one refers to *periodic boundary conditions* (see Figure 8-3).

For this 3-D-ring kind of box there are momentum eigenstate solutions to the eigenvalue problem of Equation 8-15:

$$\langle XYZ|n_1n_2n_3\rangle = \frac{1}{L^{3/2}} \exp\left(\frac{i\mathbf{P}\cdot\mathbf{R}}{\hbar}\right) \qquad (8\text{-}20)$$

where

$$\mathbf{P}\cdot\mathbf{R} \equiv \frac{n_1hX}{L} + \frac{n_2hY}{L} + \frac{n_3hZ}{L} \qquad (8\text{-}21)$$

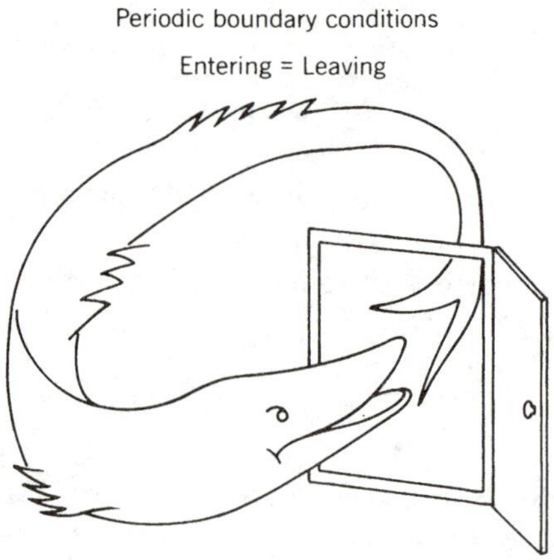

Periodic boundary conditions

Entering = Leaving

Figure 8.3 **Periodic boundary conditions: entering = leaving.**

The n, like the v, are integers, but there is this important difference: they have different domains. The n range over *all* integers, over zero and negative as well as positive ones, whereas the v are only positive integers. In state-index space the points v_1, v_2, and v_3 lie in only one of the eight quadrants formed by the v_1, v_2, and v_3 axes. In the n_1, n_2, n_3 state-index space, occupation points lie in all eight quadrants. The two cases are illustrated in Figure 8-4.

There is another difference between the physical box and the mathematical ring one: the formula for the energy. Applying the hamiltonian operator of Equation 8-15 to the states of the 3-D ring-box, Equation 8-20, we find the energies

$$E(\mathcal{W}) = \frac{h^2}{2 \mathcal{M} L^2} (n_1^2 + n_2^2 + n_3^2) \qquad (8\text{-}22)$$

This formula differs from that in (8-19) by a factor of 4.

The differences notwithstanding, one treatment is just as valid as the other in speaking of a free particle. The 3-D ring-box is just as good a model of the experimental chamber as is the material box. They are equivalent because an experimental chamber is so very large relative to anything atomic.

Put quantitatively, this means that any particle wavelength, λ, is very much smaller than the box size L. The particles—atoms of momentum $\sqrt{2 \mathcal{M} E(\mathcal{W})}$—we are considering are characterized by wavelengths $\lambda = L/|n|$ or $2L/v$ which are microscopic compared with L: only very large values of n or v characterize an experimental atomic particle in a macroscopic box.

To analyze the idea, we need examine only one of the dimensions, say X. The extension of the argument to three dimensions is transparent.

The two kinds of boxes are compared for the one-dimensional case in Figure 8-5. There we see that the traveling waves of (8-20) serve as an alternative basis with which to describe a standing wave. The pair of traveling waves $n = 50$ and $n = -50$ are

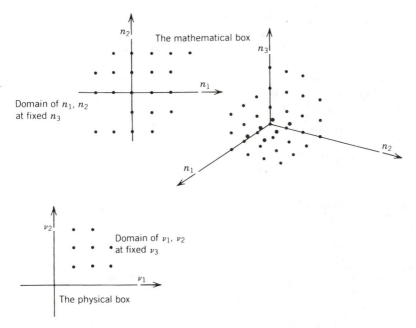

Figure 8.4 State-index spaces for the particle in two different boxes. Each index point in the space labels a state. The states of a physical box are $|\nu_1,\nu_2,\nu_3\rangle$. The states of the mathematical box are $|n_1,n_2,n_3\rangle$.

the alternative for the pair of standing wave states $\nu = 100$ and $\nu = 101$. The standing wave pair can be decomposed into the running wave pair. The almost-degenerate standing wave pair may be replaced by the precisely degenerate traveling wave pair. When $\nu (= 2|n|)$ is essentially indistinguishable from $\nu + 1 (= 2|n| + 1)$, the error is negligible. This happens when ν—or $|n|$—is large, precisely in the limit of a macroscopic box. At small ν or $|n|$, the approximation breaks down. That it does so is evident in Figure 8-5. At small ν the two boxes are *not* equivalent.

Because of the monumental mathematical simplification that results from periodic boundary conditions, we will treat free particles as if their number in the box remains constant: any that leave are replaced by an equal number entering. The box represents the experimental chamber but idealized as an abstract mathematical experimental chamber, one for which the eigenstates are traveling waves, as in Equation 8-20.

CAST THE CENTRAL FIELD HAMILTONIAN IN THE BASIS OF POLAR COORDINATES

Now we turn to the second part of the decoupled hamiltonian: that of a mass m moving in a central field pulling it toward the origin. The center from which the

The 1-D Particle-in-a-box:
two models

Standing wave states

$$\langle X|e^{-it\hat{H}/\hbar}|\nu\rangle =$$

$$\sqrt{\frac{2}{L}}\, e^{-itE_\nu/\hbar}\sin \nu\pi X/L$$

$$0 \leqslant X \leqslant L$$
$$\nu = 1, 2, 3, 4 \ldots$$

Almost degenerate pair
$$\nu = 2|n| \text{ and } 2|n| + 1$$ $$|n| = \text{large}$$

$$E_\nu = \frac{h^2}{8 \, \mathcal{M}L^2}\, \nu^2$$

Traveling wave states

$$\langle X|e^{-it\hat{H}/\hbar}|n\rangle =$$

$$\sqrt{\frac{1}{L}}\, e^{i(2\pi nX/L - E_n t/\hbar)}$$

all values of X
$$n = \ldots . -1, 0, 1, 2 \ldots$$

Precisely degenerate pair
$$n = \pm|n|$$

$$E_n = \frac{h^2}{2 \, \mathcal{M}L^2}\, n^2$$

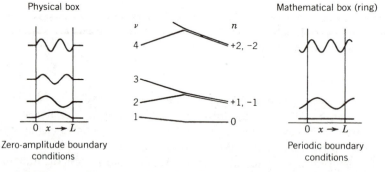

Physical box

$0 \; x \rightarrow L$

Zero-amplitude boundary
conditions

ν n
4 — — +2, −2

3 —

2 — — +1, −1
1 — — 0

Mathematical box (ring)

$0 \; x \rightarrow L$

Periodic boundary
conditions

Figure 8.5 The one-dimensional particle-in-a-box: the material box versus the mathematical ring-box.

field emanates is the position of particle 1. The position of particle 2 relative to that of particle 1 is **r**.

The energies of this system are the internal energy levels of the atom. The eigenvalue problem that yields the $E(\mathcal{m})$ is that of Equation 8-16.

We work in the basis of position because that is the eigenspace of the potential operator. Because the potential is a radial one, we expect that the coordinates r, θ, and ϕ would be a better choice than x, y, and z. The first step, therefore, is to cast the hamiltonian, $\hat{H}_{\mathcal{m}}$, of (8-16) in the language of polar coordinates: we must find $\langle r, \theta, \phi| \hat{H}_{\mathcal{m}}$.

The potential energy term is the easy one:

$$\langle r,\theta,\phi| \, V(\hat{r}) = V(r) \, \langle r,\theta,\phi| \tag{8-23}$$

The spherical coordinate position basis is specifically chosen to make this term easy. To write down the kinetic energy term in this basis requires more effort.

We first define an operator $\hat{p}_r$. We may call it the radial momentum as a mnemonic aid, but this operator is merely a computational tool. It is not a hermitian operator, and it is not to be associated with a measurable (see Chapter 5, "Procede from Meaning").

The instruction defining the operator, $\hat{p}_r$, is

$$\langle r,\theta,\phi| \hat{p}_r = -i\hbar \frac{\partial}{\partial r} \langle r,\theta,\phi| \tag{8-24}$$

Thus $\hat{p}_r$ has no θ or ϕ dependence. Because the vector operator dot product $\hat{\mathbf{r}} \cdot \hat{\mathbf{p}}$, when cast into spherical coordinates, is also a purely radial operator, there is a relationship between $\hat{p}_r$ and $\hat{\mathbf{r}} \cdot \hat{\mathbf{p}}$.

$$\hat{\mathbf{r}} \cdot \hat{\mathbf{p}} = \hat{r}\, \hat{p}_r \tag{8-25}$$

This equation merely restates the familiar identity from differential calculus that

$$x\frac{\partial}{\partial x} + y\frac{\partial}{\partial y} + z\frac{\partial}{\partial z} = r\frac{\partial}{\partial r} \tag{8-26}$$

Notice that the operator $\hat{\mathbf{p}} \cdot \hat{\mathbf{r}}$ is not $\hat{p}_r \hat{r}$ (see Problem 8.8). Rather,

$$\hat{\mathbf{p}} \cdot \hat{\mathbf{r}} = \hat{r}\hat{p}_r - 3i\hbar = \hat{p}_r \hat{r} - 2i\hbar \tag{8-27}$$

In terms of the radial momentum, the kinetic energy can be split into a purely radial operator part and a purely angular operator part: a decoupling can be effected. Particularly useful is the finding that the entire angular part is contained within one single well-known operator; the total angular momentum operator, $\hat{L}^2$. Here is the result put in a form useful for memorizing:

$$\hat{L}^2 = \hat{r}^2\hat{p}^2 - \hat{p}_r\hat{r}^2\hat{p}_r \tag{8-28}$$

It's easy to memorize because it parallels the classical expression for the angular momentum magnitude as the dot product of two cross products. The familiar classical relation is this

$$L^2 = (\mathbf{r} \times \mathbf{p}) \cdot (\mathbf{r} \times \mathbf{p}) = r^2 p^2 - (\mathbf{r} \cdot \mathbf{p})^2 \tag{8-29}$$

When writing the operator counterpart of this equation, the trick is to order things carefully: write $r^2 p^2$, not $p^2 r^2$, and convert $(\mathbf{r} \cdot \mathbf{p})^2$ into the symmetric expression $\hat{p}_r\hat{r}^2\hat{p}_r$ or $\hat{r}\hat{p}_r^2\hat{r}$. These last two are equivalent (see Problem 8.9).

That (8-29) is true does not constitute a proof that (8-28) is true: (8-29) is merely a memory aid by which you can appreciate (8-28). The reason for having such an aid is because the formal derivation of (8-28) is so tedious. It amounts to writing the ∇^2 operator in spherical coordinates. You write ∇^2 in terms of r, θ, ϕ, $\partial/\partial r$, $\partial/\partial\theta$, and $\partial/\partial\phi$. You do it by using the scheme illustrated in Figure 5-6. You calculate $\partial^2/\partial x^2 + \partial^2/\partial y^2 + \partial^2/\partial z^2$ in polar coordinates. The task is the subject of Problem 8.10. You may consult Problem 5.22 to carry it out. Here's what you will find:

$$\langle r,\theta,\phi|\,\hat{p}^2 = -\hbar^2\nabla^2\,\langle r,\theta,\phi|$$

$$= \frac{1}{r^2}\langle r,\theta,\phi|\hat{L}^2 - \frac{\hbar^2}{r^2}\frac{\partial}{\partial r}\left(r^2\frac{\partial}{\partial r}\right)\langle r,\theta,\phi| \qquad (8\text{-}30)$$

The first equality reminds us of the meaning of the $\hat{p}^2$ operator in position-space coordinates.

The second equality here is merely the representation of $-\hbar^2\nabla^2$ in polar coordinates. By the expression $\langle r,\theta,\phi|\hat{L}^2$ is meant the one displayed in Equation 5-79. The operator prescription for $\hat{L}^2$ in r, θ, ϕ space is given there. It contains only θ and ϕ, no r.

The equality of the extremes confirms the rule laid down in (8-28). Applying this rule yields the result in (8-30). Thus (8-30) proves (8-28).

This equation allows us to write the hamiltonian operator in spherical coordinates. It permits us to cast the eigenvalue problem of (8–16) in the r, θ, ϕ basis. Here's how that eigenvalue problem looks in this basis:

$$\langle r\theta\phi|\hat{H}_m\,|E(m)\ldots\rangle =$$

$$\frac{1}{2\,m\,r^2}\langle r\theta\phi|\hat{L}^2|E(m)\ldots\rangle + \left[\frac{-\hbar^2}{2\,m\,r^2}\frac{\partial}{\partial r}\left(r^2\frac{\partial}{\partial r}\right) + V(r)\right]\langle r\theta\phi|E(m)\ldots\rangle$$

$$= E(m)\,\langle r\theta\phi|E(m)\ldots\rangle \qquad (8\text{-}31)$$

8.4 ENERGY-COMPATIBLE OPERATORS LABEL THE EIGENSTATES

Equation 8-31 is a three-dimensional one. A solution contributes three to the six dimensions of the full two-particle problem. Because the bra contains three coordinates, we must expect a complete ket to contain three quantum numbers. The dots after $E(m)$ in the ket represent the other two labels that complete a description of the state.

These labels must be the eigenvalues of some operators other than $\hat{H}_m$: they correspond to observables that are simultaneous with energy. Therefore the operators yielding the remaining pair of labels must commute with $\hat{H}_m$; they must also commute with each other.

The form in which $\hat{H}_m$ appears in (8-31) makes it evident that $\hat{L}^2$ is one such operator. Because $\hat{L}^2$ operates only in the θ,ϕ domain and not in the r domain, it commutes with all r-domain operators. In addition, it commutes with itself. Thus it commutes with $\hat{H}_m$ (see Problem 8.14).

$$[\hat{L}^2, \hat{H}_m] = 0 \qquad (8\text{-}32)$$

A third eigenvalue label for the states might come from any single component of angular momentum: every component commutes with $\hat{L}^2$. And being independent of the variable r, each one also commutes with $\hat{H}_m$. The tradition is to favor the z-component: we use the eigenvalues m of the operator $\hat{L}_z$.

The set of states embracing the $E(m)$ that we use are $|E(m),\ell,m\rangle$. Each one is a simultaneous eigenstate of $\hat{H}_m$, $\hat{L}^2$, and $\hat{L}_z$. Thus

$$\hat{L}^2 |E(m),\ell,m\rangle = \ell(\ell+1)\hbar^2 |E(m),\ell,m\rangle \qquad (8\text{-}33)$$

$$\hat{L}_z |E(m),\ell,m\rangle = m\hbar |E(m),\ell,m\rangle \qquad (8\text{-}34)$$

and by virtue of (8-31) combined with (8-33),

$$\left[-\frac{\hbar^2}{2mr^2}\frac{\partial}{\partial r}\left(r^2 \frac{\partial}{\partial r} \right) + \frac{\ell(\ell+1)\hbar^2}{2mr^2} + V(r) \right] \langle r\theta\phi|E(m),\ell,m\rangle$$

$$= E(m)\,\langle r\theta\phi|E(m),\ell,m\rangle \qquad (8\text{-}35)$$

These equations are three simultaneous eigenvalue problems for the states $|E(m),\ell,m\rangle$. Happily we already know the solutions to the first two of them: ℓ must be a nonnegative integer, and m spans the integers from $-\ell$ to $+\ell$. Half-integers are not allowed because they have no position representation. Thus the atom is characterized by an orbital angular momentum quantum number, ℓ, equal to zero or one or two, and so forth. And given some value of ℓ, the solutions to (8-34) tell us that m may be any integer from $-\ell$ to $+\ell$.

The choice of $\hat{L}^2$ as a simultaneous operator has simplified the energy eigenvalue problem: Equation 8-35 is a purely radial one. It is one-dimensional in r. This equation governs only the r dependence of $\langle r\theta\phi|E(m),\ell,m\rangle$. It says nothing about the θ and ϕ dependence. But through (8-33) and (8-34), the θ and ϕ dependence is already completely prescribed. It is contained in the functions $\langle\theta,\phi|\ell,m\rangle$ given as Equation 5-87 in Chapter 5, the spherical harmonic functions $Y_{\ell,m}(\theta,\phi)$.

Thus the bracket $\langle r\theta\phi|E(m),\theta,\phi\rangle$ breaks up into a two-part product function. The purely radial part, $R(r)$, solves (8-35): it is parametrically dependent on ℓ because ℓ appears in (8-35). The other part is the spherical harmonic function $Y_{\ell,m}(\theta,\phi)$.

$$\langle r\theta\phi|E(m),\ell,m\rangle = \langle r|E(m,\ell)\rangle\,\langle\theta,\phi|\ell,m\rangle = R(r)\,Y_{\ell,m}(\theta,\phi) \qquad (8\text{-}36)$$

We need to find the functions $\langle r|E(m,\ell)\rangle = R(r)$ that solve (8-35). The energy levels—the $E(m)$ quantum numbers—depend upon it. Of course the values of $E(m)$ that come out must be related to what value of ℓ you insert in (8-35). It is for this reason that an explicit ℓ dependence is expressly displayed in the internal energy: $E(m) = E(m,\ell)$. The solutions to (8-35) answer this physical question: Given the state of orbital angular momentum characterized by ℓ and m, what values of energy are possible?

8.5   THE ELECTROSTATIC CENTRAL FIELD GOVERNS HYDROGEN

We can answer this question exactly in the case of the hydrogenlike atom; when the radial potential is the electrostatic one,

$$V(r) = -\frac{Ze^2}{r} \tag{8-37}$$

In this equation, e is the magnitude of the charge of the electron in electrostatic units, in esu. To use coulombs you must note that

$$e^2(\text{esu}) = \frac{e^2(\text{coulombs})}{4\pi\varepsilon_0} = 13.6 \ eV - \mathring{A} \tag{8-38}$$

This potential models the hydrogen atom when $Z = 1$ and the *reduced* mass m, of Equation 8-7, is that of the proton–electron combination. It models deuterium when $Z = 1$ and m is the reduced mass of the deuteron and the orbiting electron. It models the singly ionized helium atom when $Z = 2$ and m is the reduced mass of the helium nucleus and the orbiting electron. The helium nucleus, also called an alpha particle, has two protons and two neutrons—four atomic mass units of mass. The potential represents an electrostatic force of attraction between a nucleus of charge $+Ze$ and an orbiting electron of charge $-e$.

With this potential as the $V(r)$ in Equation 8-35, the critical equation to be solved— the radial equation—becomes

$$-\frac{\hbar^2}{2mr^2}\frac{d}{dr}\left(r^2\frac{dR}{dr}\right) + \frac{\hbar^2\ell(\ell+1)}{2mr^2}R - \frac{Ze^2}{r}R - E(m)R = 0 \tag{8-39}$$

BOUND STATES ARE CONFINED ONES

We are interested in the *bound-state* solutions to this equation. These solutions are the ones that are localized near the origin (see Figure 8-6). They are the ones in which the orbiting particle remains within the force field. Such solutions die away at large r; the probability vanishes for finding a large separation between the particles. A bound state is that in which the particles making up the atom remain together.

Such states are ones of negative internal energy: $E(m) < 0$, They are the ones in which the strength of the potential exceeds the kinetic energy.

To prove that all of the bound states have $E(m) < 0$, you need only inspect the very large r assymptotic form of Equation 8-39. That's the subject of Problem 8.15. You will find that there are solutions that oscillate with undiminished amplitude right out to infinity. These are the unbound ones, characterized by $E(m) \geq 0$. All of the solutions that die out assmptotically at infinity are characterized by $E(m) < 0$.

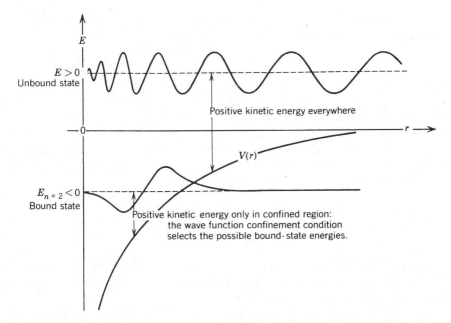

Figure 8.6 Bound-state wave functions are confined: the amplitude to find the particle is large only near the origin. Unbound-state wave functions are unconfined: even far from the origin there is an amplitude to find the particle. Unconfined states form a continuum: confined states form a discrete set.

Solutions that grow indefinitely large at infinity are, of course, not physically acceptable ones. Here we seek the bound states, the solutions for which $E(m) < 0$.

The following is a well-known tabulated equation of mathematics to which (8-39) is isomorphic:

$$\frac{1}{\rho^2}\frac{d}{dp}\left(\rho^2\frac{dR}{d\rho}\right) - \frac{\ell(\ell+1)}{\rho^2}R + \frac{n}{\rho}R - \frac{1}{4}R = 0 \qquad (8\text{-}40)$$

This equation has single-valued, continuous, and square integrable solutions for integer ℓ's only in this special case: that n be a positive integer, $n = 1, 2, 3, \ldots$ Physically acceptable solutions don't exist for arbitrary n and ℓ: for a given n the permissible values of ℓ are 0 or 1 or 2 and so on, up to $n - 1$.

Under these conditions the solutions to this equation involve a famous set of functions called associated Laguerre polynomials written as $L_{n-1-\ell}^{(2\ell+1)}(\rho)$. In terms of these, the solutions (unnormalized) are

$$R(\rho) = \rho^\ell\, e^{-\rho/2}\, L_{n-1-\ell}^{(2\ell+1)}(\rho) \qquad (8\text{-}41)$$

Information about the Laguerre polynomials may be found among the six mathematical references cited in Chapter 4.

It remains for us to implement the isomorphism between (8-40) and (8-39). It is accomplished by means of the following association.

$$r^2 = \rho^2 \frac{\hbar^2}{[-8\,m\,E(\,m\,)]} \tag{8-42}$$

Make this substitution in Equation 8-39. You will find immediately that the result looks exactly like Equation 8-40. It looks that way only if we associate the quantity $-Ze^2\sqrt{-8\,m\,E(\,m\,)}/4\hbar E(\,m\,)$ with the n of (8-40). Thus (8-39) only has physically acceptable solutions when that quantity is an integer. This integer is the energy index: it determines the values that $E(\,m\,)$ can have. These values are

$$E(\,m\,) = -\frac{Z^2 e^4\,m}{2\hbar^2 n^2} = E_n(\,m\,) \tag{8-43}$$

For the hydrogen atom, Z is 1, and the mass, m, is effectively that of the electron. For this famous case, the numbers yield that $E_{n=1} = -13.6$ electron volts. This is the ground-state energy of the hydrogen atom.

The first excited state has energy $E_2(\,m\,) = -3.4$ eV. This level is fourfold degenerate; there are four different states that have this same energy. They are $|n\ell m\rangle = |2,0,0\rangle$, $|2,1,+1\rangle$ $|2,1,0\rangle$, and $|2,1,-1\rangle$. In the ket label the energy index, n, is recorded rather than the energy itself. The first state is called the $2s$ state: the 2 is the value of n, and the letter s traditionally indicates $\ell = 0$. The remaining three states are called $2p$ states. That $\ell = 1$ is conveyed by the letter p.

The n here must not be confused with the n_1, n_2, and n_3 of Equations 8-20 and 8-21. The n_1, n_2, and n_3 index the translation states of the whole atom; n indexes the internal energy of the atom.

Here are the first six normalized radial function solutions, $R(r) = \langle r|E_n(\,m\,,\ell)\rangle = \langle r|n(\ell)\rangle$, to (8-39). Each different value of the n and ℓ in $n(\ell)$ corresponds to a different state. The symbol $n(\ell)$ merely shows that states are indexed not only by n but also parametrically by ℓ. By $n(\ell)$ in the ket is *not* meant some function of ℓ that n is but, rather, that ℓ is an auxilliary part of the ket label.

$$\langle r|n(\ell)\rangle = \langle r|1(0)\rangle = \frac{2}{a^{3/2}} e^{-r/a} \tag{8-44}$$

$$\langle r|2(0)\rangle = \frac{2}{(2a)^{3/2}} \left(1 - \frac{r}{2a}\right) \exp\left(\frac{-r}{2a}\right) \tag{8-45}$$

$$\langle r|2(1)\rangle = \frac{1}{3^{1/2}(2a)^{3/2}} \left(\frac{r}{a}\right) \exp\left(\frac{-r}{2a}\right) \tag{8-46}$$

$$\langle r|3(0)\rangle = \frac{2}{(3a)^{3/2}} \left(1 - \frac{2r}{3a} + \frac{2r^2}{27a^2}\right) \exp\left(\frac{-r}{3a}\right) \tag{8-47}$$

$$\langle r|3(1)\rangle = \frac{4\sqrt{2}}{3(3a)^{3/2}} \left(\frac{r}{a}\right) \left(1 - \frac{r}{6a}\right) \exp\left(\frac{-r}{3a}\right) \tag{8-48}$$

$$\langle r|3(2)\rangle = \frac{\sqrt{8}}{27\sqrt{5}(3a)^{3/2}} \left(\frac{r}{a}\right)^2 \exp\left(\frac{-r}{3a}\right) \tag{8-49}$$

In these expressions the parameter, a, is a characteristic length that occurs naturally in the problem. For hydrogen, this length is the first Bohr orbit radius: it is 0.5 Å for hydrogen. In general the quantity, a, is a shorthand notation for a collection of parameters that have the dimensions of length

$$a \equiv \frac{\hbar^2}{m\,e^2 Z} \tag{8-50}$$

The radial functions $\langle r|n(\ell)\rangle$, in addition to solving the eigenvalue Equation 8-39, must be properly normalized so as to serve their purpose as amplitudes for probability density. They must satisfy the certainty condition

$$1 = \langle n\ell m|n\ell m\rangle = \sum_{r,\theta,\phi} |\langle r\theta\phi|n\ell m\rangle|^2 \tag{8-51}$$

Thus

$$|\langle r\theta\phi|n\ell m\rangle|^2 = |\langle r|n(\ell)\rangle|^2 \, |\langle \theta,\phi|\ell,m\rangle|^2 \tag{8-52}$$

is a probability per unit volume of r,θ,ϕ space. The elemental volume of this spherical coordinate space is $r^2 dr \sin\theta \, d\theta \, d\phi$. Hence we must interpret the sum as an integral over the continuous variables r, θ, and ϕ as follows:

$$\sum_{r,\theta,\phi} \equiv \int_0^\infty r^2 dr \int_0^\pi \sin\theta \, d\theta \int_0^{2\pi} d\phi \tag{8-53}$$

The *summation* thus spans the entire domain of spherical coordinate measurement space.

With this understanding, and the rules governing $\langle \theta,\phi|\ell,m\rangle$ in Equations 5-94 and 5-95, it follows that the radial function normalization condition is

$$\int_0^\infty r^2 \, dr \, |\langle r|n(\ell)\rangle|^2 = 1 \tag{8-54}$$

The functions of (8-44) through (8-49) have been normalized to fit this condition.

Figure 8-7 summarizes the important features of the internal energy states of the hydrogenlike atom.

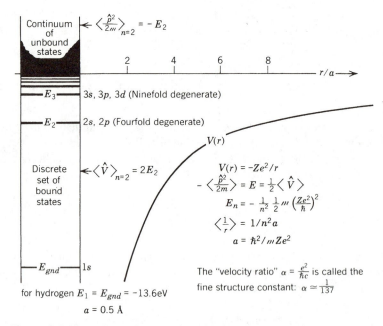

Figure 8.7 The two-body electrostatic planetary atom: key features.

8.6 THE GRAND RESULT IS A 6-D STATE

We have considered a two-particle atom: a mass m_1 attracts another mass m_2 via a radial potential. The combined system is a six-dimensional one. The stationary eigenstates of the system solve a 6-D hamiltonian eigenvalue problem.

To obtain these states we decoupled the hamiltonian so as to create six 1-D problems to replace the one 6-D problem. From these six 1-D solutions we want to compose the solution to the whole problem. We want to display the complete solution to the energy eigenvalue problem (Equation 8-5) for the 6-D hamiltonian of Equation 8-2 or 8-13.

It is best written in terms of a **K**-vector: **k** represents the quantum numbers n_1, n_2, and n_3 (see Equation 8-21):

$$k_x = \frac{2\pi n_1}{L}, \quad k_y = \frac{2\pi n_2}{L}, \quad k_z = \frac{2\pi n_3}{L} \tag{8-55}$$

Thus, the condition $k_x = 6\pi/L$ means $n_1 = 3$. (see Chapter 2, "Its Real Name Is Its Index").

In the basis of physical **R,r** space, a 6-D stationary state wave function is $\langle \mathbf{R,r} | \mathbf{K}, n, \ell, m \rangle$ where

$$\langle \mathbf{R},\mathbf{r}|\mathbf{K},n,\ell,m\rangle = \frac{1}{L^{3/2}}e^{i\mathbf{K}\cdot\mathbf{R}}\,R(r)\,Y_{\ell,m}(\theta,\phi)$$
$$= \langle \mathbf{R}|\mathbf{K}\rangle\,\langle r|n(\ell)\rangle\,\langle\theta,\phi|\ell,m\rangle \tag{8-56}$$

The vector $\mathbf{K}$ is one of the possible $\mathbf{k}$'s, and the state $|\mathbf{K},n,\ell,m\rangle$ is the eigenstate that solves the whole 6-D problem in (8-5) for two particles attracted to each other via a coulomb field.

This wave function describes a two-particle atom whose center of mass has momentum $\mathbf{P} = \hbar\mathbf{K}$, whose internal energy is at level n, whose total angular momentum index is ℓ, and whose z-component of angular momentum is $m\hbar$. To name the numbers K_x, K_y, K_z, n, ℓ, and m is to name the state (see Problem 8.20).

Equation 8-56 displays the state of the atom in the 6-D position space $\mathbf{R}$, $\mathbf{r}$. Keeping in mind the meaning, (8-55), of $\mathbf{k}$, the same state may be represented in the six-dimensional space $\mathbf{k}$, $\mathbf{r}$.

$$\langle \mathbf{k},\mathbf{r}|\mathbf{K},n,\ell,m\rangle = \langle \mathbf{k}|\mathbf{K}\rangle\,\langle\, r|n(\ell)\rangle\,\langle\theta,\phi|\ell,m\rangle \tag{8-57}$$

SIGNIFICANCE IS ELUSIVE

These results have an evident practical significance. On the two-particle atom is built the theory of multiple-particle atoms, the theory of the whole periodic table of the elements and of atomic spectra. The historical quantum theory validation process followed just this course. Through it the theory has been abundantly validated.

For a student of the subject, the practical significance of the two-particle atom calculation is that it enables one to do more calculations. The significance of any result in physics is judged by the quantity of new results generated. The immediate purpose of doing physics is to do more physics!

That it serves as a basis for increasingly complex computations is one significance of the two-particle atom study. It also tells us some remarkable things about the nature of nature.

TEST FOR A STATE:
THERE EXIST CERTAINTY EXPERIMENTS

To perceive these, we must return to the meaning of the words "to be in a state." They mean that "there exists an experiment whose outcome is certainty."

Consider a 1-D system. If it is in a state, then there will be some measurement that must show it.

Proof: Call the state $|\alpha\rangle$. It's the eigenstate of some physical operator. Call that $\hat{\alpha}$. Make the measurement corresponding to $\hat{\alpha}$. That measurement must produce

$|\alpha\rangle$ *with certainty*: the certainty measurement defines the state. It yields the label that marks the state.

In N-dimensions it takes a series of measurements to define a state. That a state exists means that there is a series of experiments whose outcome is certainty. The state is defined by a set of certainty measurements.

More fundamentally, a state is defined by its labels, each of which is the outcome of a certainty measurement.

8.7   A STATE IS DEFINED BY ITS LABELS

Now return to our two interacting particles. Armed with an operational definition of the meaning of *state*, we can explore the question: Does particle 1 have a state? We answer the question by considering it experimentally.

The system is in the stationary state $|\mathbf{K},n,\ell,m\rangle$. We measure the position $\mathbf{r}_1$ of particle 1. The outcome is surely uncertain: there is a probability distribution over x_1, y_1, and z_1. It's evident from the form of the wave function: just substitute into (8-56) for $\mathbf{R}$ and $\mathbf{r}$ their meanings from (8-8) and (8-10).

The probability of finding particle 1 at $\mathbf{r}_1$ in $d^3\mathbf{r}_1$ and particle 2 at $\mathbf{r}_2$ in $d^3\mathbf{r}_2$ when the system is in the state $|\mathbf{K}, n,\ell,m\rangle$ is $d^3\mathbf{r}_1 d^3\mathbf{r}_2 |\langle\mathbf{r}_1,\mathbf{r}_2|\mathbf{K},n,\ell,m\rangle|^2$. There is no position $\mathbf{r}_1 = \mathbf{b}$ at which you can be certain to find 1. Rather, there is a distribution of possible positions at which you might find it. Clearly particle 1 is *not* in an eigenstate of position (see Problem 8.21).

Just as clearly, it is not in an eigenstate of momentum. The amplitude $\langle\mathbf{p}_1,\mathbf{r}_2|\mathbf{K},n,\ell,m\rangle$ is not a self-space bracket.

For particle 1 to have a state, there must be some physical operator, $\hat{\Omega}_1$, that concerns particle 1 alone—like $\hat{\Omega}_1 = \Omega(\hat{\mathbf{r}}_1,\hat{\mathbf{p}}_1)$—for which $|\mathbf{K},n,\ell,m\rangle$ is an eigenstate. The state of particle 1 would then be labeled by Ω_1 where

$$\hat{\Omega}_1 |\mathbf{K},n,\ell,m\rangle \stackrel{?}{=} \Omega_1 |\mathbf{K},n,\ell,m\rangle \qquad (8\text{-}58)$$

If there exists no operator that has this property, then no measurement on particle 1 alone can produce certainty: particle 1 has no state!

But there is, indeed, no $\hat{\Omega}_1$. No such operator commutes with both $\hat{p}_1^2/2m_1$ and the interaction, $\hat{V}$, between the two particles. Thus no $\hat{\Omega}_1$ exists that commutes with the hamiltonian. For all operators, $\hat{\Omega}_1$

$$[\hat{\Omega}_1,\hat{H}] \neq 0 \qquad (8\text{-}59)$$

A state cannot be formed that is eigen to both $\hat{H}$ and $\hat{\Omega}_1$, so no $\hat{\Omega}_1$ exists that can satisfy (8-58).

What we have shown is that if a system of two interacting particles is in a stationary state, the component parts have no states. When the system as-a-whole is in a stationary state, its elemental parts lose their identity; they are not in definable states, they have no labels.

FOR AN INTERACTING PAIR STATE NEITHER PARTICLE HAS A LABEL

By contrast, notice that the two new particles, manufactured from particles 1 and 2, do have states. The compound particle of mass $\mathcal{M}$ is in an eigenstate of momentum. The measurement corresponding to $\hat{\mathbf{P}} = \hbar\hat{\mathbf{K}}$ defines it.

The light particle of mass m also has a state: $|n,\ell,m\rangle$.

Both of the new particles have states. Neither old particle has a state.

This fact may seem innocuous to some: it wasn't to Einstein. The reason is this: that something be identifiable is the essence of its physical reality. To carry no label is not to be identifiable as a physical entity. But to have no state is to carry no label!

AN ELEMENT OF PHYSICAL REALITY: SOMETHING YOU CAN LABEL

Suppose we believe that particle 1 has a reality all its own, an immutable real existence even when it is part of a larger system. Then, that it has no state appears as an inadequacy of physical theory: there is no label for a *real* thing. Quantum mechanics appears to be incomplete in that it fails to account for "an element of physical reality."

This is the substance of a famous paper—"Can Quantum-Mechanical Description of Physical Reality Be Considered Complete?"—by Einstein, Podolsky, and Rosen, published by the *Physical Review* in 1935 (*Phys. Rev. 47.* 777, 1935).

In this paper the authors explore the implications of the idea that a particle may have no state. They do it by examining a stunning thought-experiment, whose dramatic impact unveils the incredible meaning buried within the equations' mute symbols.

To analyze the EPR thought-experiment mathematically requires a few conceptual clarifications.

Stripped of extraneous detail, the mathematical essentials are rendered best by considering two 1-D interacting particles. The pair constitute a system whose state is $|\Gamma\rangle = |\alpha,\beta\rangle$. The operators $\hat{\alpha}$ and $\hat{\beta}$ refer to experiments on the system-as-a-whole. Our hydrogenlike atom provides examples: $\hat{\mathbf{P}}$ and $\hat{\mathbf{L}}^2$ illustrate whole-system operators like $\hat{\alpha}$ and $\hat{\beta}$.

Of course, because α and β both label the state, we know that

$$[\hat{\alpha},\hat{\beta}] = 0 \tag{8-60}$$

Let $\hat{1}$ define some measurable of particle 1 alone and $\hat{2}$ refer to some measurable of particle 2 alone. The operators $\hat{\mathbf{r}}_1$ and $\hat{\mathbf{r}}_2$ or $\hat{\mathbf{p}}_1$ and $\hat{\mathbf{r}}_2$ are obvious examples of $\hat{1}$ and $\hat{2}$ for a 3-D pair.

Because particle 1 is independent of particle 2, it follows that

$$[\hat{1},\hat{2}] = 0 \qquad (8\text{-}61)$$

When the system is in the state $|\Gamma\rangle$, neither one of the pair is in an identifiable state. The mathematical counterpart of this idea resides in the noncommutability of the particle operators with the whole-system operators. For example, it may be that

$$[\hat{1},\hat{\alpha}] \neq 0 \qquad \text{and} \qquad [\hat{2},\hat{\beta}] \neq 0 \qquad (8\text{-}62)$$

8.8   A PRODUCT WAVE FUNCTION SIGNIFIES DISCERNIBLE PARTICLES

Suppose, instead of the state $|\Gamma\rangle$, the pair of particles were in a state $|\psi\rangle$ that could be represented as a product.

$$\langle 1,2|\psi\rangle = \langle 1|A\rangle \langle 2|B\rangle \qquad (8\text{-}63)$$

Here the number 1 stands for any measurement result characterizing particle 1 *alone;* if $\hat{1} = \hat{x}_1$, then $1 = x_1$; if $\hat{1} = \hat{p}_1$, then by 1 is meant the measurement result p_1, and so on.

Because of this product form there exist individual particle operators to which $|\psi\rangle$ is eigen. There must be an $\hat{A}$ operator that has only to do with particle 1 and a $\hat{B}$ operator that has only to do with particle 2. These operators are the ones that produce the labels A and B. Thus

$$\langle 1,2|\hat{A}|\psi\rangle = \langle 1|\hat{A}|A\rangle \langle 2|B\rangle = A \langle 1,2|\psi\rangle \qquad (8\text{-}64)$$

and

$$\langle 1,2|\hat{B}|\psi\rangle = \langle 1|A\rangle \langle 2|\hat{B}|B\rangle = B \langle 1,2|\psi\rangle \qquad (8\text{-}65)$$

When the system-as-a-whole is in a state like $|\psi\rangle$, each component particle does have its own state. The state of particle 1 is $|A\rangle$, and that of particle 2 is $|B\rangle$.

The stationary state of the hydrogenlike atom provides an example in which a product wave function defines discernible entities.

Equation 8-56 shows that the *particle* of mass $\mathcal{M}$ behaves independently of the *particle* of mass m. These are the discernible entities. Because the wave function $\langle \mathbf{R},\mathbf{r}|\mathbf{K},n,\ell m\rangle$ appears as a product $\langle \mathbf{R}|\mathbf{K}\rangle \langle \mathbf{r}|n,\ell,m\rangle$, each of these particles is identifiable. Each has its own state.

Armed with this notion we can interpret experiments on a compound system state $|\Gamma\rangle = |\alpha,\beta\rangle$ that cannot be resolved into a product concerning particles 1 and 2. Such a state is $|K,n,\ell,m\rangle$. Although it is a product of the states of $\mathscr{N}$ and m, it cannot be resolved into a product of the states of particle 1 and particle 2.

OBSERVE ONE PARTICLE: BOTH ACQUIRE STATES

Consider the meaning of the bracket $\langle 1,2|\Gamma\rangle$: it is the amplitude governing a probability density. Given that the system is in the state $|\Gamma\rangle$, the probability to find particle 1 at 1 in $d1$ and particle 2 at 2 in $d2$ is

$$d1 \ d2 \ P(1,2|\Gamma) \ d\Gamma = d1 \ d2 \ |\langle 1,2|\Gamma\rangle|^2 \ d\Gamma \qquad (8\text{-}66)$$

You make one of the measurements first and then the other. Because $\hat{1}$ and $\hat{2}$ commute, the order is immaterial.

Suppose you make the $\hat{1}$ measurement first. You find $1 = c$: particle 1 is in the state $|c\rangle$. Having made the measurement, you have put the system in a new state. Call it $|S\rangle$. Its wave function must be of the form

$$\langle 1,2|S\rangle = \langle 1|c\rangle \langle 2|\Gamma_c\rangle \qquad (8\text{-}67)$$

This is simply an expression of the fact that 1 is known, with certainty, to be c: $\langle 1|c\rangle$ is a self-space bracket.

That the full wave function, $\langle 1,2|S\rangle$, depends on both variables means that the self-space bracket must be multiplied by some function of 2. In its form lies its substance. The wave function of the state $|S\rangle$ is of the product form. Thus the function $\langle 2|\Gamma_c\rangle$ must be the wave function of particle 2.

Equation 8-67 shows that by virtue of the measurement on particle 1, both particles acquire states! Neither had one before the measurement.

8.9 THERE EXIST CONDITIONAL STATES

The state of particle 2 is labeled $|\Gamma_c\rangle$ to indicate that it depends on two things, the original state of the system, $|\Gamma\rangle$, and the state, $|c\rangle$, of particle 1, discovered in the $\hat{1}$ measurement. Thus the state of particle 2 is a *conditional one;* the state of particle 2 depends on what is found in the measurement of particle 1!

The bracket, $\langle 2|\Gamma_c\rangle$, is the amplitude to find that a measurement corresponding to $\hat{2}$ will yield 2, subject to the condition that particle 1 is known to have $1 = c$. The probability density to find 2 is thus the conditional one $P_c(2)$, where

$$P_c(2) = P(2|\Gamma_c) = |\langle 2|\Gamma_c\rangle|^2 \qquad (8\text{-}68)$$

This conditional probability satisfies a certainty condition. To find some value of 2 among all those possible is a certainty. Thus

$$\langle \Gamma_c | \Gamma_c \rangle d\Gamma_c = \int_2 d2\, P(2|\Gamma_c)\, d\Gamma_c = \langle \Gamma_c | \Gamma_c \rangle^{-1} \sum_2 |\langle 2|\Gamma_c \rangle|^2 = 1 \qquad (8\text{-}69)$$

If the state label, Γ_c, belongs to a discrete set, then, of course, $d\Gamma_c = \langle \Gamma_c | \Gamma_c \rangle^{-1} = 1$. For the continuous case $d\Gamma_c = \langle \Gamma_c | \Gamma_c \rangle^{-1} = 0$. But in either case, if particle 2 is known to have label Γ_c, then the probability is unity to find this particle at some value of 2 within the domain of 2 values. That's what the equation says (see Chapter 2, "State Preparation is Always Differential" and also Equation 2-32).

PROCEED FROM MEANING

The mathematics of elementary probability theory tells us how the joint probability density of Equation 8-66,

$$P(1,2) = P(1,2|\Gamma) \qquad (8\text{-}70)$$

and the conditional probability density of Equation 8-68,

$$P_1(2) = P(2|\Gamma_1) \qquad (8\text{-}71)$$

are related. The relation is

$$d1\, d2\, P(1,2)\, d\Gamma = d2\, P_1(2)\, d\Gamma_1\, d1\, P(1)\, d\Gamma \qquad (8\text{-}72)$$

This says that the probability to find 1 *and* 2 is, simply, that to find 1 and then, having found 1, to find 2.

Put more precisely, the statement is this. The probability to find 1 in $d1$ *and* 2 in $d2$ is the product of two probabilities: that to find 1 in $d1$ multiplied by the probability, $d2\, P_1(2)\, d\Gamma_1$, to find 2 in $d2$ given that Γ_1 is in the $d\Gamma_1$ implied by the spread, $d1$, in the 1 measurement.

In a notation making explicit that we are always referring to an initial state of the system-as-a-whole called $|\Gamma\rangle$, the essence of Equation 8-72 may be recast this way:

$$P(1,2|\Gamma) = P(2|\Gamma_1)\, d\Gamma_1\, P(1|\Gamma) = P(2|\Gamma_1) \frac{1}{\langle \Gamma_1 | \Gamma_1 \rangle} P(1) \qquad (8\text{-}73)$$

The function $P(1) = P(c)$ or $P(1|\Gamma) = P(c|\Gamma)$ is the probability density to find $1 = c$, *regardless of what value 2 has*. It is simply

$$P(c\,|\Gamma) = P(c) \equiv \sum_2 P(c,2|\Gamma) = \sum_2 |\langle c,2|\Gamma \rangle|^2 \qquad (8\text{-}74)$$

ONE MEASUREMENT ON A BIVARIABLE STATE
CREATES A UNIVARIABLE STATE, PARAMETRIZED

Using these statements about the meaning of things it is a simple matter to deduce the wave function, $\langle 2|\Gamma_c\rangle$, of (8-67). Inserting (8-66) and (8-68) into (8-73), the deduction is that

$$\langle 2|\Gamma_c\rangle = \frac{\langle c,2|\Gamma\rangle}{\sqrt{d\Gamma_c P(c)}} = \langle c,2|\Gamma\rangle \sqrt{\frac{\langle\Gamma_c|\Gamma_c\rangle}{\sum_2 |\langle c,2|\Gamma\rangle|^2}} \qquad (8\text{-}75)$$

where $P(c)$ is the sum given by (8-74). This is the wave function for the state $|\Gamma_c\rangle$ in which particle 2 is left after a measurement, $\hat{1}$, on particle 1 yields $1 = c$.

The particle-2 conditional-state wave function is effectively the original joint state wave function parametrized by the finding for particle 1. The signature of the conditional state is contained in the joint state wave function: you find $\langle 2|\Gamma_c\rangle$ merely by inspecting $\langle c,2|\Gamma\rangle$. The dependence on the yet-unprobed variable—the 2 in (8-75)—exposes the state $|\Gamma_c\rangle$. The square root is a normalizing factor.

8.10 MEASUREMENT CREATES REALITY

The mathematical result is Equation 8-75. Its significance is hardly perceptible. Here is the example chosen by Einstein, Podolsky, and Rosen to make its significance apparent.

Two particles are in a pair state. Suppose this state were $|\Gamma\rangle = |K=0, x=b\rangle$: a state of two 1-D particles whose total momentum is zero and whose distance apart is b. This is not a stationary state, but it is a pair-system-as-a-whole state. A stationary initial state of the pair is the subject of Problem 8.26.

Consider two different experiments. Both experiments are performed on a system prepared in the state $|\Gamma\rangle = |0,b\rangle$.

In the *position experiment,* we measure the *position* of particle 1 and ask for the state in which particle 2 is left.

In the *momentum experiment,* we measure the *momentum* of particle 1 and calculate the state in which particle 2 is left.

The calculations are merely a matter of applying (8-75) to the experimental events involved. We calculate $\langle 2|\Gamma_c\rangle$ for two different c's: x_1 and p_1. To do so, we must construct representations of the original whole-system state $|\Gamma\rangle = |K=0, x=b\rangle$ in the language of the individual particles, the 1 and 2. If we choose for 2 the position, x_2, then the brackets, $\langle c,2|\Gamma\rangle$, that we need are $\langle x_1,x_2|0,b\rangle$ and $\langle p_1,x_2|0,b\rangle$.

The whole-system state $|\Gamma\rangle = |K,b\rangle$ is easily written in terms of X and x, the center-of-mass and relative position coordinates.

$$\langle X,x|K,b\rangle = \frac{1}{\sqrt{L}} \exp(iKX) \langle x|b\rangle \qquad (8\text{-}76)$$

The $\langle x|b\rangle$ is a self-space bracket, a Dirac δ-function in x-space.

From this equation we can deduce the two brackets that we need:

$$\langle x_1,x_2|K=0,b\rangle = \frac{1}{\sqrt{L}}\langle x_2-x_1|b\rangle \tag{8-77}$$

and calling $p_1 = \hbar k_1$,

$$\langle k_1,x_2|K=0,b\rangle = \frac{1}{L}\exp[-ik_1(x_2-b)] \tag{8-78}$$

The first comes from applying the prescription

$$\langle x_1,x_2|0,b\rangle = \langle X=x_1\frac{m_1}{\mathscr{M}}+x_2\frac{m_2}{\mathscr{M}}, x=x_2-x_1|0,b\rangle \tag{8-79}$$

to (8-76). It's a mere matter of coordinate transformation: the 1-D counterpart of those in Figure 8-2.

The second is obtained from the first via a ket-bra sum:

$$\langle k_1,x_2|0,b\rangle = \sum_{x_1}\langle k_1|x_1\rangle\langle x_1,x_2|0,b\rangle \tag{8-80}$$

We now consider the *position experiment*. Measuring the *position* of particle 1, you find it to be $x_1 = c$. Set $x_1 = c$ in (8-77). Because $\langle x_2-c|b\rangle = \langle x_2|c+b\rangle$, it is clear that the state in which particle 2 is left is the eigenstate of position $|\Gamma_c\rangle = |x_2=c+b\rangle$.

To complete the calculation of $\langle x_2|\Gamma_c\rangle$ is only a matter of evaluating the normalizing square root factor in (8-75). Because Γ_c represents a position, it is a continuous variable: $d\Gamma_c = d(c+b) = 1/\langle c+b|c+b\rangle$. Because of the meaning of $P(c)$ as shown in (8-74), the square of the normalizing factor is

$$d\Gamma_c\,P(c) = d(c+b)\sum_{x_2}|\langle x_1=c,x_2|0,b\rangle|^2$$

$$= d(c+b)\sum_{x_2}\frac{1}{L}|\langle x_2|c+b\rangle|^2$$

$$= \frac{1}{L}d(c+b)\langle c+b|c+b\rangle = \frac{1}{L} \tag{8-81}$$

Using this result in the formula of (8-75) shows directly that the wave function, $\langle x_2|\Gamma_c\rangle$, for the state in which particle 2 is left after a measurement of the position of 1 finds it at $x_1 = c$ is

$$\langle x_2|\Gamma_c\rangle = \langle x_2|c+b\rangle \tag{8-82}$$

Particle 2 is left in a pure position state: $|x_2=c+b\rangle$.

Now consider the *momentum experiment*. The wave function for the state in which particle 2 is left is a matter of analyzing the joint amplitude for that event: (8-78) instead of (8-77). Symbolize by $|\Gamma_\kappa\rangle$ the state in which particle 2 is left after a measurement of the momentum of particle 1 reveals $k_1 = \kappa$. What is called c in (8-75) is here dubbed κ: it is the finding for the measurement, $\hat{k}_1$, of the momentum of particle 1.

Because of its form, Equation 8-78 reveals the state in which particle 2 is left; it is a state of momentum $-\hbar\kappa$. Just substitute $k_1 = \kappa$ into the equation and inspect it. You will recognize $\langle x_2|-\kappa\rangle$. Thus $\Gamma_\kappa = -\kappa$.

To calculate the normalizing factor we take note that Γ_κ has an index: the values of momenta are quantized by the box of size L. Thus $d\Gamma_\kappa$ must be taken as unity: $d\Gamma_\kappa = 1$.

Using this thought and (8-78) in (8-74), we find

$$d\Gamma_\kappa P(\kappa) = \int_L dx_2 \frac{1}{L^2} = \frac{1}{L} \tag{8-83}$$

Thus the wave function for the state in which particle 2 is left, after a measurement of the momentum of particle 1 finds $k_1 = \kappa$, is

$$\langle x_2|\Gamma_\kappa\rangle = \frac{1}{\sqrt{L}} \exp[-i\kappa(x_2-b)] = e^{ib\kappa} \langle x_2|k_2=-\kappa\rangle \tag{8-84}$$

Particle 2 is left in a pure momentum state: $|k_2=-\kappa\rangle$.

TO HAVE NO STATE IS TO HAVE NO LABEL

You haven't made any measurement on particle 2. By virtue of one experiment you deduce that this particle is in an eigenstate of position: $|x_2=c+b\rangle$. You deduce from the other experiment that it is in an eigenstate of momentum: $|k_2=-\kappa\rangle$. The particle can't be in both of these states simultaneously; the two are incompatible.

Thus particle 2 does not have its own identifiable (labelable) state when the pair are in the whole-system state $|K=0,b\rangle$. It has no state until you look at particle 1. The two thought-experiments show that its label is manufactured entirely by the experiment you do on particle 1!

When the system-as-a-whole has a state, its parts carry no labels.

NO STATE LABEL: NO REALITY

In the EPR paper, the particles 1 and 2 are held to be *elements of reality*. That they have no labels when they are in a pair state suggests a failing of quantum mechanics: it is incomplete in that elements of reality are not accounted for.

The modern view is to accept as the truth about nature exactly what the mathematics says. In this view, particles don't have an immutable reality; they can pop in and out of reality. Particles without state labels have no marker in any measurement space: no signature of real existence. They are outside reality.

Upon measurement of a state for particle 1, the pair system disappears from reality, and the two individual particles pop into reality. The process is called particle annihilation and creation: the pair atom is annihilated, and the two new particles, 1 and 2, are created.

In this view, quantum mechanics is certainly complete: to "every element of physical reality" there is "a counterpart in physical theory"; a state label. Something that loses its label drops out of physical reality. Completeness is ensured by the meaning assigned to reality.

When two particles combine to form an atom, they *disappear from reality*. In place of two 3-D particles appears one 6-D particle. The implication is that systems in nature with a plethora of labels for their states can be split. Multiple labels are the signature of unexpressed particles. These particles are an alternative manifestation of the multiple quantum numbers of the original unit.

THE WHOLE UNIVERSE PARTAKES IN EVERY EVENT

When you make a measurement on particle 1, it acquires a state: what you know about it—its state—is obtained in the measurement. Particle 2 acquires its state without a measurement on it: particle 2 assumes its state the very instant you make a measurement on particle 1.

Because we measure only particle 1, how does particle 2 "get the message" in what state to materialize?

To ask this question is to violate the spirit of quantum mechanics. An illegitimate question can appear quite sensible; it appears so in the context of a prejudice. "How far to the ends of the earth?" is sensible in the context of an earth that has ends. Informed minds recognize the question as a relic of archaic thought. Because they perceive the world differently, new minds don't ask the questions that old minds ask.

The mathematics of quantum mechanics gives this answer to the question of communication between the two particles.

Before they enter reality, there are no particles between which communications can take place. In the question a real existence is ascribed to the particles before they do, in fact, exist. Pete and Harry can't converse before they are born.

A state occupies the whole universe instantaneously; its existence involves no communication between its parts. No messages pass. The whole universe participates in the structure of a state. The state has an amplitude everywhere: at all x_1 and x_2.

Before the measurement there is no particle pair; there is only a gigantic atom. This atom pervades all space. The experiment dematerializes the atom, and in its place two particles appear. Each materializes, as it must in the universe, so as to preserve the laws of nature.

Particles do not have immutable reality. What is preserved immutably in reality are not particles but, rather, the laws of nature. A law of nature is expressed in the EPR example in that the center-of-mass, being stationary, remains so.

Quantum mechanics teaches us that when a system materalizes, its elements dematerialize! A system is not a composite of its elements: its set of state labels acknowledges none of its elements.

DOES THOUGHT HAVE A WAVE FUNCTION?

Up to this point the picture portrayed is simply what the mathematics says. Although not supported specifically by the mathematics given here, arguments based on the picture go further.

If the whole universe partakes of any event so, too, must the observer: he is not exempted from the universe. The observer is in a conspiracy with the observed!

Like every element of the universe, the observer, too, is in some conditional state. His every decision is merely the execution, in detail, of the grand design. Thus his very decision to measure the position, and not the momentum, of particle 1 is already fixed in the universal scheme of things. Particle 2 has no need of communication about the matter: because of fixed destiny, its fate is ordained to enter reality as a particle in the state $|x_2 = c + b\rangle$. It is born as a $|x_2 = c + b\rangle$ particle.

Thus the observer's decision is related to the particles' wave function. His thoughts have wave function qualities!

NO ELEMENT OF REALITY IS REAL

No element of the universe is truly isolated. An isolated system is a conceptual fiction: there are no such systems. Every element of reality is, in fact, part of some larger element within which it interacts.

Therefore it doesn't have a state.

Because it has no state, it's not real.

No element of reality is real!

How could one ever understand reality if its elements are unreal? Maybe one can never understand it through its elements.

PROBLEMS

8.1 Solve the eigenvalue problem $\hat{H}|E\ldots\rangle = E|E\ldots\rangle$ for the case of two free particles. The hamiltonian is that of Equation 8-2 with $V = 0$; there is only kinetic energy. Show that if we decouple $\hat{H}$ into six parts,

$$\hat{H} = \hat{H}_1 + \hat{H}_2 \ldots \hat{H}_6$$

where

$$\hat{H}_1 = \frac{\hat{p}_{1x}^2}{2m_1}, \quad \hat{H}_2 = \frac{\hat{p}_{1y}^2}{2m_1}, \quad \ldots \quad \hat{H}_6 = \frac{\hat{p}_{2z}^2}{2m_2}$$

then

$$\langle \mathbf{r}_1, \mathbf{r}_2 | E \ldots \rangle = \langle x_1 | E_1 \rangle \langle y_1 | E_2 \rangle \ldots \langle z_2 | E_6 \rangle$$

where

$$\hat{H}_1 | E_1 \rangle = E_1 | E_1 \rangle, \quad \ldots \quad \hat{H}_6 | E_6 \rangle = E_6 | E_6 \rangle$$

and

$$E = E_1 + E_2 + \ldots + E_6$$

It takes six numbers to specify a state. What other five will complete the state $|E \ldots \rangle$? That is, what can the dots be? Also give an example of a measurable that *cannot* be one of the dots: name an impossible as well as a possible state that embraces energy.

Answer: Eigenstates of $\hat{H}$ can be $|n_1,n_2, \ldots ,n_6\rangle$ or $|E,n_1,n_2, \ldots ,n_5\rangle$. Impossible is $|E,x_1, \ldots \rangle$

8.2 Consider the 12 operators $\hat{x}_1$, $\hat{y}_1$, $\hat{z}_1$, $\hat{x}_2$, $\hat{y}_2$, $\hat{z}_2$, $\hat{p}_{1x}$, $\hat{p}_{1y}$, $\hat{p}_{1z}$, $\hat{p}_{2x}$, $\hat{p}_{2y}$, and $\hat{p}_{2z}$. There are 66 different commutators (excluding mere ordering interchange) that one may construct from pairs among the group. All are zero except 6.

$$[\hat{x}_1,\hat{p}_{1x}] = [\hat{x}_2,\hat{p}_{2x}] = [\hat{y}_1,\hat{p}_{1y}] = \ldots = [\hat{z}_2,\hat{p}_{2z}] = i\hbar \tag{A}$$

$$[\hat{x}_1,\hat{y}_1] = [\hat{x}_1,\hat{x}_2] = [\hat{x}_1,\hat{p}_{2x}] \ldots = 0$$

Now consider the 12 operators $\hat{X}$, $\hat{Y}$, $\hat{Z}$ $\hat{x}$, $\hat{y}$, $\hat{z}$, $\hat{P}_X$, $\hat{P}_Y$, $\hat{P}_Z$, $\hat{p}_x$, $\hat{p}_y$, and $\hat{p}_z$. Show that for these exactly the same conditions prevail. There are 6 nonzero commutators. All others of the 66 are zero.

$$[\hat{X},\hat{P}_X] = [\hat{x},\hat{p}_x] = [\hat{Y},\hat{P}_Y] = \ldots [\hat{z},\hat{p}_z] = i\hbar \tag{B}$$

$$[\hat{X},\hat{Y}] = [\hat{X},\hat{x}] = [\hat{X},\hat{p}_x] = \ldots [\hat{z},\hat{P}_Z] = 0$$

Hint: Substitute in from the definitions (8-8), (8-9), (8-10), and (8-11) and use the results in (A).

8.3 Prove Equation 8-17. It results from the combination of (8-13), (8-5), and (8-14), (8-15), and (8-16).

8.4 Show that the eigenvalue problem of (8-15) can be stated as

$$-\frac{\hbar^2}{2 \mathscr{M}} \left(\frac{\partial^2}{\partial X^2} + \frac{\partial^2}{\partial Y^2} + \frac{\partial^2}{\partial Z^2} \right) \langle XYZ|E(\mathscr{M}) \ldots \rangle = E(\mathscr{M})\langle XYZ|E(\mathscr{M}) \ldots \rangle$$

and that the solution is a product of three solutions, each one to the problem

$$-\frac{\hbar^2}{2 \mathscr{M}} \frac{d^2}{d\xi^2} \langle \xi|E\rangle = E \langle \xi|E\rangle$$

one with $\xi = X$, one with $\xi = Y$, and one with $\xi = Z$, as shown in (8-18), where, for the case of a particle physically confined within a cubical box of side L

$$\langle \xi|E\rangle = \begin{cases} 0 & -\infty < \xi \leq 0 \\ \sqrt{2/L} \, \sin(\pi\nu\xi/L) & 0 \leq \xi \leq L \\ 0 & L \leq \xi < \infty \end{cases}$$

and
$$L\sqrt{8 . //E/h^2} = \nu = \text{positive integer}$$

8.5 Solve the eigenvalue problem of Equation 8-15 for a box that is not cubical, for one whose sides are L_1, L_2, and L_3 where $L_1L_2L_3 = L^3$.

8.6 Prove that the states $\langle XYZ|\nu_1\nu_2\nu_3\rangle$ are *not* eigenstates of $\hat{P}_X$ or $\hat{P}_Y$, and so on.

You do it by examining $\langle XYZ|\hat{P}_X|\nu_1\nu_2\nu_3\rangle$. The form of the result does not fit the eigenvalue problem: it does not equal a constant times $\langle XYZ|\nu_1\nu_2\nu_3\rangle$. Is $|\nu_1\nu_2\nu_3\rangle$ an eigenstate of $\hat{P}^2$?

8.7 Calculate the energies of the material box states $\nu = 100$ and $\nu = 101$. Compare these energies with those for each of the degenerate pair of traveling wave ring-box states $n = 50$ and $n = -50$.

8.8 Prove this commutator relationship between the operator $\hat{p}_r$ and the radial coordinate operator $\hat{r}$;
$$[\hat{r},\hat{p}_r] = i\hbar$$

and then prove
$$\hat{\mathbf{p}}\cdot\hat{\mathbf{r}} \equiv \hat{p}_x\hat{x} + \hat{p}_y\hat{y} + \hat{p}_z\hat{z} = \hat{\mathbf{r}}\cdot\hat{\mathbf{p}} - 3i\hbar \neq \hat{p}_r\hat{r}$$

8.9 Prove this equivalence:
$$\hat{p}_r \hat{r}^2 \hat{p}_r = \hat{r} \hat{p}_r^2 \hat{r}$$

It is best done by using the commutator result of the previous problem. Using the two results in Problem 8.8 and the fact that $\hat{r}$ is hermitian, show that $\hat{p}_r$ is not hermitian.
$$\hat{p}_r^\dagger = \hat{p}_r - \frac{2i\hbar}{\hat{r}}$$

You can do it by using the results of Problem 8.8 in computing
$$(\hat{r} \hat{p}_r)^\dagger = (\hat{\mathbf{r}}\cdot\hat{\mathbf{p}})^\dagger$$

8.10 Prove that the ∇^2 operator in spherical coordinates can be written as
$$\nabla^2 \langle r\theta\phi| = -\frac{1}{r^2}\langle r\theta\phi|\hat{L}^2/\hbar^2 + \frac{1}{r^2}\frac{\partial}{\partial r}\left(r^2 \frac{\partial}{\partial r}\right)\langle r\theta\phi|$$

Solution aid: Use this ubiquitous textbook result: Given a function in spherical coordinates, $f(r,\theta,\phi)$, then
$$\nabla^2 f = \frac{1}{r^2}\left[\frac{1}{\sin\theta}\frac{\partial}{\partial\theta}\left(\sin\theta\frac{\partial f}{\partial\theta}\right) + \frac{1}{\sin^2\theta}\frac{\partial^2 f}{\partial\phi^2}\right] + \frac{1}{r^2}\frac{\partial}{\partial r}\left(r^2\frac{\partial f}{\partial r}\right)$$

Consult (5-79) to complete the solution.

8.11 By writing ∇^2 in cyclindrical coordinates, ρ, ϕ, and z, show that the momentum operator squared satisfies the equation
$$\hat{\rho}^2 \hat{p}^2 = (\hat{\rho}\hat{p}_\rho)^2 + \hat{L}_z^2 + \hat{\rho}^2\hat{p}_z^2$$

where the $\rho\phi z$ space instruction for $\hat{p}_\rho$ is

$$\langle \rho\phi z|\hat{p}_\rho = -i\hbar \frac{\partial}{\partial\rho} \langle \rho\phi z|$$

The operator instructions in $\rho\phi z$ space for $\hat{L}_z$ and $\hat{p}_z$ are the ones already introduced in the text

$$\langle \rho\phi z|\hat{L}_z = -i\hbar \frac{\partial}{\partial\phi} \langle \rho\phi z| \qquad \text{and} \qquad \langle \rho\phi z|\hat{p}_z = -i\hbar \frac{\partial}{\partial z} \langle \rho\phi z|$$

8.12 Do the operators $\hat{p}_\rho$, $\hat{L}_z$, and $\hat{p}_z$ commute with one another? Do $\hat{\rho}$, $\hat{L}_z$, and $\hat{p}_z$ commute?

8.13 Consider a bead in a cylindrical potential; its potential energy varies only with the ρ coordinate. It is independent of ϕ or z, so $V = V(\rho)$. Hence it is described by the hamiltonian operator

$$\hat{H} = \frac{\hat{p}^2}{2m} + V(\hat{\rho})$$

Show that $[\hat{L}_z, \hat{H}] = 0$ and $[\hat{p}_z, \hat{H}] = 0$. Pinpoint verbally what the significance of these results are. How can you use them?

Answer: States may be characterized by the L_z, p_z, and E labels. They are the invariants of the motion.

8.14 Show that for the central field hamiltonian of Equation 8-16, it is indeed, true that $\hat{L}^2$ commutes with $\hat{H}$. Use (8-28) in conjunction with (8-24) and (8-25) or just (8-30) to prove it.

8.15 Show that Equation 8-39 can be rewritten, using the variable $\ell n\, R$ instead of R, as ($a = \hbar^2/mZe^2$)

$$-\left(\frac{d(\ell n R)}{dr}\right)^2 - \frac{d^2\ell n R}{dr^2} - \frac{2}{r}\frac{d\ell n R}{dr} + \frac{\ell(\ell+1)}{r^2} - \frac{2}{a}\frac{1}{r} = \frac{2mE}{\hbar^2}$$

In this form the equation is better analyzed for its assymptotic, $r \to \infty$, solutions. You consider the case of large r ($r^{-1} \to 0$) where the relative change in the wave function per unit distance, $\dfrac{1}{R}\dfrac{dR}{dr}$, might remain finite. Then the only term on the left side that could remain finite is the first term. If this term becomes constant at large r, then its derivative, the second term, will approach zero, as do all the reciprocal r terms. The equation's validity at large r requires that $d\ell n\, R/dr$ become a finite constant at infinity; this constant remains to balance the right side, which is always a fixed constant.

From this, deduce that the $E < 0$ solutions for R die out at infinity but that the $0 \leqslant E$ solutions remain finite at large r. The former are the bound ones; the latter, the unbound.

Solution: Examine the equation in the limit that only the constant term on the left remains. The two cases arise as $\sqrt{-2m\,E/\hbar^2}$ is real or imaginary. Solutions that grow with r must be rejected as nonphysical.

8.16 What are the internal energy and the degeneracy of the second excited state of the ideal model hydrogen atom treated in the text?

Answer.: -1.5 eV, there are nine $|n=3,\ell,m\rangle$ states.

8.17 What photon frequency will be high enough just to ionize the hydrogen atom completely? It's in the X-ray spectrum.

8.18 Demonstrate the truth of the following commutator relations:

$$[\hat{r}\,\hat{p}_r,\,\hat{p}^2] = 2i\hbar\hat{p}^2 \text{ (best done as } [\hat{\mathbf{r}}\cdot\hat{\mathbf{p}},\,\hat{p}^2]\text{using rectilinear coordinates)}$$
$$[\hat{r}\,\hat{p}_r,\,\hat{r}^{-1}] = i\hbar\hat{r}^{-1} \text{ (best done in } r,\theta,\phi \text{ space)}$$
$$[\hat{r}\,\hat{p}_r,\,\hat{r}^{-2}] = 2i\hbar\hat{r}^{-2}$$
$$[\hat{r}\hat{p}_r,\,\hat{r}^{-2}\,\hat{\Omega}] = 2i\hbar\hat{r}^{-2}\hat{\Omega} \text{ if } [\hat{r}\hat{p}_r,\,\hat{\Omega}] = 0$$

and if $\hat{H} = \hat{T} + \hat{V}$ where $\hat{V}$ is the electrostatic potential energy, then

$$\frac{i}{\hbar}\,[\hat{r}\hat{p}_r,\hat{H}] = -2\hat{T} - \hat{V} \qquad (\hat{V} = \text{electrostatic})$$

but for the general central field $\hat{V} = V(\hat{r})$ where $\langle r|\hat{F} = -\partial V/\partial r\langle r|$

$$[\hat{r}\hat{p}_r,\hat{H}] = 2i\hbar\hat{T} + i\hbar\hat{r}\hat{F}$$

8.19 The stationary states of the central field hamiltonian $\hat{H}_m$, with the time dependence explicitly exhibited, are $|n\ell m\rangle \exp(-itE/\hbar)$ where $E = E_n(\ell)$. A matrix element of the operator $\hat{r}\hat{p}_r$ at the time t is thus (with $\Delta E = E' - E$)

$$\exp\left(\frac{-it\Delta E}{\hbar}\right)\langle n\ell m|\hat{r}\,\hat{p}_r|n'\ell'm'\rangle =$$

$$\langle n(\ell)|\hat{r}\,\hat{p}_r|n'(\ell')\rangle \exp\left(\frac{-it\Delta E}{\hbar}\right)\langle \ell m|\ell'm'\rangle$$

If $n,\ell,m = n',\ell',m'$, the element is a diagonal one of the $\hat{r}\hat{p}_r$ matrix. It is also the average value of the operator $\hat{r}\hat{p}_r$ for the state $|n\ell m\rangle$. This matrix element is independent of time because $\Delta E = 0$. It follows, on applying the theorem of Equation 7-23 concerning how averages evolve in time, that

$$\langle n\ell m|[\hat{r}\,\hat{p}_r,\hat{H}]|n\ell m\rangle = 0$$

Thus although $\hat{r}\hat{p}_r$ does not commute with $\hat{H}$, the *average value of its commutator with* $\hat{H}$ *must be zero.* Prove this, and deduce for the case of the electrostatic $1/r$ potential that for the central field hamiltonian, $\hat{H}_m$,

$$-\langle n\ell m|\hat{T}|n\ell m\rangle = E_n = \frac{1}{2}\langle n\ell m|\hat{V}|n\ell m\rangle$$

The magnitude of the total energy is just the average of the kinetic energy which, in turn, has a magnitude just equal to one-half of the average potential energy (see Figure 8-7).

8.20 The model hydrogen atom of this chapter is in the state $K_x=5\text{Å}^{-1}$, $K_y=6\text{Å}^{-1}$, $K_z=0$, $n=2$, $\ell=1$, $m=0$. What is the energy eigenvalue, E, of the $\hat{H}$ operator in Equation 8-5? That is, what is the energy of this state—in electron volts?

Answer: -3.3 eV.

8.21 A hydrogen atom in its ground state moves with momentum $\mathbf{P} = \hbar\mathbf{K}$. What is the probability of finding the electron within $d^3\mathbf{r}_1 = dx_1\, dy_1\, dz_1$ at a particular point $\mathbf{r}_1$ near the center of the experimental box?

Answer: $d^3\mathbf{r}_1 \sum\limits_{\mathbf{r}_2} |\langle \mathbf{r}_1, \mathbf{r}_2 | \mathbf{K}, n=1, \ell=0, m=0\rangle|^2 =$

$$\frac{d^3\mathbf{r}_1}{L^3} \int\limits_{-L/2\,=\,-\infty}^{L/2\,=\,\infty}\!\!\int\!\!\int \frac{dx_2 dy_2 dz_2}{\pi a^3}\, \exp[-2a^{-1}\sqrt{(x_2-x_1)^2 + (y_2-y_1)^2 + (z_2-z_1)^2}]$$

$$= dx_1 dy_1 dz_1 / L^3$$

We use $L = \infty$ in the limits on the integral. It's the only sensible thing to do. The scale on which the integrand is nonzero, the distance $a = 0.5\text{Å}$, is microscopically small relative to L.

8.22 A well-known equation of mathematics is

$$\frac{1}{\rho^2}\frac{d}{d\rho}\left(\rho^2 \frac{dR}{d\rho}\right) - \frac{\ell(\ell+1)}{\rho^2} R + R = 0$$

Its solutions are known as the spherical Bessel functions:

$$j_\ell(\rho) = (-\rho)^\ell \left(\frac{1}{\rho}\frac{d}{d\rho}\right)^\ell \left(\frac{\sin\rho}{\rho}\right)$$

and

$$n_\ell(\rho) = -(-\rho)^\ell \left(\frac{1}{\rho}\frac{d}{d\rho}\right)^\ell \left(\frac{\cos\rho}{\rho}\right)$$

The $n_\ell(\rho)$ functions diverge at the origin where $\rho \to 0$; the $j_\ell(\rho)$ functions remain finite there.

$$\lim_{\rho\to 0} n_\ell(\rho) = -\infty; \qquad \lim_{\rho\to 0} j_\ell(\rho) = \delta_{\ell 0}$$

For large ρ, the behavior of these functions is

$$n_\ell(\rho) \to -\frac{1}{\rho}\cos(\rho - \ell\pi/2); \qquad j_\ell(\rho) \to \frac{1}{\rho}\sin(\rho - \ell\pi/2)$$

Combinations of $j_\ell(\rho)$ and $n_\ell(\rho)$ are also solutions of the differential equation. The most useful combinations are the spherical Hankel functions, defined as

$$h_\ell^{(1)}(\rho) = j_\ell(\rho) + i\, n_\ell(\rho)$$

and

$$h_\ell^{(2)}(\rho) = j_\ell(\rho) - i\, n_\ell(\rho)$$

From their definitions, calculate and exhibit the four functions $j_\ell(\rho)$, $n_\ell(\rho)$, $h_\ell^{(1)}$, and $h_\ell^{(2)}(\rho)$ for the case of $\ell = 0$ and $\ell = 1$, and show that these eight functions do, indeed, solve the differential equation. Then sketch each of these functions as plots versus ρ so as to visualize how they look.

8.23 Show that the equation, with a sign change in the last term on the left,

$$\frac{1}{\rho^2}\frac{d}{d\rho}\left(\rho^2 \frac{dR}{d\rho}\right) - \frac{\ell(\ell+1)}{\rho^2} R - R = 0$$

is solved by the spherical Bessel or Hankel functions of imaginary argument: $R(\rho) = A j_\ell(i\rho) + B n_\ell(i\rho)$ or $A' h_\ell^{(1)}(i\rho) + B' h_\ell^{(2)}(i\rho)$. Of the four functions, $j_\ell(i\rho)$, $n_\ell(i\rho)$, $h_\ell^{(1)}(i\rho)$, and $h_\ell^{(2)}(i\rho)$, which one remains finite at large ρ?

Answer: $h_\ell^{(1)}(i\rho)$.

Is there more than one of them that remains finite as ρ goes to infinity?

Answer: No.

Key to the problem: You need to realize and use the fact that this identity is true even if θ has an imaginary part; θ may be complex:

$$e^{i\theta} = \cos\theta + i\sin\theta$$

The harmonic functions of imaginary argument are

$$\cos i\theta = \cosh\theta \quad \text{and} \quad \sin i\theta = i\sinh\theta$$

8.24 A very important problem—especially in nuclear physics—is that of the "spherical square well" potential. This is a radial potential that is a flat-bottomed well of depth B; it is constant at this depth from the origin out to a characteristic radius b. Outside this radius the potential energy is zero.

$$V(r) = \begin{cases} -B & 0 < r < b \\ 0 & b < r \end{cases}$$

Show that the radial Equation 8-35, generated by this potential, is isomorphic to the differential equation in Problem 8.22 when $0 \leq r \leq b$ and isomorphic to the differential equation in Problem 8.23 when $b \leq r$. What relationships effect the association?

Plot $V(r)$ versus r, and on this same diagram, show what the first two bound-state energy levels might be, and sketch what the radial wave functions might look like.

8.25 For each of these four statements, write down the reasons that it must be true for the bound states, $\langle r|E(\ell)\rangle$, of the particle in the "spherical square well" of Problem 8.24.

a. $\lim_{r\to\infty} \langle r|E(\ell)\rangle = 0$

b. $\lim_{r\to 0} \langle r|E(\ell)\rangle = \text{finite}$

c. $\lim_{\varepsilon\to 0} \langle r|E(\ell)\rangle|_{b-\varepsilon} = \lim_{\varepsilon\to 0} \langle r|E(\ell)\rangle|_{b+\varepsilon}$

d. $\lim_{\varepsilon\to 0} d\langle r|E(\ell)\rangle/dr|_{b-\varepsilon} = \lim_{\varepsilon\to 0} d\langle r|E(\ell)\rangle/dr|_{b+\varepsilon}$

Show that the implementation of items a through d tells us these facts about the radial part of the bound-state wave function, $\langle r|E(\ell)\rangle$ where $-B < E < 0$.

a. $\langle r|E(\ell)\rangle = A\, h_\ell^{(1)}(ir\sqrt{-2mE}/\hbar) \qquad b \leq r$

b. $\langle r|E(\ell)\rangle = C\, j_\ell(r\sqrt{2m(B+E)}/\hbar) \qquad 0 \leq r \leq b$

c. $C\, j_\ell(b\sqrt{2\,m\,(B+E)}/\hbar) \;=\; A\, h_\ell^{(1)}\,(ib\sqrt{-2\,m\,E}/\hbar)$

d. $C\, \dfrac{d}{dr}\, j_\ell(r\sqrt{2\,m\,(B+E)}/\hbar)\Big|_{r=b} \;=\; A\, \dfrac{d}{dr}\, h_\ell^{(1)}\,(ir\sqrt{-2\,m\,E}/\hbar)\Big|_{r=b}$

Deduce that finite constants A and C may be found that will satisfy both items c and d only if this condition is true:

$$\det \begin{vmatrix} j_\ell(b\sqrt{2\,m}\ (B+E)/\hbar) & h_\ell^{(1)}(ib\sqrt{-2\,m\,E}/\hbar) \\[2mm] \dfrac{d}{dr}\, j_\ell\,(r\sqrt{2\,m\,(B+E)}/\hbar)\big|_b & \dfrac{d}{dr}\, h_\ell^{(1)}\,(ir\sqrt{-2\,m\,E}/\hbar)\big|_b \end{vmatrix} = 0$$

Thus only the energies, E, for which this determinent is zero are allowed. That the determinant be zero is the condition governing what the energy levels of the system are.

8.26 In the limit $\hat{V} = 0$, where the potential of interaction is zero, there must still exist pair states of the system-as-a-whole. One such is $|K=0,\ q=Q\rangle$ where the total momentum of the system, $\hbar K$, is zero, but there is a pair relative velocity, $p/m = \hbar Q/m$ (see Equation 8-11). The simultaneous solutions to the $\hat{V} = 0$ one-dimensional counterparts of (8-15) and (8-16) are

$$\langle X,x|K,q\rangle \;=\; \frac{1}{L}\,\exp\,[i(KX+qx)]$$

among which $|K=0,q=Q\rangle$ is one state. This is a stationary state of the pair system-as-a-whole.

a. Suppose you measure the position of particle 1. In what state is particle 2 left?

Answer: The state $|k_2=Q\rangle$.

b. You measure the momentum of particle 1. Exhibit the state in which particle 2 is left.

Answer: Again the state $|k_2=Q\rangle$.

c. You measure the momentum of particle 2. Exhibit the state in which particle 1 is left.

Answer: The state $|k_1=-Q\rangle$.

d. You measure the position of 2. Exhibit particle 1's state.

Answer: The state $|k_1=-Q\rangle$.

Note:

$$\langle X,x|K=0,\ q=Q\rangle \;=\; \langle x_1|k_1=-Q\rangle\langle x_2|k_2=Q\rangle$$

This state is separable both in X,x space and in x_1,x_2 space. In this case, the compound system and its parts are simultaneously present in reality.

Chapter 9

Indistinguishable Particles:
Identical Bosons,
and Identical Fermions

ELECTRONS ARE INDISTINGUISHABLE PARTICLES

Are electrons like people? Are they like trees? You know a tree when you see one. You recognize a member of the class: a person belongs to the class of all people, a tree to the class of trees, and an electron to the class of electrons.

But among people there are differences that distinguish one from another. Bill Jones is not quite identical to Max Kitzel. Bill and Max are distinguishable, even though they are both people. Max is the ticklish one. You can distinguish Bill from Max by tickling. The observable is tickling; the measurement result distinguishes them.

What about electrons? Are two electrons distinguishable from each other? Can the electron at the point r_1 be labeled Bill and that at r_2, Max? Can we know that circling the proton is electron Bill and that the free one is Max?

The answer is no. Electrons are indistinguishable from one another. There exists no experiment by which you can distinguish one electron from another, or one proton from another, or even one atom of a species from another of the species.

INDISTINGUISHABILITY IS AN OBSERVABLE

Happily we needn't perform every conceivable experiment on electrons to know that they are indistinguishable from one another; merely assuming it theoretically and examining the consequences is enough. That electrons are indistinguishable particles has predictable experimental consequences which have been indisputably verified. We know that electrons are indistinguishable because they exhibit the behavior of indistinguishable particles and not that of distinguishable ones. The behavior difference is a measurable property: thus indistinguishability is an observable!

9.1 EXCHANGEABLE IS INDISTINGUISHABLE

Indistinguishability means that if two from among the species are exchanged, there exists no experiment by which you can detect it. A pair are indistinguishable if their exchange is undetectible by any physical experiment, if the exchange affects nothing in the universe. Among distinguishable particles such an exchange must produce some observable effect in the universe. The distinction depends upon *exchange*. To examine the logic of indistinguishability, the idea of exchange must be put mathematically.

EXCHANGING THEIR LABELS EXCHANGES PARTICLES

Consider N particles. The position coordinate of particle 1 is $\mathbf{r}_1$. This particle might have spin, and so we allow for a z-component spin index, m_1. It takes four coordinates to characterize particle 1, and together they label this particle. Particle 2 is labeled x_2, y_2, z_2, and m_2.

Contract all of the coordinate indices referring to a single particle into a single symbol. The symbol χ_1 or just plain 1 can be a shorthand notation for the four quantum indices x_1,y_1,z_1,m_1. Particle 2 is indexed by χ_2 or merely 2.

A *particle-indexed basis* is $\langle \chi_1 ... \chi_N |$. By this bra is meant $\langle x_1,y_1....z_N,m_N |$. This is one of many possible particle-indexed spaces. Another is $\langle \mathbf{p}_1,m_1...\mathbf{p}_N,m_N |$. Either one of these particle-indexed spaces is abbreviated as $\langle 1,2...N |$.

As in the previous chapter, 1 and 2 are used here as symbols, not numbers in the arithmetic sense. They index the *different particles* but always in the *same measurement space:* 1 and 2 might mean x_1 and x_2 or p_1 and p_2, but not x_1 and p_2.

A particle-indexed space is one in which the particles are labeled or enumerated as if they were distinguishable; the first is particle 1, the second particle 2, and so on. This is the space in which we think when we conceive N particles—distinguishable or not. We label them in our minds. It is generally in this space that we construct physical operators.

Among all those that may be present, we focus on a particular pair of particles. Call them particle 1 and particle 2. Our results will apply to any possible pair—particle i and particle j also. To consider these two, just relabel them; call the ith one particle 1 and the jth one particle 2.

We focus on one particular pair without disturbing any of the other particles. That is the pair we are calling 1 and 2. We want to consider the exchange of these two particles.

9.2　AN OPERATOR EFFECTS THE EXCHANGE

We make the exchange through an exchange operator. The operator $\hat{Y}_{1,2}$ exchanges particle 1 with particle 2. To consider the exchange of 3 with 5, there is available the operator $\hat{Y}_{3,5}$. Because here we contemplate only the exchange of 1 with 2, we drop the 1,2 subscript and use merely $\hat{Y} = \hat{Y}_{1,2}$.

By the same token, the dots that represent all of the undisturbed other particle-indexed coordinates 3 to N will be left out. They are suppressed in the notation but are understood to be present. The conceptual essentials are captured by thinking in terms of a two-particle system.

The instruction for the $\hat{Y}$ operator is given in particle-indexed space: in any $\langle 1,2|$ space. The instruction is to exchange the particle indices 1 and 2 of any state of the system.

Symbolically the defining instruction for the $\hat{Y}$ operator is

$$\langle 1,2|\hat{Y} = \langle 2,1| \tag{9-1}$$

Here's an example. In the case of two electrons a state might be $|S\rangle$ where

$$\langle 1,2|S\rangle = S(m_1,\mathbf{r}_1,m_2,\mathbf{r}_2) = \langle m_1,m_2|\sigma\rangle\,\psi(\mathbf{r}_1,\mathbf{r}_2) \tag{9-2}$$

The left-hand equality says that to each of the sixfold-infinity-plus-four points of $x_1y_1z_1m_1x_2y_2z_2m_2$ space there corresponds a complex number S. To have S for every such point is to have the state of the system in $x_1...m_2$ (or 1,2) representation.

On the right-hand side $\psi(\mathbf{r}_1,\mathbf{r}_2)$ is a wave function, a function of the 6-D position space coordinates only. To every point of the six-dimensional space $x_1...z_2$ there corresponds a complex number called ψ.

For electrons where $s_1 = s_2 = 1/2$, there are only four points in the two-dimensional m_1m_2 space. To each of these there corresponds a complex number called $\langle m_1,m_2|\sigma\rangle$. The product of these two complex numbers is the amplitude $\langle 1,2|S\rangle$.

THE EXCHANGE OPERATOR BRACKETED: THE AMPLITUDE FOR THE EXCHANGED EVENT

What Equation 9-1 says is that the $\hat{Y}$ operator manufactures from $\langle 1,2|S\rangle = \langle \chi_1,\chi_2|S\rangle$ the bracket $\langle \chi_2,\chi_1|S\rangle = \langle 2,1|S\rangle$.

$$\langle 1,2|\hat{Y}|S\rangle = \langle m_2,m_1|\sigma\rangle\,\psi(\mathbf{r}_2,\mathbf{r}_1) \tag{9-3}$$

The exchange operator produces the amplitude for the exchanged event. It manufactures a new function of χ_1 and χ_2 from the old one; the new one is the old function with the index 1 everywhere exchanged with the index 2.

In Figure 9-1 the two particles are shown as a circle and a square. A circle at

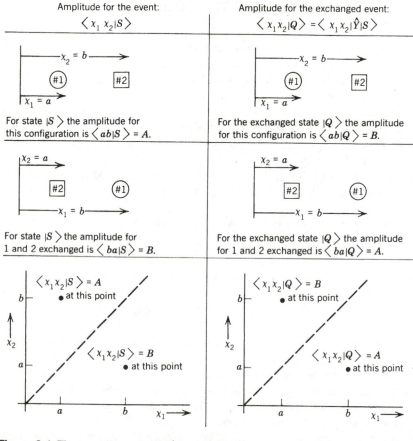

**Figure 9.1 The meaning of <1,2|Q> = <2,1|S> for two spinless 1-D particles.
The exchange operator produces |Q> from |S>. The exchanged state |Q> of 1
and 2 is the state |S> of 1 exchanged with 2. $\langle 1,2|Q \rangle = \langle 1,2|\hat{Y}|S \rangle = \langle 2,1|S \rangle$**

position a and a square at position b are the bra configuration of one possible event
result. The exchanged configuration, a circle at b and a square at a, is the other event
result.

THE EXCHANGED STATE IS THE STATE EXCHANGED

Like all operators, $\hat{Y}$ produces a new ket from an old one. The old one is $|S\rangle$;
call the new one $|Q\rangle$. Thus (see Problem 9.1)

$$|Q\rangle = \hat{Y}|S\rangle \tag{9-4}$$

The instruction equation, (9-1), tells us what this new ket is when represented in particle-indexed space.

$$\langle 1,2|Q\rangle = Q(\chi_1,\chi_2) = S(\chi_2,\chi_1) = \langle 2,1|S\rangle \qquad (9\text{-}5)$$

The new state is whatever function of χ_1 and χ_2 the old one was, but with the subscripts 1 and 2 interchanged.

The situation is illustrated in Figure 9-1 for the simplest possible case: two spinless 1-D particles. The effect of the exchange operator is to replace the function $\langle x_1,x_2|S\rangle$ by the new function $\langle x_2,x_1|S\rangle = \langle x_1,x_2|Q\rangle$. The complex number formerly associated with the *point* $x_1,x_2 = 3\text{Å},5\text{Å}$ is now to be associated with the *point* 5Å,3Å, and vice versa.

The exchange operator has this remarkable feature: it is both hermitian and unitary! Thus

$$\hat{Y} = \hat{Y}\dagger = \hat{Y}^{-1} \qquad (9\text{-}6)$$

The substance of Problems 9.2 and 9.3 is the proof of these equations.

THE EXCHANGE OPERATOR
MAKES EXCHANGED OPERATORS

The idea of exchange gives meaning to the *order* of the particles in the bra: $\langle 2,1| \neq \langle 1,2|$. Via the $\hat{Y}$ operator a significance is attached to this order. In general, the ordering of the labels in a bra is without significance. For example, no distinction is made between $\langle \mathbf{r},m|$ and $\langle m,\mathbf{r}|$ for a single particle. There is no physical basis for any distinction. But among particles a reordering does have significance: it means to exchange them.

Now consider some general multiparticle operator of physics: $\hat{\Omega} = \Omega(\hat{1},\hat{2},...)$. This operator is particle indexed. An example of what $\hat{\Omega}$ might be is the two-particle hamiltonian of Chapter 8 describing a pair of interacting particles, Equation 8-2.

Given an operator $\hat{\Omega} = \Omega(\hat{1},\hat{2})$, one can manufacture an exchanged operator $\underline{\hat{\Omega}}$ by

$$\underline{\hat{\Omega}} \equiv \Omega(\hat{2},\hat{1}) \qquad (9\text{-}7)$$

You construct $\underline{\hat{\Omega}}$ by exchanging the labels 1 and 2 on the indexed particle operators wherever they occur in $\hat{\Omega}$. The operator $\underline{\hat{\Omega}}$ is defined by how you get it from $\hat{\Omega}$. If by $\hat{\Omega}$ is meant $\hat{x}_1\hat{p}_2$, then by $\underline{\hat{\Omega}}$ is meant $\hat{x}_2\hat{p}_1$. Notice that $\underline{\hat{\Omega}}$ does *not* mean $\hat{p}_2\hat{x}_1$; you exchange the labels, not the operators.

We know the meaning of $\langle 2,1|S\rangle$, but we must be sure that we understand the meaning of $\langle 2,1|\hat{\Omega}|S\rangle$. It is this

$$\langle 2,1|\hat{\Omega}|S\rangle = \langle 1,2|\underline{\hat{\Omega}}|Q\rangle \qquad (9\text{-}8)$$

This equation ensures internal self-consistency, it guarantees that $\langle 2,1|\hat{\Omega}|S\rangle$ is what you get when you form the state $\hat{\Omega}|S\rangle$ in normal particle-indexed space and *then* exchange the particle indices. You get the same thing as the state $\underline{\hat{\Omega}}|Q\rangle$ in $\langle 1,2|$ space.

Equation 9-8 preserves the meaning already assigned to $\langle 2,1|$: an exchange of coordinates after having the state in $\langle 1,2|$ coordinates. To calculate $\langle 2,1|\hat{\Omega}|S\rangle$ you first get $\langle 1,2|\hat{\Omega}|S\rangle$; and then you exchange the coordinates. The procedure offered in the equation accomplishes this task. In Problem 9.5 a simple example makes the general truth of (9-8) transparent. The problem also demonstrates that operating on an exchanged state, $\langle 1,2|\hat{\Omega}\hat{Y}|S\rangle$, is *not the same* as exchanging after the operation, $\langle 1,2|\hat{Y}\hat{\Omega}|S\rangle$.

Because $|S\rangle$ and $|Q\rangle$ are just the exchanged states related through $\hat{Y}$ (see Equation 9-4) and because the bras in (9-8) are also connected by the $\hat{Y}$ operator (9-1), the content of (9-8) can be put in operator form:

$$\hat{Y}\Omega(\hat{1},\hat{2}) = \Omega(\hat{2},\hat{1})\,\hat{Y} \qquad (9\text{-}9)$$

Equation 9-8 is just the bracketed form of Equation 9-9: $\langle 1,2|$ on the left and $|S\rangle$ on the right.

This very important equation is the critical one governing the study of multiple-particle systems. It is the mathematical statement of the effect of particle interchange on physical observations.

INDISTINGUISHABILITY: EXCHANGE AFFECTS NO PHYSICAL OPERATOR

If the two particles, 1 and 2, are indistinguishable, then *any* physical operator, $\hat{\Omega}$, governing their behavior has the property

$$\Omega(\hat{2},\hat{1}) = \Omega(\hat{1},\hat{2}) \qquad (9\text{-}10)$$

This is the mathematical statement of indistinguishability: for an indistinguishable pair no physical operator exists for which this equation is not true.

If this equation were not true for some operator, the pair would be distinguishable; the inequality would guarantee it. The operator for which (9-10) is not true furnishes the experiment you should perform to distinguish one particle from the other! If $\hat{\Omega}$ differs from $\hat{\Omega}$, then some *state* of $\hat{\Omega}$ must differ in some way from those of $\hat{\Omega}$: this state difference defines the distinction between the particles.

If there is no distinguishing experiment, then there will be no physical operator that distinguishes the two. Thus it is that for indistinguishable particles all physical operators must have the property of Equation 9-10.

The operator $\hat{x}_1\hat{p}_2$ does not have this property and so could not refer to a pair of indistinguishable particles. Its eigenvalues are the position of particle 1 and the momentum of particle 2. This presumes that you could measure the position of 1 and be sure that it is not that of 2, that you could distinguish 1 from 2.

An example of an operator that does have the property of (9-10) is $\hat{x}_1\hat{p}_2 + \hat{x}_2\hat{p}_1$.

INDISTINGUISHABLE-PARTICLE OPERATORS ALL COMMUTE WITH EXCHANGE

The indistinguishability condition, (9-10), when combined with (9-9), can be recast as an operator equation:

$$[\hat{Y},\hat{\Omega}] = 0 \tag{9-11}$$

Indistinguishable particles have the property that $\hat{Y}$ commutes with every possible physical operator involving the pair.

It follows that $\hat{Y}$ must commute with every conceivable hamiltonian involving the pair; with all possible hamiltonians. Regarding an indistinguishable pair, $\hat{Y}$ commutes with the grand hamiltonian of the universe.

Operators that commute with the hamiltonian produce the invariants of the system (see Chapter 7, especially Equation 7-23). Thus, indistinguishability is an invariant, a conserved quantity in nature. Particles that are indistinguishable from one another forever remain so.

9.3 THE EIGENVALUES OF EXCHANGE: BOSONS AND FERMIONS

Like other operators, $\hat{Y}$ has eigenvalues. Call them Y. The values that Y can have label the eigenstates of the $\hat{Y}$ operator. Such an eigenstate is $|Y...\rangle$.

Because $\hat{Y}$ commutes with all physical operators concerning any particular indistinguishable pair, it follows that a value of Y labels every physical state of the pair. The dots remind us that there are other labels present which complete the characterization of the state. I'll suppress the dots when appropriate, when the other labels are not relevant to the exposition.

You deduce the values of Y by letting $\hat{Y}$ operate twice.

$$\langle 1,2|Y\rangle = \langle 1,2|\hat{Y}^2|Y\rangle = Y^2 \langle 1,2|Y\rangle \tag{9-12}$$

The left-hand equality expresses this thought: because of its meaning, $\hat{Y}^2$, operating on any state, eigen or not, simply returns the state back again. A double exchange equals no exchange.

The right-hand side says that the effect of $\hat{Y}^2$ on an eigenstate of $\hat{Y}$ must be the same state back again multiplied by Y^2.

The equality of the extremes tell us that $Y = +1$ and $Y = -1$ are the eigenvalues of the $\hat{Y}$ operator! Those are the only values of Y that guarantee the truth of Equation 9-12.

There are, thus, two kinds of indistinguishable-particle states. One kind has the property that the wave function for the exchanged configuration is exactly the same as that for the original configuration. In the other kind the exchanged configuration wave function is the negative of the original one. There are, thus, two types of indistinguishable particles.

The b-type, called Bose–Einstein particles, or bosons, have $Y = +1$. Boson states are even upon exchanging an indistinguishable pair.

$$\langle 2,1|b \rangle = \langle 1,2|b \rangle \tag{9-13}$$

When $\hat{Y}$ operates on states with this property, it produces the original state back again multiplied by $+1$.

The f-type, called Fermi–Dirac particles, or fermions, belong to the class with $Y = -1$. Fermions are those groups of indistinguishable particles whose states are always odd in the interchange.

$$\langle 2,1|f \rangle = - \langle 1,2|f \rangle \tag{9-14}$$

The operator $\hat{Y}$ produces $Y = -1$ on states with this property.

In either case, the probability of the configuration is completely unaffected by exchanging the particles.

$$|\langle 2,1|Y \rangle|^2 = |\langle 1,2|Y \rangle|^2 \tag{9-15}$$

This equation is true whether $Y = +1$ (bosons) or $Y = -1$ (fermions). That this must be so is evident physically: the exchange, being undetectable, cannot affect the probability—which is detectable.

Y-NESS IS CONSERVED

Fermions can never change into bosons, nor bosons into fermions: Y-ness is preserved. That is because $\hat{Y}$ commutes with all hamiltonians for these particles. If at one time in the history of the universe, a pair were in one or the other of the eigenstates of $\hat{Y}$, the pair must forever remain in that eigenstate. No force in the universe can change their Y-ness property. Thus the Y-ness property is a way to classify the particles of nature.

All indistinguishable particles of half-integral spin fall into the fermion class. The great physicist Wolfgang Pauli demonstrated that eigenstates of $\hat{S}^2$ and $\hat{Y}$ must be related in this way. The class of fermions include the electron, the proton, and the neutron.

All indistinguishable particles of zero or integral spin fall into the boson class.

These include the photon, the π meson (pion), the deuteron (bound neutron–proton pair), alpha particles (helium atom nuclei), and all even-mass-number nuclei.

TO EXCHANGE THE WHOLE IS TO EXCHANGE ALL PARTS

A *deuteron* is a compound particle; a neutron and a proton bound together. It is the nucleus of an unstable isotope of the hydrogen atom called *deuterium*.

The deuteron is a pair of fermions bound together, a pair of *distinguishable* fermions.

All neutrons are fermions, indistinguishable from one another but quite distinguishable from protons. Protons are also indistinguishable fermions, indistinguishable from one another but not indistinguishable from neutrons. A deuteron is a proton–neutron pair that remains bound together acting as an independent unit (see Figure 9-2).

An ensemble of deuterons behaves like indistinguishable bosons. An exchange consideration shows it.

Consider a particular pair of deuterons: call one of the pair A and the other B. If the state of the system of all deuterons is $|S\rangle$, then the representation of this state in deuteron-indexed space is $\langle A,B...|S\rangle$.

The Y-class of the deuteron is fixed by what happens to this wave function when

---

Two bound pairs of fermions equals one pair of bosons

The exchange of A with B
equals
The double exchange: 1 with 3 and 2 with 4

$$\langle AB|\hat{Y}_{AB}|S\rangle = \langle 1234|\hat{Y}_{24}\hat{Y}_{13}|S\rangle$$
$$= -\langle 1234|\hat{Y}_{24}|S\rangle$$
$$= \langle 1234|S\rangle = \langle AB|S\rangle$$

Compound particle A　　　Compound particle B

Figure 9.2 The exchange property of a compound particle equals the net exchange property of all of its constituents.

two deuterons are exchanged: according to whether $\langle B,A...|S \rangle$ is the negative of $\langle A,B...|S \rangle$ or its equal, the deuteron is a fermion or a boson.

But the exchange of the two deuterons corresponds to a double exchange among the composite particles. Exchanging A with B is equivalent to the exchange of the A-proton with the B-proton plus the exchange of the A-neutron with the B-neutron. Because exchange among the deuterons corresponds to a pair of fermion exchanges, the net effect is the boson property. The exchanged wave function is just equal to the original one. Thus deuterons are bosons.

The quantitative rendering of this argument is displayed in Figure 9-2.

From the example the generalization is evident: a compound particle is a boson or a fermion according to the net exchange property of all of its constituents.

Thus the hydrogen ion—a proton—is a fermion, but the hydrogen atom—proton and electron bound together—is a boson. Any electrically neutral atom of atomic number Z and atomic mass number A consists of $A + Z$ fermions. Hence such atoms are fermions or bosons according to whether $A + Z$ is an odd or an even number. The nitrogen atom is a fermion; the oxygen atom a boson.

INDISTINGUISHABILITY HAS CONSEQUENCES

The physical consequences of indistinguishability are apparent with two particles, and so we shall confine ourselves to a two-particle system. To spotlight the indistinguishable particle effects unencumbered by irrelevant complexity, we resort to the simplest system possible: the two particles can move in only one dimension, and they interact via a stylized ideal 1-D potential.

A pair of particles, whose masses, m, are equal, interact via the potential, $U(x_2 - x_1)$. It is an infinite square well of width $2b$, as shown in Figure 9-3.

The two particles are bound together: they can't get out of each other's range. The *range* of interaction is b: each particle confines the other to remain no greater than a distance b away from it as if they were held together by a string. The force is only there when the string becomes taut. The $x = x_2 - x_1$ of Figure 9-3 is a relative coordinate: it locates particle 2 with respect to particle 1.

The hamiltonian for the interacting pair is

$$\hat{H} = \frac{\hat{p}_1^2}{2m} + \frac{\hat{p}_2^2}{2m} + U(\hat{x}_2 - \hat{x}_1) \tag{9-16}$$

In terms of the center-of-mass (X, P) and relative (x, p) coordinates, this becomes (see Figure 8-2)

$$\hat{H} = \frac{\hat{P}^2}{4m} + \frac{\hat{p}^2}{m} + U(\hat{x}) \tag{9-17}$$

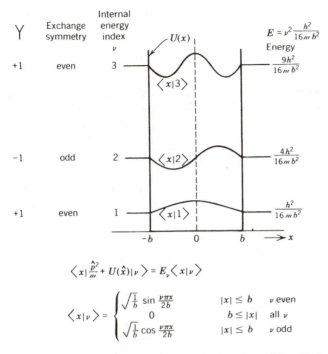

$$\left\langle x\left|\frac{\hat{p}^2}{m} + U(\hat{x})\right|\nu\right\rangle = E_\nu\langle x|\nu\rangle$$

$$\langle x|\nu\rangle = \begin{cases} \sqrt{\frac{1}{b}}\ \sin\frac{\nu\pi x}{2b} & |x| \le b \quad \nu \text{ even} \\ 0 & b \le |x| \quad \text{all } \nu \\ \sqrt{\frac{1}{b}}\ \cos\frac{\nu\pi x}{2b} & |x| \le b \quad \nu \text{ odd} \end{cases}$$

Figure 9.3 A particle is bound to its neighbor by a string of length b. The energy eigenstates of the relative coordinate hamiltonian: $x = x_2 - x_1$.

The relationships between the particle-indexed coordinates and the center-of-mass/relative coordinates are those given in Chapter 8 when applied in one dimension.

The hamiltonian of Equation 9-17 is the sum of two decoupled parts. The energy eigenvalue problem for the center-of-mass part generates the quantum number K; it indexes the kinetic energy of translation of the whole atom. The state $|K\rangle$ is a solution of

$$\frac{\hat{P}^2}{4m} |K\rangle = \frac{\hbar^2 K^2}{4m} |K\rangle \tag{9-18}$$

It characterizes the compound particle in the macroscopic experimental chamber: a box of extent $L \gg b$. The experimental chamber is the mathematical box which folds back on itself as a ring.

Instead of the integer index label $KL/2\pi$, we use K to label these states. The values of K are limited by

$$\frac{KL}{2\pi} = \dots -2, -1, 0, 1, 2, \dots \tag{9-19}$$

The representation of the state $|K\rangle$ in X-space is

$$\langle X|K\rangle = \frac{1}{\sqrt{L}} \exp(iKX) \tag{9-20}$$

The same state represented in P-space is the self-space bracket $\langle P|K\rangle$: unless the value of P is equal to $\hbar K$, the state doesn't exist, and the bracket is zero.

The other decoupled part of the hamiltonian in (9-17) concerns the bound-state internal energy levels of the pair. These are indexed by v, the set of all positive integers that characterize the solutions to

$$\langle x| \frac{\hat{p}^2}{m} + U(\hat{x})|v\rangle = v^2 \left(\frac{h^2}{16\,m\,b^2}\right) \langle x|v\rangle \tag{9-21}$$

See Figure 9-3.

The wave functions $\langle x|v\rangle$ are

$$\langle x|v\rangle = \begin{cases} \dfrac{1}{\sqrt{b}} \sin v\pi x/2b & |x| < b, \; v = \text{even} > 0 \\[2mm] 0 & b < |x|, \; \text{all } v \\[2mm] \dfrac{1}{\sqrt{b}} \cos v\pi x/2b & |x| < b, \; v = \text{odd} > 0 \end{cases} \tag{9-22}$$

Using these states we can construct the solutions to the two-particle problem. The problem

$$\hat{H}|K,v\rangle = E|K,v\rangle \tag{9-23}$$

has as its eigenvalues the total energy, E, which is the sum of two parts.

$$E = \frac{\hbar^2 K^2}{4\,m} + \frac{v^2 h^2}{16\,m\,b^2} \tag{9-24}$$

The state $|K,v\rangle$, from which this result comes, has an X, x representation that is the product of (9-20) and (9-22).

$$\langle X,x|K,v\rangle = \langle X|K\rangle \langle x|v\rangle \tag{9-25}$$

A more useful representation for this same state is that of P and x:

$$\langle P,x|K,v\rangle = \langle P|K\rangle \langle x|v\rangle \tag{9-26}$$

CLASSIFY THE STATES BY Y

We now examine the exchange property of these states.

The $\hat{Y}$ operator exchanges coordinates x_1 with x_2. In X,x-space this exchange appears as a reflection in x with no change at all in X. This is apparent from the transformation equations of Figure 8-2. The center-of-mass coordinate, X, is not affected by the exchange of 1 with 2. The value of x is affected: its sign is changed.

Mirror symmetry in x-space corresponds to exchange symmetry in particle-index space.

Thus in x,X-space the instruction for the $\hat{Y}$ operator is

$$\langle X,x|\hat{Y} = \langle X, -x| \tag{9-27}$$

See Problem 9.6.

Keeping this in mind, it becomes evident that the states $|K,\nu\rangle$ resolve themselves naturally into Y-ness categories. All states in which the quantum number ν is odd are symmetric states. They are characterized by $Y = +1$. All states in which ν is even are antisymmetric states. They're characterized by $Y = -1$ (see Problem 9.7).

Now we can examine the difference in energy-level structure that indistinguishability makes.

If the two particles are distinguishable, then the state produced by exchanging the particles can be distinguished from that without the exchange: both are valid states. A distinguishable pair has the level structure indicated in Figure 9-4, it is independent of the Y property.

If the pair are indistinguishable bosons, then states with $Y = -1$ will be forbidden to the system. Because the levels $\nu = 2,4, \ldots$ have $Y = -1$, an indistinguishable boson pair can never be in any of these states. You see in the figure that the $\nu = 2$ level is missing.

9.4 THERE ARE NO SPINLESS FERMIONS

To examine the level structure of a fermion pair, we must consider spin. There are no spinless fermions; all fermions have half integral spins.

We consider the simplest case, that of two indistinguishable fermions with the lowest possible half-integral spin, $s = 1/2$. They might be two neutrons.

For any one of these particles, a purely spatial state is not a complete description. The spin part is lacking. States of the pair system require a four-dimensional space like $\langle x_1,m_1,x_2,m_2|$ or $\langle x,X,m_1, m_2|$.

The hamiltonian for the system contains no spin dependence. Hence the energy eigenstates of the pair must be of the product form: a spin part multiplied by a space part. Any state of the form

$$\langle x_1,m_1,x_2,m_2|E,\ldots\rangle = \langle m_1,m_2|\sigma\rangle \langle X,x|K,\nu\rangle \tag{9-28}$$

will solve the eigenvalue problem for the hamiltonian of Equation 9-16 or 9-17.

The space part, $\langle X,x|K,\nu\rangle$, is what we have already calculated, Equation 9-25.

The state $|\sigma\rangle$ refers only to the spin condition of the pair: σ must be a two-label function in the spin z-component space, m_1,m_2.

An evident example for σ are the values of m_1 and m_2 themselves. That choice generates self-space brackets: the bracket

$$\langle m_1,m_2|m_1=1/2,m_2=1/2\rangle = \langle m_1|1/2\rangle \langle m_2|1/2\rangle \tag{9-29}$$

is zero unless both particles have spin up, in which case it is one.

There are four possible states of the pair in this space: $|1/2,1/2\rangle$, $|-1/2,-1/2\rangle$, $|1/2,-1/2\rangle$, and $|-1/2,1/2\rangle$. Two of them cannot refer to indistinguishable particles; the last two. They are not eigenstates of the $\hat{Y}$ operator. Such states cannot characterize an indistinguishable pair.

CLASSIFY SPIN STATES BY Y

One can recombine these four states to form a set of four new states which *are* eigenstates of $\hat{Y}$. Why numbers called s and m are chosen to label these states will be explained shortly. The states are

$$\langle m_1,m_2|s=1,m=1\rangle = \langle m_1|1/2\rangle \langle m_2|1/2\rangle \tag{9-30}$$

$$\langle m_1,m_2|1,0\rangle = \frac{1}{\sqrt{2}}[\langle m_1|1/2\rangle \langle m_2|-1/2\rangle + \langle m_2|1/2\rangle \langle m_1|-1/2\rangle] \tag{9-31}$$

$$\langle m_1,m_2|1,-1\rangle = \langle m_1|-1/2\rangle \langle m_2|-1/2\rangle \tag{9-32}$$

$$\langle m_1,m_2|0,0\rangle = \frac{1}{\sqrt{2}}[\langle m_1|1/2\rangle \langle m_2|-1/2\rangle - \langle m_2|1/2\rangle \langle m_1|-1/2\rangle] \tag{9-33}$$

The first three of these are symmetric states. If you exchange 1 and 2, you will get the state back again. Hence these states are characterized by $Y = +1$. All the states carrying the label $s = 1$ are symmetric (see Problem 9.8).

The last state is antisymmetric. The exchange produces a minus sign, so for this state, $Y = -1$. The state labeled $s = 0$ is antisymmetric (see Problem 9.9).

Now consider the spin-spatial product that results when one of these four states multiplies the spatial function $\langle X,x|K,\nu\rangle$, as in Equation 9-28. This product carries a Y-label: it is always either odd or even upon interchange of 1 with 2. Thus, as examples (see Problem 9.15),

$$\hat{Y}|K,v=1,s=1,m=1\rangle = |K,v=1,s=1,m=1\rangle \qquad (9\text{-}34)$$

but

$$\hat{Y}|K,v=2,s=1,m=0\rangle = -|K,v=2,s=1,m=0\rangle \qquad (9\text{-}35)$$

Every state of the form $|K,v,s,m\rangle$ is a simultaneous eigenstate of both $\hat{H}$ and $\hat{Y}$.

$$|E,Y,\ldots\rangle = |K,v,s,m\rangle \qquad (9\text{-}36)$$

Depending on the values of the quantum numbers v and s, the Y-value can be either $+1$ or -1.

The $Y = +1$ states are forbidden to a pair of indistinguishable fermions: those states are not part of its energy spectrum. Only $Y = -1$ states are allowed.

Hence of the four conceivable product states with energy E_{gnd} ($v=1$), only one can characterize a pair of indistinguishable fermions: the state $|K,v=1,s=0,m=0\rangle$. This is the only one with $Y = -1$.

Because the spatial wave function of the first excited state, $v = 2$, is antisymmetric, only products formed with symmetric spin states have a net antisymmetry. Of the four spin states, three are symmetric. Thus the first excited state is threefold degenerate, a *triplet* state.

The last of the three energy-level diagrams in Figure 9-4 shows the energy-level structure for two bound spin-1/2 indistinguishable fermions. Evident differences in energy-level structure characterize the three different classes of particles. Indistinguishability has quite observable experimental consequences.

9.5 FROM TWO SPINS, A NET SPIN

The four spin-state pair combinations in Equations 9-30, 9-31, 9-32, and 9-33 are the eigenstates of two operators. Two operators generate two labels.

The quantum numbers s and m are the eigenvalue indices for the pair of operators $\hat{S}^2$ and $\hat{S}_z$ where

$$\hat{S}^2 |s,m\rangle = s(s+1)\hbar^2 |s,m\rangle \qquad (9\text{-}37)$$

and

$$\hat{S}_z |s,m\rangle = m\hbar |s,m\rangle \qquad (9\text{-}38)$$

These two operators characterize a vector operator $\hat{\mathbf{S}}$ which has all the properties of a vector angular momentum operator whose magnitude squared is $\hat{S}^2$ and whose z-component is $\hat{S}_z$. They obey the fundamental rules of angular momentum, Equations

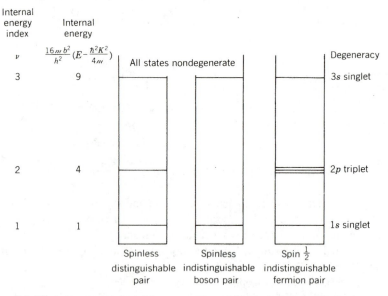

Figure 9.4 **Three energy-level diagrams for two interacting particles: a distinguishable pair, indistinguishable spinless bosons, and indistinguishable spin-1/2 fermions.**

5-48 and 5-49, which generate an s and an m. This vector operator is the operator vector sum of the particle-indexed vector spin operators $\hat{\mathbf{S}}_1$ and $\hat{\mathbf{S}}_2$.

$$\hat{\mathbf{S}} = \hat{\mathbf{S}}_1 + \hat{\mathbf{S}}_2 \tag{9-39}$$

The states that solve (9-37) and (9-38) are precisely those of (9-30), (9-31), (9-32), and (9-33).

Proof: You simply write $\hat{S}_z$ and $\hat{S}^2$ in terms of $\hat{S}_{1z}$, $\hat{S}_{2z}$, $\hat{S}_{1x}$, and so forth, as (9-39) directs. That allows you to carry out the operations of (9-37) and (9-38) in $\langle m_1, m_2 |$ space: you calculate $\langle m_1, m_2 | \hat{S}^2 | s, m \rangle$ and $\langle m_1, m_2 | \hat{S}_z | s, m \rangle$. This process is the subject of Problems 9.10 through 9.14. You find that (9-37) and (9-38) are, indeed, satisfied.

The significance of these results is in the idea of *net spin*.

The pair of fermions may be viewed as having a *net spin angular momentum* or a *total spin angular momentum*. The quantum index number for the magnitude of this net spin is s, so that the magnitude of the total angular momentum is $\hbar\sqrt{s(s+1)}$.

The solution to (9-37) for a pair of spin-1/2 particles yields that s is either zero or one. Thus two spin-1/2 particles can exhibit a net spin state of zero or one unit of angular momentum.

The number m is the quantum index for the z-component of this net or total angular momentum. A pair of half-unit spins can show a net z-component of $+1, 0,$ or -1.

In s, m space the single-particle spin state combinations separate out quite naturally into three even ones and an odd one. The single (singlet) $s = 0$ state is odd ($Y = -1$). All three (triplet) $s = 1$ states are even ($Y = +1$).

States of total angular momentum of the pair and its z-component are automatically Y-classified. The values of s signal the Y classification. For two particles the highest value of s always indicates an exchange symmetric $Y = +1$ state. The symmetry alternates with each unit of decreasing s (see Figure 9-5).

9.6 THE IDEAL GAS: NONINTERACTING PARTICLES

Consider a box of noninteracting 1-D particles, the proverbial ideal gas. That the particles don't interact means that the hamiltonian is already particle decoupled.

$$\hat{H} = H(\hat{1}, \hat{2}...) = H(\hat{1}) + H(\hat{2}) + \ldots \qquad (9\text{-}40)$$

From $\|m_1, m_2\rangle$	Comes $m = m_1 + m_2$	Forming s with $Y = +1$ $Y = -1$	$s_1 + s_2 = 1$ $s = 1$	$s_2 - s_1 = 0$ $s = 0$
$\|+\frac{1}{2}, +\frac{1}{2}\rangle$	1	1 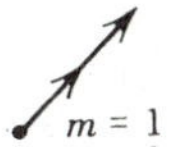	$\nearrow$ $m = 1$ $s = 1$	
$\|+\frac{1}{2}, -\frac{1}{2}\rangle$ and $\|-\frac{1}{2}, +\frac{1}{2}\rangle$	0	1 0	 $m = 0$ $\rightarrow$ $s = 1$	 $m = 0$ $\leftrightarrows$ $s = 0$
$\|-\frac{1}{2}, -\frac{1}{2}\rangle$	-1	1	$s = 1$ $m = -1$ $\searrow$	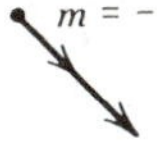
		Even Odd		

Figure 9.5 The states $\| s_1, s_2, s, m \rangle$ are linear combinations of the states $\| s_1, s_2, m_1, m_2 \rangle$.

The total energy of the system is E, the eigenvalue in the problem

$$\hat{H} \,|E,...\rangle = E\,|E,...\rangle \tag{9-41}$$

We shall examine the case in which $H(\hat{1})$ and $H(\hat{2})$ and so on all have exactly the same form: $H(\hat{1} = \hat{\chi}) = H(\hat{2} = \hat{\chi}) = \ldots$. The hamiltonian does not distinguish among the particles.

Suppose the solutions to the single-particle eigenvalue problem are $|v\rangle$ where, if by $\hat{j}$ is meant $\hat{1}$ or $\hat{2}$ or $\hat{3}$, and so on, then

$$H(\hat{j})\,|v\rangle = E(v)\,|v\rangle \tag{9-42}$$

The total energy is then the sum of single-particle energies.

$$E = E(v_1) + E(v_2) + \ldots \tag{9-43}$$

The v-value of the jth particle is v_j.

An energy eigenstate of the whole system of particles is a product of single-particle energy eigenstates.

$$\langle 1,2,...|E...\rangle = \langle 1,2,...|v_1,v_2,...\rangle = \langle 1|v_1\rangle \langle 2|v_2\rangle \ldots \tag{9-44}$$

Any product of this form will solve (9-41) to produce the E of (9-43).

The traditional case of particles-in-a-box offers the simplest example. For this case the hamiltonian for the jth particle is

$$H(\hat{j}) = \frac{\hat{p}_j^{\,2}}{2m} \tag{9-45}$$

so that

$$E(v) = \frac{v^2 h^2}{8mL^2} \tag{9-46}$$

The positive integer v is the index governing the wave function $\langle j|v\rangle$.

$$\langle j|v\rangle = \begin{cases} 0 & x_j < 0,\, L < x_j \\[2mm] \sqrt{2/L}\,\sin\dfrac{v\pi x_j}{L} & 0 < x_j < L \end{cases} \tag{9-47}$$

The state of (9-44) is a product of these brackets. For each different $j = 1, 2, 3$, and so forth, there is a v-value; v_1, v_2, v_3, and so forth. The total energy of (9-43) is a sum of energies from (9-46), each for a different j-value.

The key features of this result are (1) Many of the energy levels are degenerate; that is, the same value of E may result from many different product states. (2) Not all such product states are simultaneously eigenstates of the exchange operator.

These ideas are easily exemplified in a spinless two-particle ideal gas. The two particles are a distinguishable pair. Particle 1 is in state v_1, and particle 2 is in state v_2. The representative of the eigenstate of energy $|E,d\rangle = |v_1 v_2\rangle$ in the basis x_1, x_2 is

$$\langle x_1, x_2 | E, d \rangle = \langle x_1 | v_1 \rangle \langle x_2 | v_2 \rangle \tag{9-48}$$

The energy of the state is (see Problem 9.16)

$$E = (v_1^2 + v_2^2) \frac{h^2}{8\,m\,L^2} \tag{9-49}$$

NONDEGENERATE $\hat{H}$ EIGENSTATES MUST ALSO BE $\hat{Y}$ EIGENSTATES

The state $v_1 = 3 = v_2$ is nondegenerate; there is only this one state that yields the energy $E = 18h^2/8\,m\,L^2$.

Imagine an interchange of the particles in this state. It results in the same state back again. The state $|v_1 = 3 = v_2\rangle$ of (9-48) is, thus, an eigenstate of the $\hat{Y}$-operator.

$$\hat{Y} |v_1 = 3, v_2 = 3\rangle = +1 |v_1 = 3, v_2 = 3\rangle \tag{9-50}$$

The nondegenerate energy eigenstate $|v_1 = 3 = v_2\rangle$ is a simultaneous eigenstate of the exchange operator.

Any nondegenerate state of the pair must be a simultaneous eigenstate of exchange. Because $\hat{Y}$ and $\hat{H}$ commute, Y and E are permissible simultaneous labels. If to a value of E there corresponds only one state, it must accept a Y label.

DEGENERATE $\hat{H}$ EIGENSTATES NEED NOT BE $\hat{Y}$ EIGENSTATES

If v_1 and v_2 are not equal, then the state in Equation 9-48 is one of a pair of degenerate ones. For example, suppose $v_1 = 4$ and $v_2 = 7$. The state $v_1 = 7, v_2 = 4$ has exactly the same energy; $65\ h^2/8\,m\,L^2$. This new state is quite distinct from the old one. It is the old one exchanged, $\hat{Y}|E,d\rangle$.

$$\langle x_1, x_2 | \hat{Y} | E, d \rangle = \langle x_2, x_1 | v_1, v_2 \rangle = \langle x_2 | v_1 \rangle \langle x_1 | v_2 \rangle \tag{9-51}$$

The two distinctly different states $|E,d\rangle$ and $\hat{Y}|E,d\rangle$ are degenerate in energy (see Problem 9.17).

The two states $|E,d\rangle$ and $\hat{Y}|E,d\rangle$ are energy eigenstates of the system: eigenstates of $\hat{H}$. However, neither of these are eigenstates of $\hat{Y}$.

Proof: When $\hat{Y}$ operates on either, it produces the other, not the same, state.

FURTHER MEASUREMENT BREAKS THE DEGENERACY

The two states have the same energy but are distinguishable from each other. They can be distinguished by further measurement.

There must be some operator that commutes with the hamiltonian that will expose the degeneracy: otherwise, the particles wouldn't, in fact, be distinguishable. By the measurement corresponding to this operator you can distinguish the state $|\nu_1=4,\nu_2=7\rangle$ from the state $|\nu_1=7,\nu_2=4\rangle$.

Here's a simple example. As an auxilliary measurement to the finding that $E=65h^2/8\,m^2L^2$, measure the energy of particle 1 alone. Because $H(\hat{1})$ commutes with the total system hamiltonian, this measurement doesn't destroy the state label. If you find $16h^2/8\,mL^2$, then the system state is $|4,7\rangle$ and not $|7,4\rangle$. You have broken the degeneracy.

The energy-level diagram for a pair of distinguishable noninteracting particles-in-a-box is displayed in Figure 9-6.

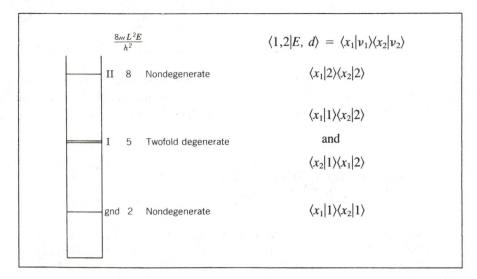

Figure 9.6 The energy-level structure for two distinguishable noninteracting particles-in-a-box.

FROM DEGENERATE $\hat{H}$ EIGENSTATES CONSTRUCT $\hat{Y}$ EIGENSTATES

The state $|v_1 = 4, v_2 = 7\rangle$ has the same energy as does the state $|v_1 = 7, v_2 = 4\rangle$, but neither of these is an eigenstate of $\hat{Y}$. Therefore they cannot describe indistinguishable particles. However, $\hat{Y}$ and $\hat{H}$ commute. So, if to a single value of E there corresponds many states, then combinations of them must accept Y-labels. Because the labels E and Y are always possible, simultaneous eigenstates of $\hat{Y}$ may always be constructed. One constructs eigenstates of $\hat{Y}$ from degenerate $\hat{H}$ eigenstates. Here's an example.

$$\langle 1,2 | E = 65h^2/8 \, m L^2, Y = +1 \rangle = \frac{1}{\sqrt{2}} |\langle 1|4 \rangle \langle 2|7 \rangle + \langle 2|4 \rangle \langle 1|7 \rangle| \qquad (9\text{-}52)$$

$$\langle 1,2 | E = 65h^2/8 \, m L^2, Y = -1 \rangle = \frac{1}{\sqrt{2}} |\langle 1|4 \rangle \langle 2|7 \rangle - \langle 2|4 \rangle \langle 1|7 \rangle| \qquad (9\text{-}53)$$

These states have the dual property that they are eigenstates of $\hat{Y}$ and yield the same energy. They are simultaneous eigenstates of both $\hat{Y}$ and $\hat{H}$. To characterize indistinguishable particles we require such states.

The only spinless indistinguishable particles are bosons.

If $v_2 \neq v_1$, the only boson state with these two numbers having the energy E of Equation 9-49 is $|E, b\rangle$ where

$$\langle x_1, x_2 | E, b \rangle = \frac{1}{\sqrt{2}} |\langle x_1|v_1 \rangle \langle x_2|v_2 \rangle + \langle x_1|v_2 \rangle \langle x_2|v_1 \rangle| \qquad (9\text{-}54)$$

That's because this state is symmetric in 1 and 2. Its companion, the evident generalization of (9-53), is antisymmetric and therefore forbidden to bosons. Hence, for a pair of spinless bosons the energy level, $65h^2/8 \, m L^2$, is nondegenerate. There is only one state with this energy; that of (9-52).

Figure 9-7 shows the energy-level structure for an indistinguishable pair of noninteracting bosons-in-a-box.

WHAT YOU KNOW IS A DISTRIBUTION

The only decisive feature of the state $|E, b\rangle$ is what values of v are *occupied*.

One of the indistinguishable boson pair is in the state $v = 4$ and the other is in state $v = 7$: that is a complete description of the state whose wave function is given in (9-52). The equation holds no more information than this word description does.

This description is the verbal rendering of a distribution, $n(v)$. The particular

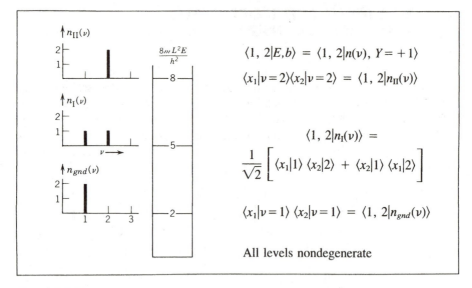

Figure 9.7 The energy level structure for two indistinguishable noninteracting spinless bosons-in-a-box.

distribution portrayed in (9-52) is $n(v) = \delta_{v,4} + \delta_{v,7}$. States $v = 4$ and 7 are *occupied*, each by one of a pair of indistinguishable particles, and no particles occupy any other v-states. A distribution describes state occupation.

The notion of *state occupation* releases us from thinking in particle-indexed space. It focuses on the critical feature of indistinguishability; that the only thing one may know about the indistinguishable particles of the gas is how many of them are in any one single-particle state. "Which particle is in which state?" constitutes an unacceptable consideration: to pose it is to expect an answer to a meaningless question.

THE DISTRIBUTION GOES IN THE KET

Occupation or distribution states are a natural expression of the dictum: "Put What You Know in the Ket." What we know is the *occupation distribution* $n(v)$. This is the label that we put in the ket. It designates what is known of a noninteracting indistinguishable multiparticle state.

The distribution state of one boson occupying $v = 4$ and the other occupying $v = 7$ is

$$|n(v) = \delta_{v,4} + \delta_{v,7}, Y = +1\rangle$$

The state $v_1 = 3 = v_2$ also has a distribution representation. The distribution is $n(v \neq 3) = 0$, while $n(v = 3) = 2$: all of the particles occupy $v = 3$, and so no other v-value is occupied. The pair are in the state $|n(v) = 2\delta_{v,3}, Y = +1\rangle$. The 1,2 space wave function for this state is the simple product in (9-48) with $v_1 = v_2 = 3$.

The distributions $n_{gnd}(v)$, $n_I(v)$ and $n_{II}(v)$ for the ground state and the next two excited states of a pair of noninteracting indistinguishable spinless bosons are depicted in Figure 9-7.

9.7 AN INDISTINGUISHABLE PARTICLE GAS: THINK IN DISTRIBUTION SPACE

Consider, instead of two, three noninteracting indistinguishable bosons-in-a-box. Their energy eigenstate label, $n(v)$, is a three-particle distribution function. In $n(v)$ all three particles must be accounted for. Hence, for every state-of-the-system $|n(v)\rangle$, the sum of $n(v)$ over all v must be three.

For N particles

$$\sum_v n(v) = N \tag{9-55}$$

Consider the energy, E, of some given multiparticle state. This state is always some combination of distinguishable multiparticle states. Distinguishable multiparticle states are ones for which you assign particle 1 to one single-particle state and particle 2 to another, and so on. Distinguishable multiparticle states are the simple products of (9-44).

It is from these that the $\hat{Y}$ eigenstates are constructed. All of the simple product states that contribute to the construction are necessarily degenerate in energy. The system energy, E, is the same for an indistinguishable multiparticle distribution state as it is for any one of the simple product states composing it. That total system energy is the sum, shown in (9-43), of single-particle state energies, $E(v)$.

In this sum there will be, for any given value of $v = \mu$, $n(\mu)$ terms with exactly the same energy $E(\mu)$: each value of energy $E(\mu)$ will appear $n(\mu)$ times in the sum in (9-43). The same energy $E(\mu)$ appears every time one of the particles is in a state $v_j = \mu$. Hence the total energy given in (9-43) may be alternatively written as

$$\sum_v E(v)n(v) = E \tag{9-56}$$

This expression facilitates thinking in distribution state language. It confirms the idea that the essential attribute of any state of the system is in the label $n(v)$; to give this distribution is to designate the state and its energy.

IN PARTICLE-INDEXED SPACE, CUMBERSOME DISTRIBUTION STATES

We rely on old languages to describe new perceptions. The formulas for distribution states in particle-indexed space exemplify the point. The matrix elements

$\langle 1,2..|n(v)\rangle$ are both cumbersome and not generally useful computationally. Nevertheless we want to see them. They are a traditional point of reference. With them we can prove the very theorems that release us from thinking in particle-indexed space.

Equation 9-54 shows the two-particle distribution $\langle 1,2|n(v)\rangle$ where, if we call $v_1 = a$ and $v_2 = b \neq a$, then $n(v) = \delta_{v,a} + \delta_{v,b}$. When you set $v_1 = v_2 = 3$, the distribution $n(v) = 2\delta_{v,3}$ is represented by (9-48).

As a three-particle example consider this one: three different v-states are occupied. Call them a, b, and c. The symmetric state is constructed by summing over all the permutations of the three single-particle states and dividing by $\sqrt{3!}$. Thus

$$\sqrt{3!}\ \langle 1,2,3|n(v) = \delta_{va} + \delta_{vb} + \delta_{vc}, Y = +1\rangle =$$

$$\langle 1|a\rangle\langle 2|b\rangle\langle 3|c\rangle + \langle 1|a\rangle\langle 3|b\rangle\langle 2|c\rangle + \langle 2|a\rangle\langle 1|b\rangle\langle 3|c\rangle +$$
$$\langle 2|a\rangle\langle 3|b\rangle\langle 1|c\rangle + \langle 3|a\rangle\langle 1|b\rangle\langle 2|c\rangle + \langle 3|a\rangle\langle 2|b\rangle\langle 1|c\rangle \qquad (9\text{-}57)$$

Notice that the form in this equation is appropriate only to cases where $a \neq b \neq c$. If $a = b$, the form will be different (see Problem 9.18).

The cumbersome sum of distinguishable-particle-state products means what is put compactly by the occupation state bracket: that each of the three states $v = a$, b, and c are occupied by a particle.

The extension to bosons in N different states is evident: sum the N-term product distinguishable-multiparticle wave function over all possible permutations of the particle index and divide by $\sqrt{N!}$. The exchange of any pair will leave such a state unchanged: it is an eigenstate of $\hat{Y}$ with $Y = +1$ for all possible exchanges. Any of the exchange operators, $\hat{Y}_{12}$, $\hat{Y}_{13}$, $\hat{Y}_{23}$, and so on yield $+1$. And because each product term in the sum is itself a state of energy E, so is the sum. The state is eigen to all $\hat{Y}$'s and $\hat{H}$.

FERMIONS: $Y = -1$, HALF-INTEGRAL SPIN

An ideal gas of indistinguishable fermions can be found only in states labeled by $Y = -1$. Only states that are antisymmetric upon particle exchange are allowed in the spectrum.

Fermions always have spin. We'll consider the simplest case, $s = 1/2$.

The complete spin-1/2 fermion single-particle eigenstate for the jth particle hamiltonian, $H(j)$, of Equation 9-42 or 9-45, is the spin-state space-state product

$$\langle j|v,M\rangle = \langle x,m|v,M\rangle = \langle x|v\rangle\langle m|M\rangle \qquad (9\text{-}58)$$

Here j represents not only the particle's spatial position $x_j = x$ but also its spin z-component $m_j = m$. By the bracket $\langle x_j|v\rangle = \langle x|v\rangle$ is meant just that given in (9-47).

The subscript j is omitted when its inclusion carries no conceptual thrust: any isolated particle is the jth.

A fermion with spin up is in the state $|M = +1/2\rangle$ whose representative in spin component space is $\langle m|+1/2\rangle$. Both by $|m\rangle$ and by $|M\rangle$ are meant eigenstates of the operator $\hat{S}_z$:

$$\hat{S}_z |m\rangle = m\hbar |m\rangle \qquad \text{and} \qquad S_z |M\rangle = M\hbar |M\rangle \qquad (9\text{-}59)$$

Thus $\langle m|+1/2\rangle$ is a self-space bracket in m-space. It expresses the spin z-component $+1/2$ state.

A single fermion in the ground state of the box with spin up is characterized by $v = 1$ and $M = 1/2$: its representation in x,m space is $\langle x|1\rangle \langle m|1/2\rangle$. A ground-state spin-down fermion is in the state $|v=1,M=-1/2\rangle$.

THE SELF-SPACE BRACKET IS A SINGLE-PARTICLE DISTRIBUTION

We need not think in x,m space. We would do better to think in v,m space. Let the j of (9-58) represent the single-particle space $v_j = v$, $m_j = m$. In this space the state $|v=1,M=-1/2\rangle$ appears as

$$\langle j|1,-1/2\rangle = \langle v,m|v=1,M=-1/2\rangle = \langle v|1\rangle \langle m|-1/2\rangle \qquad (9\text{-}60)$$

It is a product of self-space brackets. Only at $v = 1$ and $m = -1/2$ is a point in v,m space occupied.

We see from this equation that the state of a single particle represented in its home state basis is a distribution portrayal of the state! States of one particle alone may be described in the language of distribution.

For our fermion each state of the system needs two labels, a value of v, say $v = \mu$, and a value of m, say $m = M$. The distribution among states is a two-dimensional function: $n = n(v,m)$. The distribution $n(v,m)$ that represents the state $|v=\mu,m=M\rangle$ is this one: the point μ,M of v, m space is occupied, and no other point of v,m space is occupied. Precisely this same statement describes the self-space bracket $\langle v,m|\mu,M\rangle$. Hence

$$n(v,m) = \langle v,m|\mu,M\rangle = \langle v|\mu\rangle \langle m|M\rangle \qquad (9\text{-}61)$$

A single-particle distribution is simply its self-space bracket.

Using the boldface symbol $\mathbf{v}$ to represent both v and m combined and contracting μ and M into the boldface $\boldsymbol{\mu}$ the relation may be condensed into

$$n(\mathbf{v}) = \langle \mathbf{v}|\boldsymbol{\mu}\rangle \qquad (9\text{-}62)$$

9.8 THE SLATER DETERMINANT CONSTRUCTS FERMI STATES

There is a general prescription for fabricating an indistinguishable-multiparticle fermion state from single-particle states.

You construct a determinant called the Slater determinant. Its elements, $D_{j\mu}$, are the amplitudes, $\langle j|\mu\rangle$, which portray the jth particle in the μth single-particle state.

The single-particle state wave function is $\langle j|\mu\rangle$, where j is any set of coordinates with which to mark particle j. An example of $\langle j|\mu\rangle$ is just the bracket of (9-58), and so is (9-60).

When there are N particles to be considered, the antisymmetric (fermi) state is an $N \times N$ determinant divided by $\sqrt{N!}$. Here's how it looks for $N = 3$ when the distribution is one particle each at $\mu = a$, b, and c:

$$\sqrt{3!} \, \langle 1,2,3|n(\pmb{\nu}) = \delta_{\pmb{\nu}a} + \delta_{\pmb{\nu}b} + \delta_{\pmb{\nu}c}, \, Y = -1\rangle = \begin{vmatrix} \langle 1|a\rangle & \langle 1|b\rangle & \langle 1|c\rangle \\ \langle 2|a\rangle & \langle 2|b\rangle & \langle 2|c\rangle \\ \langle 3|a\rangle & \langle 3|b\rangle & \langle 3|c\rangle \end{vmatrix} \quad (9\text{-}63)$$

9.9 NEVER TWO INDISTINGUISHABLE FERMIONS IN THE SAME SINGLE-PARTICLE STATE

Consider the fermion state $|f\rangle$. In particle-indexed space its representative is $\langle 1,2|f\rangle$.

Now consider this event: that the two particles are found to have the same coordinates, that $1 = \chi = 2$. An example is for both fermions to have spin up at position $x = 73\text{Å}$.

Being fermions, the exchanged state, $\langle 2,1|f\rangle$, must be equal to the negative of the original; Equation 9-14. Hence the amplitude, $\langle \chi,\chi|f\rangle$, for the event must satisfy $\langle \chi,\chi|f\rangle = -\langle \chi,\chi|f\rangle$.

The only solution is zero. Thus the probability of the event is zero. No two electrons with the same spin component can occupy the same position. Neither can two electrons be in the same single-particle energy state with the same spin. That the χ can be any single-particle index at all leads to the Pauli exclusion principle. No two indistinguishable fermions may ever be found, both of which bear exactly the same single-particle indices.

All noninteracting multiparticle distributions derive from the single-particle behavior and only this principle. As an example of its effect, the first two energy levels for a pair of noninteracting indistinguishable spin-1/2 fermions are displayed in Figure 9-8.

The Slater determinant exemplifies the Pauli exclusion principle. Its form guarantees it. A property of determinants is that if two columns are the same, then the determinant will be zero. If two of the states—say a and b in (9-63)—were the same, then two columns would be equal. Thus the Slater determinant yields zero, as it must, for the amplitude that two electrons occupy the single-particle state $a = b$.

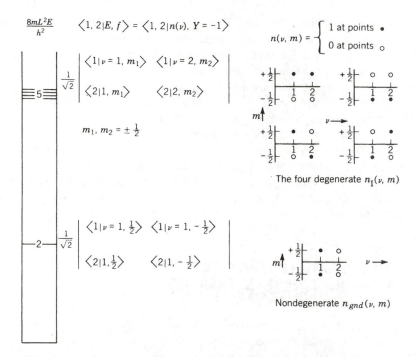

Figure 9.8 The energy-level structure for two noninteracting spin 1/2 indistinguishable fermions-in-a-box: the Slater determinant.

ODD AND EVEN SPIN STATES
EVEN AND ODD SPATIAL STATES

There is an alternative way to construct a fermion-multiparticle state from single-particle states; the Slater determinant is not the only way. The other way is to fabricate spin × spatial product states whose net effect is odd.

From purely spatial single-particle states fabricate even and odd combinations. For a single pair of particles, one in state $\nu = \alpha$ and the other in state $\nu = \beta \neq \alpha$, Equation 9-54 gives the even combination. If $\nu = \alpha = \beta$, then the simple product state of (9-48) is even. An odd combination only occurs when $\alpha \neq \beta$. It is

$$\langle x_1, x_2 | n(\nu) = \delta_{\nu,\alpha} + \delta_{\nu,\beta}, Y = -1 \rangle = \frac{1}{\sqrt{2}} [\langle x_1|\alpha\rangle \langle x_2|\beta\rangle - \langle x_1|\beta\rangle \langle x_2|\alpha\rangle] \quad (9\text{-}64)$$

This is a pure spatial state, one of physical position only. It is odd or antisymmetric in the particle-indexed spatial coordinates (see Problem 9.20).

Consider the spin-spatial product state formed when (9-64) is multiplied by $\langle m_1, m_2 | s = 1, m \rangle$. A purely spatial state of odd symmetry is multiplied by an even-symmetry pure-spin state. The product state is an eigenstate of $\hat{H}$ and has the overall odd symmetry proper to fermions. An exchange of all particle coordinates—spin and space—results in a sign change: the net effect of *odd* multiplied by *even* is *odd*.

A fourth state of the same energy is the even spatial state multiplied by the odd ($s = 0$) spin state. This product also has a net $Y = -1$. It is a valid two-fermion state.

Figure 9-9 shows an example: the first excited state energy is fourfold degenerate.

This energy was also found to be fourfold degenerate by the Slater determinant. In each case there are four states with the same energy, $5h^2/8\,mL^2$. What is different is the way of classifying these states. The spin-space product classifies the states by the pair of quantum numbers s and m; the Slater determinant by $n(\nu)$.

The ground state is nondegenerate. It is the same as that obtained from the Slater determinant. This must be so, as different descriptions can't produce different states if there is only one state possible.

MULTIPARTICLE LEVELS ARE DISTRIBUTION STATE ENERGIES

As a practical matter, the form of the state in single-particle-indexed space plays no role in constructing the multiparticle states of the system. It is only the distribution, $n(\nu)$, that dictates the system's energies. Bose distributions allow any number of particles in a state, and fermi distributions allow only one.

The scheme is best conveyed by an example. Suppose that the energy levels for a single particle in a particullar well is this simple one:

$$E(\nu=1) = 1\text{ eV}, \quad E(\nu=2) = E(\nu=3) = 3\text{ eV}, \quad \text{and} \quad 6\text{ eV} < E(\nu \geq 4)$$

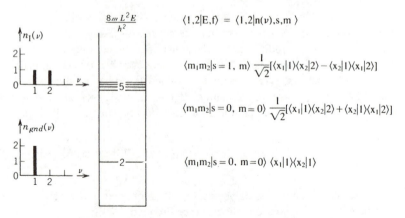

Figure 9.9 The energy-level structure for two noninteracting spin-1/2 indistinguishable fermions in a box: even × odd spin-space scheme.

By these levels are meant precisely those (9-42) which solve the single-particle energy-eigenvalue problem. The present well is not the box for which the levels are the nondegenerate ones of (9-46). In this well the states $v = 2$ and $v = 3$ are degenerate. The single-particle energy-level diagram for this well is illustrated in Figure 9-10.

Now consider the multiparticle states of three noninteracting indistinguishable spin zero bosons locked into the well with this single-particle structure. For this system of three, the total hamiltonian is

$$\hat{H} = H(\hat{1}) + H(\hat{2}) + H(\hat{3}) \tag{9-65}$$

where the eigenstates of each $H(\hat{j})$ are the same: the $|v\rangle$ whose energies are the given $E(v)$.

We first consider the ground state of the system. We want to deduce E_{gnd} and its degeneracy.

The ground state has the lowest possible energy of the three-boson system. The distribution that will yield the lowest energy is $n(v) = 3\langle v|1\rangle$. When all three bosons have $v = 1$, the lowest energy possible is achieved. This energy level is nondegenerate. Only the one distribution $3\langle v|1\rangle$ achieves $E_{gnd} = 3$ eV.

Now consider the next possible energy: the first excited energy E_I. It is evidently $E_I = 1\text{eV} + 1\text{eV} + 3\text{eV} = 5\text{eV}$. Two distinctly different distributions achieve it: $n(v) = 2\langle v|1\rangle + \langle v|2\rangle$ and $2\langle v|1\rangle + \langle v|3\rangle$. Thus the energy $E_I = 5\text{eV}$ is twofold degenerate.

The first two levels of the three-boson system are shown in Figure 9-10.

The case of three noninteracting indistinguishable spin 1/2 fermions proceeds in a parallel way. The governing hamiltonian, (9-65), does not act on the spins. But the distribution is in the space of spin z-component as well as v: $n = n(v,m)$. The critical constraint on the distribution is that no two fermions may have the same v,m value: no two may be in the same $|v,m\rangle$ state.

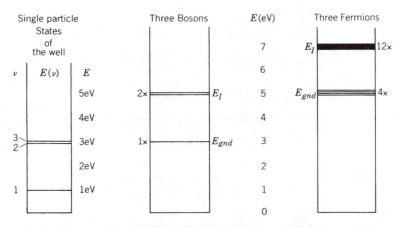

Figure 9.10 Three noninteracting indistinguishable particles in a well. Bose or fermi distribution statistics determine the energy level diagrams.

The lowest energy achievable under this constraint is $E_{gnd} = 1\text{eV} + 1\text{eV} + 3\text{eV}$ $= 5\text{eV}$. There are four distinct distributions that give it:

$$n(v,m) - \langle v,m|1, +1/2\rangle - \langle v,m|1, -1/2\rangle = \langle v,m|2, +1/2\rangle \qquad \text{or}$$
$$= \langle v,m|2, -1/2\rangle \qquad \text{or}$$
$$= \langle v,m|3, +1/2\rangle \qquad \text{or}$$
$$= \langle v,m|3, -1/2\rangle$$

Thus the ground-energy level is fourfold degenerate. A direct extension of this argument shows that the first excited energy level, E_I, is at 7eV and that it is twelvefold degenerate, as shown in Figure 9-10.

PROBLEMS

9.1 In general an operator produces a new ket multiplied by a constant. Prove that if $\hat{Y}|S\rangle = c|Q\rangle$ then $c = 1$. It is so because $\langle Q|Q\rangle = \langle S|S\rangle$ and $\hat{Y}\dagger\hat{Y} = 1$.

Solution: $\langle S|S\rangle = \langle S|\hat{Y}\dagger\hat{Y}|S\rangle = c^*c\langle Q|Q\rangle$

9.2 Prove that $\hat{Y}\dagger = \hat{Y}$.

Solution key:
$$\langle\psi|\hat{Y}\dagger|\phi\rangle = \langle\phi|\hat{Y}|\psi\rangle*$$
$$= \sum_{1,2} \langle\phi|1,2\rangle* \langle1,2|\hat{Y}|\psi\rangle*$$
$$= \sum_{1,2} \langle2,1|\hat{Y}|\phi\rangle \langle\psi|2,1\rangle$$
$$= \langle\psi|\hat{Y}|\phi\rangle$$

9.3 Prove that $\hat{Y} = \hat{Y}^{-1}$

Solution key:
$$\langle1,2|\hat{Y}\hat{Y}|\psi\rangle = \langle1,2|\psi\rangle.$$

Be sure you can complete the argument connecting the solution key to the result.

9.4 Imagine that in the neighborhood of the region shown in Figure 9-1 a good approximation to the wave function $\langle x_1,x_2|\psi\rangle$ is $x_1/15\text{Å}^2 + x_2/30\text{Å}^2$. Calculate the amplitudes A and B shown in the figure. Consider the case that the ball is at 3 Å and the cube is at 5 Å. Then consider the exchange of ball and cube.

9.5 Suppose
$$\hat{\Omega} = \hat{x}_1\hat{p}_2$$
so that

$$\hat{\Omega} = \hat{x}_2\hat{p}_1$$

and suppose that

$$\langle 1,2|S\rangle = \psi(x_1,x_2)$$

As an example, envision $\psi(x_1,x_2)$ to be $\psi(x_1,x_2) = \sin x_1 \cos x_2$ in some region of space.

   a. Exhibit $\langle 1,2|\hat{\Omega}|S\rangle$.

   b. Exhibit $\langle 2,1|\hat{\Omega}|S\rangle$.

   c. Exhibit $\langle 1,2|Q\rangle$.

   d. Exhibit $\langle 1,2|\hat{\Omega}|Q\rangle$.

Answers:

   a. $-i\hbar x_1 \, \partial\psi(x_1,x_2)/\partial x_2 \; [i\hbar x_1 \sin x_1 \sin x_2]$

   b. $-i\hbar x_2\partial\psi(x_2,x_1)/\partial x_1 \; [i\hbar x_2 \sin x_2 \sin x_1]$

   c. $\psi(x_2,x_1) \; [\; \sin x_2 \cos x_1]$

   d. $-i\hbar x_2\partial\psi(x_2,x_1)/\partial x_1 \; [i\hbar x_2 \sin x_2 \sin x_1)$

Notice that $\langle 1,2|\hat{\Omega}|Q\rangle = \langle 2,1|\hat{\Omega}|S\rangle$. Show that $\langle 1,2|\hat{\Omega}|S\rangle = \langle 2,1|\hat{\Omega}|Q\rangle$. Notice that $\langle 2,1|\hat{x}_1\hat{p}_2|S\rangle \neq -i\hbar x_1\partial\psi(x_2,x_1)/\partial x_2$: exchanging-after-operating does not equal operating-after-exchanging.

9.6 Prove (9-27).

   It's a matter of showing that

$$\psi[x_2(X,x), x_1(X,x)] = \psi[x_1(X, -x), x_2(X, -x)]$$

9.7 Prove that the states $|K,v\rangle$ of (9-25) can be classified by Y where

$$|K,v=\text{odd}\rangle = |K,v,Y=+1\rangle$$

and

$$|K,v=\text{even}\rangle = |K,v,Y=-1\rangle$$

   Solution: Apply $\hat{Y}$ to each of the states calculating in the $\langle X,x|$ basis. Use (9-20), (9-22), (9-25), and (9-27).

9.8 The three $s = 1$ spin states represented in particle-indexed spin z-component space are given in Equations 9-30, 9-31, and 9-32 as

$$\langle m_1m_2|s=1,m=1\rangle = \langle m_1|1/2\rangle \langle m_2|1/2\rangle$$

$$\langle m_1m_2|1,0\rangle = \frac{1}{\sqrt{2}}\{\langle m_1|1/2\rangle \langle m_2|-1/2\rangle + \langle m_2|1/2\rangle \langle m_1|-1/2\rangle\}$$

$$\langle m_1m_2|1,-1\rangle = \langle m_1|-1/2\rangle \langle m_2|-1/2\rangle$$

Show that each of these states are eigen to $\hat{Y}$ with $Y = +1$: they all are symmetric (even) states. And show that for each the certainty condition is satisfied.

Solution: Show $\langle m_1 m_2 | \hat{Y} | s = 1, m \rangle = \langle m_1 m_2 | s = 1, m \rangle$ and that $\langle s = 1, m | s = 1, M \rangle = \delta_{mM}$

9.9 The spin state, $|s = 0, m = 0\rangle$, represented in particle-indexed spin space is

$$\langle m_1 m_2 | s = 0, m = 0 \rangle = \frac{1}{\sqrt{2}} \{\langle m_1 | 1/2 \rangle \langle m_2 | - 1/2 \rangle - \langle m_2 | 1/2 \rangle \langle m_1 | - 1/2 \rangle\}$$

Prove that this is an odd (antisymmetrical) state against exchange; that it is an eigenstate of $\hat{Y}$ with $Y = -1$. Show that the certainty condition, $\langle s = 0, m = 0 | s = 0, m = 0 \rangle = 1$, is satisfied.

Solution: Show $\langle m_1, m_2 | \hat{Y} | s = 0, m = 0 \rangle = -\langle m_1, m_2 | s = 0, m = 0 \rangle$

9.10 Consider this product spin state:

$$\langle m_1 m_2 | \sigma \rangle = \langle m_1 | 1/2 \rangle \langle m_2 | - 1/2 \rangle$$

What does $\hat{S}_z = \hat{S}_{1z} + \hat{S}_{2z}$ do to it?

Answer:
$$\langle m_1 m_2 | \hat{S}_z | \sigma \rangle = \langle m_1 | \hat{S}_{1z} | 1/2 \rangle \langle m_2 | - 1/2 \rangle + \langle m_1 | 1/2 \rangle \langle m_2 | \hat{S}_{2z} | - 1/2 \rangle = 0$$

9.11 What does the operator $\hat{S}_{1-} \hat{S}_{2+}$ do to the state $|\sigma\rangle$ of the previous problem?

Answer:
$$\langle m_1 m_2 | \hat{S}_{1-} \hat{S}_{2+} | \sigma \rangle = \langle m_1 | \hat{S}_{1-} | 1/2 \rangle \langle m_2 | \hat{S}_{2+} | - 1/2 \rangle = \hbar^2 \langle m_1 | - 1/2 \rangle \langle m_2 | + 1/2 \rangle$$

9.12 Show that the state in Problem 9.9 is eigen to both the operator

$$\hat{S}_z = \hat{S}_{1z} + \hat{S}_{2z}$$

and the operator

$$\hat{S}^2 = (\hat{S}_{1x} + \hat{S}_{2x})^2 + (\hat{S}_{1y} + \hat{S}_{2y})^2 + (\hat{S}_{1z} + \hat{S}_{2z})^2$$

and find the eigenvalues. (Both are zero.)

Solution key: Use the result that the operator $\hat{S}^2$ may be written as

$$\hat{S}^2 = \frac{1}{2} (\hat{S}_+ \hat{S}_- + \hat{S}_- \hat{S}_+) + \hat{S}_z^2$$

where by $\hat{S}_z$, $\hat{S}_+$, and $\hat{S}_-$ are meant

$$\hat{S}_z = \hat{S}_{1z} + \hat{S}_{2z}$$
$$\hat{S}_+ = \hat{S}_{1+} + \hat{S}_{2+}$$

and

$$\hat{S}_- = \hat{S}_{1-} + \hat{S}_{2-}$$

9.13 Find out what $\hat{S}_{1+} + \hat{S}_{2+}$ produces when operating on each of the states in Problem 9.8. Calculate the kets resulting from the operation $\hat{S}_{1-} + \hat{S}_{2-}$ on each of the three states in Problem 9.8. And then exhibit the three kets resulting from the operation $\hat{S}_{1z} + \hat{S}_{2z}$ on the three states of Problem 9.8.

9.14 Show that if we call $\hat{S}_+ = \hat{S}_{1+} + \hat{S}_{2+}$, $\hat{S}_- = \hat{S}_{1-} + \hat{S}_{2-}$, and $\hat{S}_z = \hat{S}_{1z} + \hat{S}_{2z}$, then the nine results of Problem 9.13 can be summarized as

$$\hat{S}_+|s=1,m\rangle = \hbar\sqrt{2-m(m+1)}|1,m+1\rangle$$

$$\hat{S}_-|1,m\rangle = \hbar\sqrt{2-m(m-1)}|1,m-1\rangle$$

$$S_z|1,m\rangle = m\hbar|1,m\rangle$$

where $m = -1, 0,$ or $+1$.

Significance: The three spin states of Problem 9.8 behave exactly as would a triplet of angular momentum states of net spin quantum number $s = 1$.

9.15 Show that (9-34) and (9-35) are true.
You do it by examining

$$\langle X,x,m_1,m_2|\hat{Y}|K,v,s,m\rangle = \langle X, -x,m_2,m_1|K,v,s,m\rangle$$

in each case.

9.16 Show that the compound wave function of (9-48) is indeed a solution of (9-41) in x_1,x_2 representation with the energy given in Equation 9-49.

Solution key: Note that a one-dimensional operator affects, in a multidimensional state, only what is represented in its own one-dimensional space, so

$$\langle x_1x_2|\hat{H}|E,d\rangle = \langle x_1|H(\hat{1})|v_1\rangle \langle x_2|v_2\rangle + \langle x_1|v_1\rangle \langle x_2|H(\hat{2})|v_2\rangle$$

$$= [E(v_1) + E(v_2)] \langle x_1x_1|E,d\rangle$$

9.17 The hamiltonian of the system commutes with $\hat{Y}$; $[\hat{Y},\hat{H}] = 0$. Show that if $|E,d\rangle$ is some energy eigenstate of the system, it must be true that the state $\hat{Y}|E,d\rangle$ is also an energy eigenstate of the system with exactly the same energy as the original state, $|E,d\rangle$.

Solution: Prove $\hat{H}(\hat{Y}|E,d\rangle) = E(\hat{Y}|E,d\rangle)$.

9.18 Exhibit the particle-indexed basis form of the three-indistinguishable-boson-state of distribution $n(v) = 2\delta_{vb} + \delta_{vc}$.

Answer:

$$\langle 1,2,3|n(v)=2\delta_{vb}+\delta_{vc},Y=+1\rangle\sqrt{3} =$$
$$\langle 1|b\rangle \langle 2|b\rangle \langle 3|c\rangle + \langle 1|b\rangle \langle 2|c\rangle \langle 3|b\rangle + \langle 1|c\rangle \langle 2b\rangle \langle 3b\rangle$$

Notice that here the factor is $\sqrt{3}$ instead of the $\sqrt{3!} = \sqrt{6}$ of (9-57).

9.19 Prove that the state in Equation 9-57 is indeed eigen to all $\hat{Y}$'s with $Y = +1$ and simultaneously eigen to $\hat{H}$. Show that the energy of this state of three spinless bosons in a 1-D box is $E(a) + E(b) + E(c)$.

9.20 Prove that the representative of the spatial state with $\nu = 1,2$; see (9-54)

$$\frac{1}{\sqrt{2}} \{\langle x_1|1\rangle \langle x_2|2\rangle - \langle x_1|2\rangle \langle x_2|1\rangle\}$$

is an eigenstate of $\hat{Y}$ with $Y = -1$. Prove that this is a simultaneous eigenstate of $\hat{H}$ with energy $5h^2/8\,m\,L^2$. Show that this state is properly normalized; that the certainty condition is satisfied.

9.21 Two noninteracting particles are in a 1-D box and are known to have energy $5h^2/8\,m\,L^2$. Each of their positions are measured to within $dx = 0.01L$.

   a. One is a neutron with spin up in its ground state, the other a spin up proton in its first excited state. What is the probability of finding the neutron with spin up a distance $L/6$ removed from the left wall and the proton with spin up at $L/4$ from that wall?

   b. Both particles are spinless helium atoms. What is the probability of finding one distant $L/6$ from the left wall and the other distant $L/4$ from that wall?

   c. Both particles are neutrons. What is the probability of finding one at $L/6$ from the left wall with spin up and the other $L/4$ from that wall with spin up?

   Answers:
   a. 10^{-4}
   b. $(5 + 2\sqrt{6})10^{-4}/4$
   c. $(5 - 2\sqrt{6})10^{-4}/4$

9.22 The ground-state energy of a single isolated particle in the 1-D box is E_1. Show that the ground state of three noninteracting helium atoms in the box has energy $3E_1$ and that the ground state of three noninteracting electrons has energy $6E_1$. Show that four helium atoms have ground-state energy, $4E_1$, and four electrons, $10E_1$.

Solution: For the helium atom bosons $n_{gnd}(\nu) = 3\delta_{\nu 1}$. This distribution has energy $3E_1$, the lowest possible for which $\hat{Y}_{ij}$ produces $+1$ for any i,j.

   For spin 1/2 electrons the Pauli exclusion principle applies: no two may have the same ν,m values. Thus

$$n_{gnd}(\nu) = \delta_{\nu 1} (\delta_{m, 1/2} + \delta_{m, -1/2}) + \delta_{\nu, 2} \delta_{m,M}$$

The electrons are in the three lowest-energy *different* single-particle states available. This three-electron state is twofold degenerate, has an energy $E = E_1 + E_1 + 4E_1$; and $\hat{Y}_{ij}$ yields -1 for all relevant i's and j's.

9.23 A single particle bound in a 2-D harmonic oscillator well has the following energy-level scheme: The lowest state has energy $\hbar\omega$ and is nondegenerate. The next state has energy $2\hbar\omega$ and is twofold degenerate. The next has energy $3\hbar\omega$ and is threefold degenerate, and so on.

Construct the energy-level diagram showing the ground state and first excited states and their degeneracies for a system of three particles in this well. Consider the cases of distinguishable particles, indistinguishable bosons, and indistinguishable spin 1/2 fermions.

Chapter 10

Stationary-State
Perturbation Theory

10.1 THE ANSWER COMES
FROM NEARLY-THE-ANSWER

Most practical problems in physics are so complex that it is impossible to do exact calculations. One can't obtain exact eigenvalues for all hamiltonians. Some are prohibitively complicated. Perturbation theory is a procedure for making an approximate calculation.

Perturbation theory answers the question: If I can solve part of the problem, how can I estimate the solution to the total problem?

Suppose that the hamiltonian for which I want solutions is $\hat{H}$. And suppose that I know the solutions to the eigenvalue problem for the operator $\hat{H}_0$: I know its eigenstates and eigenvalues. If an eigenvalue of $\hat{H}$ does not differ much from selected $\hat{H}_0$ eigenvalues, then perturbation theory will tell us how to get this unknown $\hat{H}$-eigenvalue using the known properties of $\hat{H}_0$.

The accuracy of a perturbation calculation measures our cleverness in assessing and ordering observables. We order them according to their contribution to the effect we wish to calculate.

The greatest contribution to the energy of the system comes from $\hat{H}_0$; the difference, $\hat{H} - \hat{H}_0$, produces a perturbation. To get reasonably valid solutions for $\hat{H}$ from those available from $\hat{H}_0$, we had better choose $\hat{H}_0$ cleverly. The $\hat{H}_0$ must account for almost all of the phenomenon being examined, leaving only a small remainder to make up the rest of $\hat{H}$.

Small remainder means that the differences between eigenvalues are small compared with the values themselves. Figure 10-1 illustrates the situation.

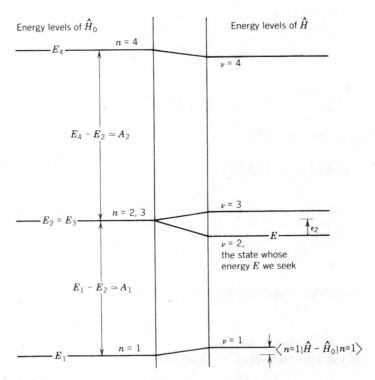

Figure 10.1 **The system governed by $\hat{H}_0$ has energy levels E_n. The perturbation $\hat{H}-\hat{H}_0$ changes the energy-level structure.**

The energy levels generated by the dominant operator, $\hat{H}_0$, are indexed by n, the E_n. Thus

$$\hat{H}_0 \, |n\rangle \; = \; E_n \, |n\rangle \tag{10-1}$$

We know the eigenstates $|n\rangle$ and the energies E_n. We don't know the eigenstates of $\hat{H}$. The vth such state is $|v\rangle$. We focus on this one single particular state of $\hat{H}$, whose energy is E. It is E that we are after. E is the energy of the vth state $|v\rangle$ of $\hat{H}$, so

$$\hat{H} \, |v\rangle \; = \; E \, |v\rangle \tag{10-2}$$

Of all the energies $E_{n=1}, E_{n=2}, E_3 \ldots$ in the spectrum of $\hat{H}_0$, only one, or at most a few, are close to the energy E we are seeking. This is illustrated in Figure 10-1. The state whose energy we seek is shown as $v = 2$. Thus the differences $E_1 - E$ and $E_4 = E$ and $E_5 - E$, and so on are relatively large numbers.

The energy E is close only to the degenerate energy level $E_2 = E_3$: only the energy differences shown as $\varepsilon_2 = E_{n=2} - E$ and $\varepsilon_3 = E_{n=3} - E$ are small.

Of course if the energy we sought were that of the state $v = 1$, then only one energy of the unperturbed system would be close to it. Only $E_{n=1} - E$ would be small; all other energy differences $E_{n \neq 1} - E$ would be large by comparison.

Perturbation theory is built upon the condition that among all the differences $E_n - E$, only a few are small. In the figure the small differences are called ε_n, the large ones A_n. The ratio of any ε to any A is small, and although the number of A's may be indefinitely large, there are only a few ε's.

We shall examine the deductions from perturbation theory before deriving them. An informal proof of these deductions will be outlined later.

Summarized here are the central results of perturbation theory, the results that one uses in practical calculations.

10.2 FIRST-ORDER NONDEGENERATE RESULT: THE NEW OPERATOR BETWEEN OLD STATES

Suppose E is close only to the nondegenerate energy E_N corresponding to the single state $|n=N\rangle$. Figure 10-1 shows such a case at $N = 1$. First-order perturbation theory yields for the energy E

$$E = \langle N|\hat{H}|N\rangle \tag{10-3}$$

This same result may be cast in terms of differences. Exactly equivalent to (10-3) is

$$E - E_N = \langle N|\hat{H} - \hat{H}_0|N\rangle \tag{10-4}$$

Here is what these equations say. Get the matrix element of the full hamiltonian $\hat{H}$, using the basis states of the approximate hamiltonian, $\hat{H}_0$. Having ourselves chosen $\hat{H}_0$, we know these basis states. Evaluate the diagonal matrix element for the particular level $n = N$. If E_N is nondegenerate, this diagonal matrix element will be a good first approximation to the energy E for the state $|v=N\rangle$ of the system characterized by the full hamiltonian.

Here is a simple example. Suppose a one-dimensional harmonic oscillator like that of Chapter 4 has a modified potential energy term: instead of the Hookes' law potential $m\omega^2 x^2/2$ this oscillator feels the potential $m\omega^2 x^2/2 + \beta x^4$. Thus

$$\hat{H} = \hat{p}^2/2m + m\omega^2 \hat{x}^2/2 + \beta \hat{x}^4 \tag{10-5}$$

We may take as $\hat{H}_0$ the traditional harmonic oscillator hamiltonian (Equation 4-3) as long as the operator $\hat{H} - \hat{H}_0 = \beta \hat{x}^4$ produces only perturbations; something small compared to what $\hat{H}_0$ produces. Under this condition there is an energy level of $\hat{H}$ not far different from the Nth level of $\hat{H}_0$. By virtue of (10-3) the energy of this level is

$$E = (N + 1/2)\hbar\omega + \beta\langle N|\hat{x}^4|N\rangle$$

$$= (N + 1/2)\hbar\omega + [6N(N+1) + 3]\,\beta(\hbar/2\,m\,\omega)^2 \qquad (10\text{-}6)$$

The matrix element in this equation has been evaluated by writing $\hat{x}$ in terms of the raising and lowering operators $\hat{a}^\dagger$ and $\hat{a}$, of Equations 4-50 and 4-51 in Chapter 4 (see also Problems 10.1 and 10.2).

Equation 10-6 gives us the first-order perturbation theory approximation to the $v = N$ energy level of $\hat{H}$. This same theory yields a formula for the eigenstate $|v = N\rangle$ itself. In the basis of the $\hat{H}_0$ eigenstates, the formula is

$$\langle n|v = N\rangle = \begin{cases} 1 & \text{if } n = N \\ \langle n|\hat{H}|n = N\rangle/(E - \langle n|\hat{H}|n\rangle) & \text{if } n \neq N \end{cases} \qquad (10\text{-}7)$$

This equation says that $\langle n|v\rangle$ is almost a self-space bracket. In first order $\langle n = N|v\rangle = 1$, but for all $n \neq N$ the bracket is not zero; it is merely small. In the state $|v = N\rangle$ of the hamiltonian $\hat{H}$, the probability of finding $n = N$ is almost a certainty; there are nonzero but small probabilities to find $n \neq N$.

In Problem 10.3 this formula is applied to our perturbed harmonic oscillator example.

10.3 A NULL RESULT IN FIRST ORDER: SUM OVER INTERMEDIATE STATES

Suppose that the first-order calculation yields no result, that $\langle N|\hat{H} - \hat{H}_0|N\rangle = 0$. If, as before, E is close only to the nondegenerate energy E_N corresponding to the single state $|n = N\rangle$, then the energy E will be given by

$$E - E_N = \sum_{\text{all } n \neq N} \frac{\langle N|\hat{H}|n\rangle\,\langle n|\hat{H}|N\rangle}{E_N - E_n} \qquad (10\text{-}8)$$

The sum on the right is over all the states, $|n\rangle$, of $\hat{H}_0$ except the one, $n = N$, being examined. All the denominators are large. The matrix elements in the numerator are, again, those of $\hat{H}$ between states of $\hat{H}_0$. But now all off-diagonal matrix elements connected to N contribute.

As before, the change in energy is given in terms only of known quantities. Both $\hat{H}$ and $\hat{H}_0$ are known, as are all the $|n\rangle$. What are not known are the eigenvalues E_v, of which E is one, and the eigenstates $|v\rangle$.

Equation 10-8 is sometimes called the "*second-order*" correction to E. But it is not, in fact, used in the traditional sense of second order—as a correction to the first order. Rather, it is used when the first-order calculation yields zero, when a null result is obtained in first order. Equation 10-8 is the second first-order correction.

Again a perturbed harmonic oscillator provides a simple example. The perturbation term typical of a mass carrying charge e in an electric field $\mathscr{E}$ is $-e\mathscr{E}\,x$. The shift in an atomic-energy level induced by a laboratory electric field is called the Stark effect. In our oscillator this effect is modeled by the hamiltonian

$$\hat{H} = \hat{p}^2/2\,m + m\,\omega^2\hat{x}^2/2 - \gamma\hat{x} \tag{10-9}$$

Although this hamiltonian may be recast so as to extract exact solutions, it may equally well be treated in perturbation theory if the $-\gamma\hat{x}$ term has a small effect. Problem 10.4 concerns the exact solutions.

The perturbation theory first trial energy correction is $\langle N|\hat{H}-\hat{H}_0|N\rangle$. It is zero for this example. Using the raising and lowering operators, Equations 4-50 and 4-51, you will find that all diagonal matrix elements of $\hat{x}$ are zero.

$$\langle n|\hat{x}|N\rangle = i\sqrt{\hbar/2\,m\,\omega}\,[\sqrt{N}\,\langle n|N-1\rangle - \sqrt{N+1}\,\langle n|N+1\rangle] \tag{10-10}$$

No first-order correction is obtained. We must therefore use the second first-order correction formula, Equation 10-8.

For the hamiltonian $\hat{H}$ of Equation 10-9 in relation to the unperturbed hamiltonian $\hat{H}_0$ of Equation 4-3, a typical term in the sum of (10-8) is

$$\frac{|\langle N|\hat{H}|n\rangle|^2}{E_N - E_n} = [\sqrt{N}\,\langle n|N-1\rangle - \sqrt{N+1}\,\langle n|N+1\rangle]^2\,\frac{\gamma^2}{2\,m\,\omega^2(N-n)} \tag{10-11}$$

The sum over n is easily performed because, for a given N, there are only two nonzero terms: when $n = N-1$ and when $n = N+1$. Carrying out the summation, the result from (10-8) is

$$E = (N+1/2)\hbar\omega - \frac{\gamma^2}{2\,m\,\omega^2} \tag{10-12}$$

The energy of each state is lowered in proportion to the square of the electric field strength.

10.4 DEGENERATE STATES, FIRST ORDER: SET THE DETERMINANT TO ZERO

The energy E_N to which E is close may come from a group of degenerate states $|n=N\rangle$, $|n=N+1\rangle$, $|n=N+2\rangle$, . . . $|n=N+\Gamma\rangle$ where $E_N = E_{N+1} = \ldots = E_{N+\Gamma}$. The degeneracy is $\Gamma+1$-fold. Figure 10-1 shows a twofold degenerate case, $N = 2$, $N + \Gamma = 3$. Because E_2 equals E_3, the energy E is close to both of them.

For this case there is not a single value of E; there are many of them. The $\Gamma+1$

old states can give rise to $\Gamma+1$ values of E. In the figure the two states $|n=2\rangle$ and $|n=3\rangle$, which have the same energy, give rise to two new states $|v=2\rangle$ and $|v=3\rangle$, which have different energies. The perturbation *splits the degeneracy*.

All of the possible values of E are given by setting the following determinant to zero:

$$
\begin{vmatrix}
\langle N|\hat{H}|N\rangle - E & \langle N|\hat{H}|N+1\rangle & \cdot\ \cdot & \langle N|\hat{H}|N+\Gamma\rangle \\
\langle N+1|\hat{H}|N\rangle & \langle N+1|\hat{H}|N+1\rangle - E & \cdot\ \cdot & \cdot \\
\langle N+2|\hat{H}|N\rangle & \langle N+2|\hat{H}|N+1\rangle & & \cdot \\
\cdot & \cdot & & \cdot \\
\cdot & \cdot & & \cdot \\
\cdot & \cdot & & \cdot \\
\langle N+\Gamma|\hat{H}|N\rangle & \cdot & & \langle N+\Gamma|\hat{H}|N+\Gamma\rangle - E
\end{vmatrix}
\tag{10-13}
$$

You set this determinant equal to zero and solve for E. There will be $\Gamma+1$ values of E, because this equation is of $(\Gamma+1)$th order in E. Each value of E corresponds to a state $|v\rangle$. And each such state is a linear combination of all the states $|n=N\rangle$, $|n=N+1\rangle$, ... $|n=N+\Gamma\rangle$ from which the original energy came.

Notice that, except for E, every term in the determinant is known. They are simply matrix elements of $\hat{H}$ in the basis $|n\rangle$. And in the limiting case of no degeneracy, which means that $\Gamma=0$, Equation 10-13 reduces to 10-3, as it must.

For an example to illustrate degenerate-state first-order perturbation theory, we consider a 2-D harmonic oscillator. The unperturbed hamiltonian, $\hat{H}_0$, is

$$
\hat{H}_0 = \hat{p}_x^2/2m + m\omega^2\hat{x}^2/2 + \hat{p}_y^2/2m + m\omega^2\hat{y}^2/2
\tag{10-14}
$$

The eigenstates of this hamiltonian are labeled by two nonnegative integers n_1 and n_2 where

$$
\hat{H}_0|n_1,n_2\rangle = (n_1 + n_2 + 1)\hbar\omega\ |n_1,n_2\rangle
\tag{10-15}
$$

Each energy is merely the sum of two orthogonal 1-D oscillator energies. In x,y-space the eigenstate wave functions are simple products of the Hermite polynomial wave functions of Equation 4-10 (see Problem 10.5).

$$
\langle x,y|n_1,n_2\rangle = \langle x|n_1\rangle\ \langle y|n_2\rangle
\tag{10-16}
$$

The ground state of this 2-D oscillator is nondegenerate. Only one state of the system, $|n_1=0,n_2=0\rangle$, yields the energy $\hbar\omega$.

The first excited state, however, is twofold degenerate. Both the states $|n_1=0,n_2=1\rangle$ and $|n_1=1,n_2=0\rangle$ yield the same energy: $2\hbar\omega$ (see Figure 10-2).

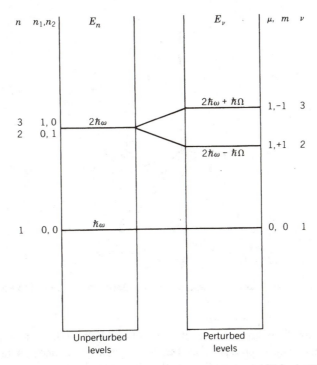

Figure 10.2 **Energy levels of the charged 2-D harmonic oscillator with magnetic field off and on.**

A CHARGE WITH ANGULAR MOMENTUM INTERACTS WITH MAGNETIC FIELD

The 2-D oscillator is a mass moving in the x-y plane under the action of a central force field attracting it to the origin. It can have z-direction angular momentum. Measurements corresponding to $\hat{L}_z$ may be made on it. The results are the L_z spectrum.

Suppose this oscillator mass carried a charge. If a z-direction magnetic field crosses its confining plane, then one direction of angular momentum will be energetically favored over the other. That is because the associated magnetic moment of the oscillator interacts with the magnetic field. The $\mathbf{v} \times \mathbf{B}$ force perceived by the charge manifests itself in the hamiltonian as the magnetic moment $-\mathbf{M} \cdot \mathbf{B}$ interaction. This interaction is the subject of Chapter 6—(compare Figure 10-3 with Figure 6-1). Our charged harmonic oscillator in such a field has the energy

$$\hat{H} = \hat{H}_0 - \Omega\hat{L}_z \tag{10-17}$$

See Equations 6-1 and 6-3 or 7-8 with $\hat{S}_z$ replaced by $\hat{L}_z$.

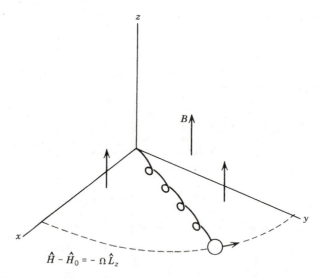

$$\hat{H} - \hat{H}_0 = -\,\Omega\,\hat{L}_z$$

Figure 10.2 Charged-mass 2-D oscillator in a magnetic field.

The magnetic field strength is contained within Ω. It produces a perturbation via the angular momentum possessed by the 2-D harmonic oscillator. A magnetic field splits the degeneracy at $E_{1,0} = E_{0,1} = 2\hbar\omega$. Splitting induced by a magnetic field is called a Zeeman effect.

To find the new energies, we must evaluate the elements $\langle n_1,n_2| - \Omega\hat{L}_z|n_1',n_2'\rangle$ so as to construct the determinant of Equation 10-13. Because we are interested in the levels only near the two $n_1,n_2 = 0,1$ and $1,0$, there are four elements to calculate. Organized in matrix form, here they are.

$$\begin{pmatrix} \langle 0,1|\hat{L}_z|0,1\rangle & \langle 0,1|\hat{L}_z|1,0\rangle \\ \langle 1,0|\hat{L}_z|0,1\rangle & \langle 1,0|\hat{L}_z|1,0\rangle \end{pmatrix} = \begin{pmatrix} 0 & i\hbar \\ -i\hbar & 0 \end{pmatrix} \tag{10-18}$$

The calculation leading to this result is the subject of Problem 10.6. It is best done by expressing the operator $\hat{L}_z$ in terms of the raising and lowering operators for n_1 and n_2.

Having these elements, use them in Equation 10-13 to find

$$\begin{vmatrix} 2\hbar\omega - E & -i\hbar\Omega \\ i\hbar\Omega & 2\hbar\omega - E \end{vmatrix} = 0 \tag{10-19}$$

or

$$E = 2\hbar\omega \pm \hbar\Omega \tag{10-20}$$

This result displays *splitting*. The two original states which yielded one single energy become states of differing energy as a result of the perturbation. The effect is illustrated in Figure 10-1.

Having used perturbation theory to find it, we can be assured of the validity of our result only if $\Omega \ll \omega$.

SPLITTING THE DEGENERACY MAKES NEW EIGENSTATES FROM OLD ONES

The $\Gamma + 1$ eigenstates of $\hat{H}$ (the $|v\rangle$) that result from the $\Gamma + 1$ degenerate eigenstates of $\hat{H}_0$ (the $|n\rangle$) are the elements of the unitary transformation matrix which solves the eigenvalue problem leading to the determinant of (10-13). Having solved that determinant, use any of the E's obtained to solve for the $\langle n|v\rangle$ in the following:

$$
\begin{pmatrix}
\langle N|\hat{H}|N\rangle & \langle N|\hat{H}|N+1\rangle & \cdots & \langle N|\hat{H}|N+\Gamma\rangle \\
\langle N+1|\hat{H}|N\rangle & \langle N+1|\hat{H}|N+1\rangle & & \cdots \\
\cdot & \cdot & \cdots & \cdots \\
\cdot & \cdot & & \cdots \\
\langle N+\Gamma|\hat{H}|N\rangle & \cdot & \cdots & \langle N+\Gamma|\hat{H}|N+\Gamma\rangle
\end{pmatrix}
\begin{pmatrix}
\langle N|v\rangle \\
\downarrow \\
n \\
\downarrow \\
\langle N+\Gamma|v\rangle
\end{pmatrix}
$$

$$
= E
\begin{pmatrix}
\langle N|v\rangle \\
\downarrow \\
n \\
\downarrow \\
\langle N+\Gamma|v\rangle
\end{pmatrix}
\tag{10-21}
$$

Different values of E are enumerated by assigning a value of v to each one. Thus $|v=N\rangle$ corresponds to the lowest E; $|v=N+1\rangle$ corresponds to the next E; and so forth.

Here is how it works in the case of our 2-D harmonic oscillator with magnetic moment. The unitary matrix of elements $\langle n_1,n_2|\mu,m\rangle = \langle n|v\rangle$ that diagonalizes $\langle n_1,n_2|\hat{H}|n_1',n_2'\rangle$ where

$$
\left(\langle n_1\underset{\downarrow}{,}n_2|\hat{H}|\overset{\rightarrow}{n_1'},n_2'\rangle\right) = \begin{pmatrix} 2\hbar\omega & -i\hbar\Omega \\ i\hbar\Omega & 2\hbar\omega \end{pmatrix}
\tag{10-22}
$$

is

$$
\begin{array}{cc}
 & \begin{array}{cc} 1,+1 & \quad 1,-1 \end{array} \\
\begin{array}{c} 0,1 \\[20pt] 1,0 \end{array} &
\begin{pmatrix}
 & \overset{\rightarrow}{} & \\
 & \langle n_1,n_2|\mu,m\rangle & \\
 & \downarrow &
\end{pmatrix}
\end{array}
= \begin{pmatrix} i/\sqrt{2} & -i/\sqrt{2} \\ 1\sqrt{2} & 1/\sqrt{2} \end{pmatrix}
\tag{10-23}
$$

A SINGLE INTEGER ENUMERATES STATES

The states of the unperturbed hamiltonian are labeled by $|n\rangle$ in the formal exposition. As is evident from the example, this single integer represents a group of labels; $|n\rangle = |n_1,n_2\rangle$. The single integer enumerates the states, thus

by the first state $|n=1\rangle$ we mean $|n_1=0,n_2=0\rangle$ and

by the second state $|n=2\rangle$ we mean $|n_1=0,n_2=1\rangle$ and

by $|n=3\rangle$ we mean $|n_1=1,n_2=0\rangle$, and so on.

A multiple-integer label can be set into a one-to-one correspondence with a single-integer label. One integer alone is all that is needed, so in general proofs only one is exhibited, although, in practice, multiple-index labels arise.

States of the new hamiltonian, $\hat{H}$, are enumerated by ν in the formal exposition. This label too can be decomposed into multiple ones.

In our example two degenerate states $n = 2$ and $n = 3$ split into two new states. The one with the lower energy, $E = 2\hbar\omega - \hbar\Omega$, we label $\nu = 2$, and the one with the higher energy, $E = 2\hbar\omega + \hbar\Omega$, we label $\nu = 3$.

In Equation 10-23, a twofold label is shown. The lower-energy $\nu = 2$ state is called $|\mu=1,m=+1\rangle$, and instead of $\nu = 3$ the label $\mu=1, m=-1$ is used.

The reason is simple: the pair of quantum numbers μ and m is physically meaningful, and the ν are not. The pair of integers index the eigenvalues of $\hat{H}_0$—$(\mu + 1)\hbar\omega$—and of $\hat{L}_z$—$m\hbar$—simultaneously. A simultaneous eigenstate of both the unperturbed hamiltonian and of the perturbation is necessarily an eigenstate of the full hamiltonian.

In this particular example a physical labeling of the states can be found to replace the ν-label of mere enumeration. A physical labeling is desirable if one is available. The perturbation result, however, does not depend on finding one. The enumeration label is always available if no other presents itself.

With the understanding that by $|\mu=1,m=+1\rangle$ is meant $|\nu=2\rangle$ and that by $|\mu=1,m=-1\rangle$ is meant $|\nu=3\rangle$, we see that the unitary transformation matrix of Equation 10-23 does, indeed, give us the new states: it gives us the new states in the language of the unperturbed old states.

10.5 Derivation of the Perturbation Theory Results

The practical results of stationary-state perturbation theory are embodied in five equations (10-3), (10-7), (10-8), (10-13), and (10-21). These five equations are the useful ones for practical calculations. Having surveyed these results and examples of their uses, we can now turn to their derivation.

MARK THE ORDER OF THE TERMS

A particularly graphic derivation applies when we limit ourselves to eigenvalue spaces that are finite and denumerable, for then we can visually display the matrices involved. We can use the methods of matrix mechanics in Chapter 6 for calculation. This method of proof doesn't apply to denumerably infinite spectra, but the results obtained turn out to be valid even in those cases.

Consider the eigenvalue equation that we want to solve: (10-2). Write it in the basis n so that it appears in matrix form (see Equation 6-33).

$$\left(\langle n|\hat{H}|\overrightarrow{m}\rangle \right)\left(\langle m|v\rangle \right) = E \left(\langle n|v\rangle \right) \tag{10-24}$$

The $|m\rangle$ and $|n\rangle$ are states in the basis generated by $\hat{H}_0$.

Every element of the $\hat{H}$ matrix is a known quantity. The solutions, E, to this eigenvalue problem are given by setting a determinant equal to zero. Call this determinant D. It contains all the states $|n\rangle$, not just a small subset of degenerate states as in (10-13) (see also Equation 6-37).

$$D = \begin{vmatrix} \langle 1|\hat{H}|1\rangle - E & \langle 1|\hat{H}|2\rangle & \cdots \\ \langle 2|\hat{H}|1\rangle & \langle 2|\hat{H}|2\rangle - E & \cdots \\ \cdot & \cdot & \cdots \\ \cdot & \cdot & \cdots \\ \cdot & \cdot & \cdots \end{vmatrix} = 0 \tag{10-25}$$

If we could solve this determinant exactly, we would have the exact eigenvalues, E. But we can't, and so we must be content to solve it approximately. To do so we must take careful note of what is small and what is large. The justification for the approximation process rests on three key conditions regarding the magnitudes of the terms in the determinant.

1. Compared with the energies E_n and E_m themselves, the off-diagonal elements, $\langle n|\hat{H}|m\rangle$, where $n \neq m$, are *always small*. That is because if we replace $\hat{H}$ by $\hat{H}_0$, we will get dead zero. Using $\hat{H}$ gives something near zero—something small.

2. The matrix element $\langle n|\hat{H}|n\rangle$ is always of the order E_n. Again, using $\hat{H}_0$ instead of $\hat{H}$ yields E_n exactly, and so $\hat{H}$ must produce something near E_n. Hence the diagonal elements of the determinant, $\langle n|\hat{H}|n\rangle - E$, are *almost always large*. They are generally of the order E or E_n.

This is illustrated in Figure 10-1. The distance labeled A_1 is an example: $A_1 = \langle 1|\hat{H}|1\rangle - E \cong E_1 - E_2$, large because the E we are seeking, though near E_2, is far from E_1.

3. The only time that $\langle n|\hat{H}|n\rangle - E$ is not large is for those few states $n = N$, $n = N+1, \ldots, n = N+\Gamma$, for which the energy E_n is near the E we are seeking.

In this case the diagonal element $\langle n|\hat{H}|n\rangle - E$ is labeled by ε to mark it as being small.

What we must do is to solve (10-25) keeping track of the relative magnitude of things. We can do it pictorially.

Consider this schematic depiction of the determinant we want to evaluate:

$$
D = \begin{vmatrix}
\varepsilon & \kappa & \kappa & \kappa & \cdots \\
\kappa & A & \kappa & \kappa & \cdots \\
\kappa & \kappa & A & \kappa & \cdots \\
\kappa & \kappa & \kappa & A & \cdots \\
\cdot & \cdot & \cdot & \cdot & \cdots
\end{vmatrix}
\tag{10-26}
$$

The position indices on the elements are suppressed, but what each element represents is evident from its position. The κ in position 1,2 represents $\langle 1|\hat{H}|2\rangle$; that in position 1,3 represents $\langle 1|\hat{H}|3\rangle$, and so on.

Thus both the determinant of (10-25) and this one are the same; both are D. The form in (10-26) focuses our attention on the relative magnitudes of the elements rather than on their individuality. All Greek alphabet lowercase elements are small compared with Latin uppercase elements.

The capital A's are the large differences $\langle n|\hat{H}|n\rangle - E$ along the diagonal of (10-25). If the E we seek is near E_1, the small ε will go in the 1,1 position. That is the case illustrated. The lower case κ's are all the off-diagonal elements of (10-25). They are expected to be of the order of ε, small compared with any A.

Now we shall make a diagrammatic evaluation of this $M \times M$ determinant.

Imagine that you have M stepping-stones. You build a path by laying stepping-stones on selected elements of the determinant. Put the first stone in the top row at some column head. Put the next stepping-stone anywhere in row 2 other than in the same column as the first stone. The next stone, 3, goes in row 3, but in a column not already occupied by either of the first two stones: it gets its own private column, as do all the stones. No two stones occupy the same column or the same row.

Each particular arrangement of M stepping-stones constitutes a path traversing the entire matrix from top to bottom. There are $M!$ such paths possible.

All the possible paths can be generated from a single one. All paths are variations on the main diagonal path. The diagonal is the elemental one from which other paths may be built.

Each stone can be placed anywhere in its row. Its column can be changed, but it cannot be moved out of its row. Every exchange of stones among columns builds another path.

Figure 10-4 illustrates it for the case $M = 4$.

Path 2 is manufactured from the diagonal path (path 1) by exchanging the columns in which the first and second stones are placed. You place the first stone in column 2 and the second in column 1 instead of the other way round.

Path 3 derives from the diagonal if the first and second columns are exchanged and then the first and fourth. It is *two exchanges removed* from the diagonal path.

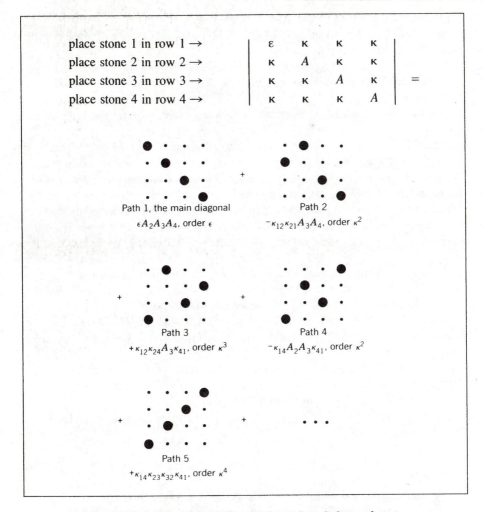

place stone 1 in row 1 →

place stone 2 in row 2 →

place stone 3 in row 3 →

place stone 4 in row 4 →

$$\begin{vmatrix} \varepsilon & \kappa & \kappa & \kappa \\ \kappa & A & \kappa & \kappa \\ \kappa & \kappa & A & \kappa \\ \kappa & \kappa & \kappa & A \end{vmatrix} =$$

Path 1, the main diagonal

$\varepsilon A_2 A_3 A_4$, order ε

Path 2

$-\kappa_{12}\kappa_{21}A_3A_4$, order κ^2

Path 3

$+\kappa_{12}\kappa_{24}A_3\kappa_{41}$, order κ^3

Path 4

$-\kappa_{14}A_2A_3\kappa_{41}$, order κ^2

Path 5

$+\kappa_{14}\kappa_{23}\kappa_{32}\kappa_{41}$, order κ^4

Figure 10.4 Five of the 24 paths for a 4 × 4 determinant.

The fourth path in the figure is simply the column interchange of the first and fourth stone in the diagonal array.

By means of a series of exchanges of the stones among columns, all of the paths may be built from the diagonal one.

In any one of these paths there are M steps: M stepping-stones are used. Multiply together all of the matrix elements covered by stepping-stones marking a particular path. To each path there corresponds a matrix element product with a sign in front of it.

The sign (+ or −) of the product is fixed by the number of column exchanges necessary to manufacture the path from the diagonal. If it takes an even number of exchanges, use a plus sign; if it takes an odd number, use a minus sign.

Examples are given in Figure 10-4.

Path 1 uses stones covering ε, A_2, A_3, and A_4 so this path gives rise to the product $+ \varepsilon A_2 A_3 A_4$: a plus sign for no exchanges.

The product corresponding to path 2 is $-\kappa_{12}\kappa_{21}A_3A_4$: a minus sign for one exchange.

To produce path 3 from the diagonal requires two exchanges. Its product is $+ \kappa_{12}\kappa_{24}A_3\kappa_{41}$.

A mathematical quantity is assigned to each path. Using these, the value of the determinant is easily expressed. It is the sum over all paths of each of these path-quantities! It is the sum of all the product terms fabricated from the paths. That sum coincides exactly with the one generated by the prescribed standard method for evaluating a determinant. We have merely given the sum a diagramatic interpretation. Diagrams make it easy to gather together terms by their order. You can do it visually.

The strength of any one product term in the sum is entirely governed by the number of A's it contains. The greater the number is of A's in the path, the lower the order will be. The orders ascend in powers of κ according to how few A's there are in the term.

The diagonal path gives the first-order term: the lowest-order term. Only paths containing all-the-A's-but-one contribute to the next order. The more the path deviates from the main diagonal, the higher the order will be.

Figure 10-4 illustrates the idea. The lowest-order term possible is that from the diagonal path, 1: the only path that embraces all three A's. It yields the first-order term, order ε or κ.

Path 2 contains one less A. Its associated product is of order κ^2, second order. There are only three paths that contain two A's. Path 4 also contains two A's.

Path 5 contains no A's. It produces a term of order κ^4.

For the general $M \times M$ determinant, in first-order approximation, its value is the diagonal term.

$$D \text{ (nondegenerate, first order)} = \varepsilon A_2 A_3 ... A_M = \varepsilon \Pi \qquad (10\text{-}27)$$

The product of A's includes all $M - 1$ of them. This product we call Π.

Setting this evaluation of the determinant equal to zero produces the nondegenerate first-order perturbation theory result set down as Equation 10-3. This equation is the statement that $\varepsilon = 0$.

Because it is the diagonal path alone that fixes this order, the result is valid for an ε in any position on the diagonal. It is good for $\varepsilon = \langle N|\hat{H}|N\rangle - E$ and not only for the case illustrated in (10-26) where $N = 1$.

We now turn to the next order (see Figure 10-5). By inspecting typical paths that include all but one of the A's, it is easy to see that they follow a pattern: each one has the recipe $- \kappa_{Nn}\kappa_{nN}\Pi/A_n$.

This is just the formula for exchanging the column of the nth stone covering an A in the diagonal with that of the Nth stone, the one covering ε. That is the only way of creating a path with all-but-one-A in it. The one A that is removed leaves the product, and so instead of Π we get Π/A_n.

The second-order contribution to the evaluation of the determinant is the sum over

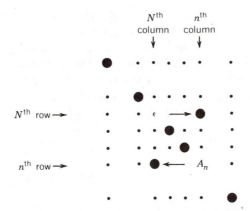

Figure 10-5 The structure of paths where stones cover all the A's but one. Stones that formerly covered ε and A_n on the diagonal now cover κ_{Nn} and κ_{nN}. The remaining stones cover all the A's but A_n. They remain undisturbed. The path product is $-\kappa_{Nn}\kappa_{nN}\Pi/A_n$.

all possible all-but-one-A paths. A different path corresponds to every n between 1 and M, excluding the n marking the ε; $n \neq N$.

Now we can portray the determinant out to the second order. It is

$$D \text{ (nondegenerate, second order)} = [\varepsilon - \sum_{n \neq N} (\kappa_{Nn}\kappa_{nN}/A_n)]\Pi \qquad (10\text{-}28)$$

That this D must equal zero is just what the second first-order perturbation theory formula, given in (10-8), expresses.

We return now to the perturbation approximation for the states of the system. We must first assess the relative orders of magnitude of matrix elements, just as we did for the determinant elements.

In zeroth order, $\langle n|v = N\rangle$ is precisely a self-space bracket: $\langle N|N\rangle = 1$, $\langle n \neq N|N\rangle = 0$. Because it must vanish when κ does, we expect in first approximation that $\langle n \neq N|v = N\rangle$ is of order κ or higher. From this fact together with the certainty condition, it follows that $\langle n = N|v\rangle = 1 - \text{order } \kappa^2$. Hence $\langle n = N|v = N\rangle$ remains at unity to order κ.

With this in mind, the new states $|v\rangle$ follow from this observation: the form for the determinant D in Equation 10-28 is generated by this matrix equation:

$$\begin{pmatrix} \varepsilon & \kappa & \kappa & \kappa & \kappa \dots \\ \kappa & A & 0 & 0 & 0 \dots \\ \kappa & 0 & A & 0 & 0 \dots \\ \kappa & 0 & 0 & A & 0 \dots \\ \cdot & \cdot & \cdot & \cdot & \cdot \\ \cdot & \cdot & \cdot & \cdot & \cdot \end{pmatrix} \begin{pmatrix} 1 \\ \langle n|v = 1\rangle \\ \big| \\ \big\downarrow \end{pmatrix} = 0 \qquad (10\text{-}29)$$

Again the ε is shown at position $N = 1$ in the illustration, but it could be at any N. The row and column of κ's intersect it wherever it is.

Equation 10-29 is simply that of (10-24) recast so that only the terms that could contribute out to second order are displayed. Only stones laid on the nonzero positions shown generate paths valid out to second order. To this order the matrix equation of (10-29) yields the correct result, precisely Equation 10-28.

Equation 10-29 allows us to calculate immediately each of the amplitudes $\langle n|v \rangle$. You just carry out the matrix multiplication; set the nth element of the resulting column matrix to zero. For the case $N = 1$ illustrated in the equation, you find

$$(n \neq 1) \qquad \kappa_{n1} + A_n \langle n|v=1 \rangle = 0 \qquad\qquad (10\text{-}30)$$

For the general case the result is

$$(n \neq N) \qquad \kappa_{nN} + A_n \langle n|v=N \rangle = 0 \qquad\qquad (10\text{-}31)$$

Putting in the meaning of κ_{nN} and A_n by comparing (10-26) with (10-25), you arrive precisely at the formal result, Equation 10-7.

When degeneracy is present, the matrix equation that portrays the lowest order is like this one:

$$
\begin{pmatrix}
\varepsilon_1 & \kappa & \kappa & 0 & 0 & \cdots \\
\kappa & \varepsilon_2 & \kappa & 0 & 0 & \cdots \\
\kappa & \kappa & \varepsilon_3 & 0 & 0 & \cdots \\
0 & 0 & 0 & A & 0 & \cdots \\
0 & 0 & 0 & 0 & A & \cdots \\
\vdots & \vdots & \vdots & \vdots & \vdots &
\end{pmatrix}
\begin{pmatrix}
\langle 1|v \rangle \\
\langle 2|v \rangle \\
\langle 3|v \rangle \\
0 \\
0 \\
\vdots
\end{pmatrix} = 0 \qquad (10\text{-}32)
$$

A threefold degeneracy at positions 1, 2, and 3 is shown. But no matter what the degeneracy nor where it occurs along the diagonal this rule applies: The matrix has nonzero terms only along the diagonal and in a square of terms surrounding the ε's. These mark the states $n = N, N+1, \ldots N+\Gamma$ for which all the energies are equal and therefore near E.

You prove that (10-32) is (10-24) portrayed correctly to first-order by considering the paths that contribute to the first-order evaluation of the associated determinant. Every such path must include all of the A's. Only paths with stones covering the elements shown as nonzero in (10-32) can contribute. All other elements may be taken as zero, as they never get covered by stones in the first-order calculation; no first-order path traverses them.

The solution to this equation involves only the matrix pertaining to the degenerate states. Diagonalizing that part alone diagonalizes the whole matrix. The principles enunciated in (10-13) and (10-21) amount just to this diagonalization procedure.

PROBLEMS

10.1 Using the harmonic oscillator raising and lowering operators, Equations 4-50 and 4-51, calculate the matrix element $\langle N|\hat{x}^4|N\rangle$ and thus establish the result shown as Equation 10-6 of the text.

10.2 Perturbation theory results have only a limited domain of validity: $\varepsilon/A \ll 1$. Show that for the perturbed harmonic oscillator of the previous problem, the result (10-6) is only valid for low-lying levels, ones where

$$[6N(N+1)+3]\beta \ll \frac{4\omega^3 m^2}{\hbar}$$

10.3 Apply Equation 10-7 to calculate the state $|v=N\rangle$ to first order for the perturbed harmonic oscillator of the previous two problems.

Answer (partial):

$$\langle n \neq N|v=N\rangle = \frac{\beta\langle n|\hat{x}^4|N\rangle}{E - E_n} = 0 \text{ unless } n = N\pm2, N \pm 4$$

From the result in the previous problem, the denominator

$$A_n = E - \langle n|\hat{H}|n\rangle \cong E - E_n \cong E_N - E_n$$

as is shown in Figure 10-1.

10.4 Solve the eigenvalue problem for the hamiltonian of Equation 10-9 exactly. To do it, argue that this hamiltonian may be recast as

$$\hat{H} = \frac{\hat{p}^2}{2m} + \frac{m\omega^2}{2}\left(\hat{x}-\frac{\gamma}{m\omega^2}\right)^2 - \frac{\gamma^2}{2m\omega^2}$$

Then show that this operator has eigenvalues

$$E = \left(n + \frac{1}{2}\right)\hbar\omega - \frac{\gamma^2}{2m\omega^2}$$

It follows because just as $[\hat{x},\hat{p}] = i\hbar$, so does $[\hat{x} - \gamma/m\omega^2, \hat{p}] = i\hbar$; a displacement of the origin's position doesn't affect the energy levels of the unperturbed simple harmonic oscillator.

10.5 If

$$\langle x|\frac{\hat{p}_x^2}{2m} + \frac{m\omega^2}{2}\hat{x}^2 |n_1\rangle = \left(n_1 +\frac{1}{2}\right)\hbar\omega\, \langle x|n_1\rangle$$

and

$$\langle y|\frac{\hat{p}_y^2}{2m} + \frac{m\omega^2}{2}\hat{y}^2 |n_2\rangle = \left(n_2 +\frac{1}{2}\right)\hbar\omega\, \langle y|n_2\rangle$$

show that the product state of (10-16) solves (10-15) to produce the energy spectrum shown there.

10.6 The operators $\hat{x}$ and $\hat{p}_x$ may be written in terms of the raising and lowering operators $\hat{a}^\dagger$ and $\hat{a}$. There must be operators $\hat{b}^\dagger$ and $\hat{b}$ which are the raising and lowering operators in terms of which $\hat{y}$ and $\hat{p}_y$ may be written.

$$\sqrt{\hbar\omega}\ \hat{b} = \frac{\hat{p}_y}{\sqrt{2\,m}} - i\omega\sqrt{\frac{m}{2}}\ \hat{y} \qquad \text{and so on}$$

Keeping in mind that both $\hat{a}$ and $\hat{a}^\dagger$ commute with $\hat{b}$ and $\hat{b}^\dagger$, show that the 2-D harmonic oscillator hamiltonian of (10-14) may be written in terms of the $\hat{a}$ and $\hat{b}$ operators as

$$\hat{H}_0 = \hbar\omega\,[\hat{a}^\dagger\hat{a} + \hat{b}^\dagger\hat{b} + 1\,]$$

and that the angular momentum operator, $\hat{L}_z = \hat{x}\hat{p}_y - \hat{y}\hat{p}_x$, in terms of these same operators, comes out to be

$$\hat{L}_z = i\hbar[\,\hat{a}\hat{b}^\dagger - \hat{a}^\dagger\hat{b}]$$

Using this result, calculate the matrix of elements $\langle n_1,n_2|\hat{L}_z|n_1', n_2'\rangle$ as shown in Equation 10-18.

Epilogue

So What?

The entire machinery of quantum mechanics resides in a few principles and rules: they are summarized inside the cover of this book. In the pages between the covers these rules are explained and illustrated by application to examples. To use them competently is to have learned quantum mechanics. So what?

CALCULATE THE RESULTS OF EXPERIMENTS

The most immediate answer is the practical one. With quantum mechanics you calculate the results expected from physical experiments. Calculations in physics are founded on quantum mechanics.

WRITE DOWN THE ANSWER, THEN DO THE COMPUTATION

Quantum mechanical calculations follow a simple two-step pattern: you *write down the answer*, and then you *do the computation*.

The reason is that the theoretical structure captures the entirety of physical measurements on the elements of nature! The questions one asks of nature are matters of experimental measurement; these can all be cast as Dirac brackets. The experimental preparations are in the ket on the right; the experimental results are in the bra on the left. The bra-ket is the amplitude for the experimental event, the answer.

When doing a problem in quantum mechanics you just write down the answer. The answer is a Dirac bracket, a unitary transformation-matrix element.

THE OPERATOR AND THE BRA-KETS: MAKE THEM COMPUTATIONALLY FRIENDLY

Having written down the answer, you must compute it. An indispensable step in this process is the evaluation of matrix elements: the evaluation of an operator bracketed between states. You must process this quantity so that you can compute it. The key hurdle is to make the states in the brackets and the bracketed operator *computationally friendly*.

For any operator you know its effect on some states: those are what I mean by the computationally friendly ones. For any state, you know the effect of at least some operators on it: the computationally friendly ones.

In exploring the quantum mechanical description of nature one acquires a reference catalogue of operator–state relationships: all of these are the computationally friendly ones. The equations in this book are entries in that reference catalogue.

To calculate a bracketed operator matrix element, you recast it so as to dip into your reference catalogue of known relationships. Out of these you compose an answer.

There are only two ways of doing the recasting: (1) via a ket-bra sum, change the states into new ones computationally friendly with the operators; or (2) via operator relationships, change the operator into one that is computationally friendly to the states. You change the states to suit the operator or change the operator to suit the states. They are suited to each other when you can refer to your reference catalogue for the answers, when everything is computationally friendly.

THE KET-BRA SUM RECASTS THE STATES

To suit the states to the operator you recast the states with a ket-bra sum. This corresponds to viewing the calculation in another language. You change your reference basis to a new one in which the operator's effects are familiar. You alter your measurement space viewpoint to accommodate the experimental apparatus.

A simple illustration is provided by the matrix element of the operator, $\hat{S}_y$, bracketed between states of $\hat{S}_z$.

$$\langle m = 1/2 | \hat{S}_y | m = 1/2 \rangle = \sum_{\mu} \mu \hbar \langle m = 1/2 | \mu \rangle \langle \mu | m = 1/2 \rangle = 0 \qquad \text{(A-1)}$$

The ket-bra sum recasts the states into $\hat{S}_y$-friendly ones: you view the matrix element in the space of the $\hat{S}_y$ measurement operator. Using the $\langle m = 1/2 | \mu \rangle$ of (6-30) you will find that the sum is zero.

Another illustration is provided by the harmonic oscillator matrix element $\langle n = 1 | \hat{x} | n = 2 \rangle$. A ket-bra sum recasts the states into $\hat{x}$-operator friendly ones.

$$\langle n=1|\hat{x}|n=2\rangle \; = \; \int_{-\infty}^{+\infty} dx \; x \; \langle x|1\rangle^* \; \langle x|2\rangle \; = \; \text{hard work} \qquad \text{(A-2)}$$

The functions $\langle x|n\rangle$ are those given in Equation 4-10. Using these the computation is a matter of carrying out the integration.

OPERATOR RELATIONS RECAST OPERATORS

To suit the operator to the states you recast the operator. To do it requires connections among operators; you must know how the bracketed operator relates to other operators computationally friendly to the bra-kets.

The same two illustrative matrix elements used above, when evaluated in this way, are:

$$\langle m=1/2|\hat{S}_y|m=1/2\rangle \; = \; \frac{1}{2i} \langle m=1/2|\hat{S}_+ - \hat{S}_-|m=1/2\rangle \; = \; 0 \qquad \text{(A-3)}$$

and

$$\langle n=1|\hat{x}|n=2\rangle \; = \; \langle n=1|\hat{a} - \hat{a}\dagger|n=2\rangle i\sqrt{\hbar/2\,m\,\omega} \; = \; i\sqrt{\hbar/\,m\,\omega} \qquad \text{(A-4)}$$

The operators are recast instead of the states. The computation is the application of reference catalogue items: Equations 4-66, 4-67, 6-11, 6-12, and 6-13.

EVERY OPERATOR MUST MESH WITH THE ENTIRETY OF PHYSICS

Operator relations are the fundamental substance of physics. They connect the outcomes of different experiments. They test the logic of physical understanding. The discovery of formerly unsuspected connections among operators is the most honored accomplishment in physics.

The foremost example of such a connection concerns the hamiltonian operator, $\hat{H}$, the one that must produce, for any system, its energy spectrum. How $\hat{H}$ is connected to the operators $\hat{r}$ and $\hat{p}$, and $\hat{M}$ are statements of physical law. Without these connections, energy measurement results could not be analyzed; they would have no meaning.

Another example of such a connection is that the magnetic moment of the electron.

A measurable assigned the operator $\hat{\mathbf{M}}$ is connected to an angular-momentum-like operator $\hat{\mathbf{S}}$.

The Stern–Gerlach apparatus confirms the relation; it measures M_z directly. But every physical observable in which $\hat{\mathbf{M}}$ partakes is related to $\hat{\mathbf{S}}$. The relation affects a myriad of physical effects. All atomic energy levels depend on the form for $\hat{H}$ which, in turn, contains the $\hat{\mathbf{M}}, \hat{\mathbf{S}}$ proportionality. That this relationship is found to be consistent with the great variety of recorded measurements is the reason that we accept it.

Each operator must mesh with the entirety of physics.

THE MEASURE OF UNDERSTANDING: TO EMBRACE MUCH WITH LITTLE

The quantitative explanation for the whole periodic table of the elements, for all of conventional chemistry, and for much of contemporary physics involves only the operators studied in this book. Great recent advances in physics are concerned with the operators that describe a coherent order in the melange of subatomic particles found in nature. The symmetries of nature expressed by this order are described by operator connections. The spectrum of particles seen is the space of eigenvalues generated by operators.

All advances in physics are completely consonant with the structure of quantum mechanics. No evidence exists that puts this structure in any doubt. New operator connections do not test the validity of quantum mechanics. Only ones that confirm its validity are accepted into physics.

Quantum mechanics describes nature correctly and incontrovertibly. It fits all the experimental results ever compiled! Its mathematical structure gives all the right answers. We must accept that it is an accurate portrayal of nature. The image portrayed derives from its mathematical structure. Critical to that image is this idea: that all we can know of reality is contained in measurement data.

WHAT YOU KNOW: MEASUREMENT DATA

We began this study with the dictum, "What You Measure is What You Know." This was an underlying theme until Chapter 8. There we found that from a measurement on one part of a system, you can deduce, without measuring it, the state of another part. Thus what you know is more than what you measure. A component of what you know are things deduced; you need not have personally performed all the measurements.

IN THE KET IS WHAT YOU KNOW

Nevertheless, what you know is a set of measurement data. What resides in the ket is what you know, the known values of a set of measurables. The values recorded in the ket are points in a set of measurement spaces.

The sum total of our verifiable perceptions are points in a compendium of measurement spaces.

ONLY IN STATES OF POSITION HAS IT POSITION

Quantum mechanics teaches us to view the universe as measurement spaces, not physical position space. Position space is not sacred among measurement spaces; it is only a profane one among many of them.

Only if it is in a state of position does something have a position. If it is in the state $|p\rangle$, then it exists, not in x-space, but only in p-space! It is to be found at a point in p-space—with certainty. Its ket label, $|p=P\rangle$, is its essential property, its only identity in reality.

Measurement data are the tracks of a system in the physical world. In the ket they define the condition of the system; they record its tracks on reality.

MAKING A MEASUREMENT ANNIHILATES A UNIVERSE

To make the measurement, $\hat{x}$, forces a result, x. It annihilates the occupation point in p-space and creates one in x-space. The universe of p-space measurements disappears. A new reality replaces it, one of position. The universe is changed by what you choose to examine about it!

The cover of this book provides us with an analogue. You can view it as a white design on a black background. Now change your point of view. See it as black ink impressed upon white paper. When viewed in this new way, the old way is automatically destroyed. The two views are incompatible. It is impossible to view the design in both ways at once!

So is it impossible to view a particle in both x-space and p-space at once. These are also incompatible.

The analogy is not complete, however. When you revert to the white-on-black point of view, you see the design again as it was. But a return to a momentum space measurement will not find the particle again with its former momentum!

IT'S A VECTOR REGARDLESS OF ITS REPRESENTATION

The state of a system is often called its *state vector*. It is a unit vector whose direction is a basis space axis. It defines a basis space: its own basis space is its home space.

The analogy between vectors and states is illustrated in Figure E-1.

A state vector may be represented in many different foreign bases. Dirac brackets correspond to projections of the vector onto axes of the basis you choose. The state vector can be represented in any suitable basis. However, its existence is not a matter of the basis in which you choose to view it. In the premeasurement lower circle of the figure, the state vector is $|r\rangle$. This is the state vector independently of whether or not you choose to describe $|r\rangle$ in the q-space basis.

To each different basis corresponds a measurement operator. It generates the unit vectors along each of the axes in that basis.

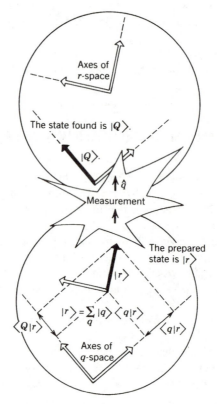

Figure E.1 A vector has reality, regardless of its representation.

Making the measurement destroys the original state vector and produces a new state vector. In Figure E-1, after making a measurement corresponding to the operator $\hat{q}$, the original state vector, $|r\rangle$, is destroyed, and a new state vector, $|q = Q\rangle$, is created. It is shown in the upper circle. Like $|r\rangle$, the state $|Q\rangle$ is a unit vector. It is a unit vector along one of the axes in the home space of the measurement-just-made-operator, $\hat{q}$.

IN THE BRA IS WHAT YOU SAY

In the ket is *what we know*. In the bra is its representation; *how we speak of what we know*. The bracket is the ket-state described in bra-words. But the labels used in the bra are the same kind as those in the ket: measurement data. Thus the words in the languages we use to describe perceptions (states) themselves issue from perceptions. We speak measurement language.

One's spoken language contains a multitude of measurement spaces. Word language grows in tandem with new perceptions. Each new finding introduces a new spectrum of words: a new basis space is generated by the new finding's operator.

THE SIGNAL OF UNDERSTANDING: PERCEIVING NATURE AS NATURAL

It is possible to feel uncomfortable with the picture of reality that issues from quantum mechanics. Much of it appears to contradict the intuition of direct physical experience: it seems to defy understanding. Yet the purpose of studying the physical world is surely to gain understanding.

The quest for understanding hangs on the meaning of the word *understanding*.

One understands when one perceives a satisfying structure. For a scientist, a satisfying structure is one that, with few axioms, accounts for and matches the behavior of nature.

Understanding is the perception of nature as natural.

You have understood something when its behavior is just what you expect it to be, when it appears that *what is, ought to be*.

Understanding is lacking when things seem to be different from what they ought to be. That incontrovertible facts appear to make no sense reveals a faulty structure of understanding. The structure is no longer satisfying.

Under these conditions to achieve understanding requires a revision in the conceptual structure. Recognizing that they are also your prejudices, you question your axioms. You cast off some old ones and take on some new ones. You alter your philosophy. To achieve understanding is to go through a conceptual metamorphosis.

UNDERSTANDING MATERIALIZES
BY CONCEPTUAL METAMORPHOSIS

The history of physics provides striking examples of the process.

Consider the introduction of gravity, that all masses pull on one another, even from afar. Newton called it an action-at-a-distance force. The phrase echoes the incomprehensibility of the concept to people of Newton's time. Only between touching bodies could forces be envisioned. That forces could reach out invisibly from one body to pull on another seemed beyond comprehension. It was not understood.

A century passed before the term *force field* was introduced to describe action-at-a-distance forces. Today every child is familiar with force fields. What in Newton's time puzzled the minds of sages is now the comfortable self-evident perception of little boys and girls.

What happened to the profound questions perceived by the sages?

They lost their meaning; they became meaningless. No one now expects that forces require touching. We recognize the expectation to be an unfounded prejudice. The problem of how forces could act at a distance was never resolved. It was dissolved. A conceptual metamorphosis dissolved it.

Understanding consisted in perceiving the question to be meaningless.

According to quantum mechanics, one cannot know both position and momentum simultaneously. Why are we denied this knowledge?

To perceive this question as inherently meaningless is what quantum mechanics teaches us.

What has a definite momentum does not have a position. To demand both together is like demanding dry wetness. It is denied us not by nature's perversity but, rather, because the demand is meaningless.

When the picture of nature that Newton presented proved to be indisputably correct, the sages of his time had to undergo a conceptual metamorphosis. Modern physics confronts us with the same task: to embrace the picture of nature presented by quantum mechanics.

It has been my aim in this book to paint that picture.

Index

A CATALOG OF SELECTED
DOVER BOOKS
IN SCIENCE AND MATHEMATICS

Astronomy

BURNHAM'S CELESTIAL HANDBOOK, Robert Burnham, Jr. Thorough guide to the stars beyond our solar system. Exhaustive treatment. Alphabetical by constellation: Andromeda to Cetus in Vol. 1; Chamaeleon to Orion in Vol. 2; and Pavo to Vulpecula in Vol. 3. Hundreds of illustrations. Index in Vol. 3. 2,000pp. 6⅛ x 9¼.
Vol. I: 0-486-23567-X
Vol. II: 0-486-23568-8
Vol. III: 0-486-23673-0

EXPLORING THE MOON THROUGH BINOCULARS AND SMALL TELESCOPES, Ernest H. Cherrington, Jr. Informative, profusely illustrated guide to locating and identifying craters, rills, seas, mountains, other lunar features. Newly revised and updated with special section of new photos. Over 100 photos and diagrams. 240pp. 8¼ x 11. 0-486-24491-1

THE EXTRATERRESTRIAL LIFE DEBATE, 1750–1900, Michael J. Crowe. First detailed, scholarly study in English of the many ideas that developed from 1750 to 1900 regarding the existence of intelligent extraterrestrial life. Examines ideas of Kant, Herschel, Voltaire, Percival Lowell, many other scientists and thinkers. 16 illustrations. 704pp. 5⅜ x 8½. 0-486-40675-X

THEORIES OF THE WORLD FROM ANTIQUITY TO THE COPERNICAN REVOLUTION, Michael J. Crowe. Newly revised edition of an accessible, enlightening book recreates the change from an earth-centered to a sun-centered conception of the solar system. 242pp. 5⅜ x 8½. 0-486-41444-2

A HISTORY OF ASTRONOMY, A. Pannekoek. Well-balanced, carefully reasoned study covers such topics as Ptolemaic theory, work of Copernicus, Kepler, Newton, Eddington's work on stars, much more. Illustrated. References. 521pp. 5⅜ x 8½.
0-486-65994-1

A COMPLETE MANUAL OF AMATEUR ASTRONOMY: TOOLS AND TECHNIQUES FOR ASTRONOMICAL OBSERVATIONS, P. Clay Sherrod with Thomas L. Koed. Concise, highly readable book discusses: selecting, setting up and maintaining a telescope; amateur studies of the sun; lunar topography and occultations; observations of Mars, Jupiter, Saturn, the minor planets and the stars; an introduction to photoelectric photometry; more. 1981 ed. 124 figures. 25 halftones. 37 tables. 335pp. 6½ x 9¼. 0-486-40675-X

AMATEUR ASTRONOMER'S HANDBOOK, J. B. Sidgwick. Timeless, comprehensive coverage of telescopes, mirrors, lenses, mountings, telescope drives, micrometers, spectroscopes, more. 189 illustrations. 576pp. 5⅝ x 8¼. (Available in U.S. only.)
0-486-24034-7

STARS AND RELATIVITY, Ya. B. Zel'dovich and I. D. Novikov. Vol. 1 of *Relativistic Astrophysics* by famed Russian scientists. General relativity, properties of matter under astrophysical conditions, stars, and stellar systems. Deep physical insights, clear presentation. 1971 edition. References. 544pp. 5⅝ x 8¼. 0-486-69424-0

Chemistry

THE SCEPTICAL CHYMIST: THE CLASSIC 1661 TEXT, Robert Boyle. Boyle defines the term "element," asserting that all natural phenomena can be explained by the motion and organization of primary particles. 1911 ed. viii+232pp. 5⅜ x 8½.
0-486-42825-7

RADIOACTIVE SUBSTANCES, Marie Curie. Here is the celebrated scientist's doctoral thesis, the prelude to her receipt of the 1903 Nobel Prize. Curie discusses establishing atomic character of radioactivity found in compounds of uranium and thorium; extraction from pitchblende of polonium and radium; isolation of pure radium chloride; determination of atomic weight of radium; plus electric, photographic, luminous, heat, color effects of radioactivity. ii+94pp. 5⅜ x 8½.   0-486-42550-9

CHEMICAL MAGIC, Leonard A. Ford. Second Edition, Revised by E. Winston Grundmeier. Over 100 unusual stunts demonstrating cold fire, dust explosions, much more. Text explains scientific principles and stresses safety precautions. 128pp. 5⅜ x 8½.   0-486-67628-5

THE DEVELOPMENT OF MODERN CHEMISTRY, Aaron J. Ihde. Authoritative history of chemistry from ancient Greek theory to 20th-century innovation. Covers major chemists and their discoveries. 209 illustrations. 14 tables. Bibliographies. Indices. Appendices. 851pp. 5⅜ x 8½.   0-486-64235-6

CATALYSIS IN CHEMISTRY AND ENZYMOLOGY, William P. Jencks. Exceptionally clear coverage of mechanisms for catalysis, forces in aqueous solution, carbonyl- and acyl-group reactions, practical kinetics, more. 864pp. 5⅜ x 8½.
0-486-65460-5

ELEMENTS OF CHEMISTRY, Antoine Lavoisier. Monumental classic by founder of modern chemistry in remarkable reprint of rare 1790 Kerr translation. A must for every student of chemistry or the history of science. 539pp. 5⅜ x 8½.   0-486-64624-6

THE HISTORICAL BACKGROUND OF CHEMISTRY, Henry M. Leicester. Evolution of ideas, not individual biography. Concentrates on formulation of a coherent set of chemical laws. 260pp. 5⅜ x 8½.   0-486-61053-5

A SHORT HISTORY OF CHEMISTRY, J. R. Partington. Classic exposition explores origins of chemistry, alchemy, early medical chemistry, nature of atmosphere, theory of valency, laws and structure of atomic theory, much more. 428pp. 5⅜ x 8½. (Available in U.S. only.)   0-486-65977-1

GENERAL CHEMISTRY, Linus Pauling. Revised 3rd edition of classic first-year text by Nobel laureate. Atomic and molecular structure, quantum mechanics, statistical mechanics, thermodynamics correlated with descriptive chemistry. Problems. 992pp. 5⅜ x 8½.   0-486-65622-5

FROM ALCHEMY TO CHEMISTRY, John Read. Broad, humanistic treatment focuses on great figures of chemistry and ideas that revolutionized the science. 50 illustrations. 240pp. 5⅜ x 8½.   0-486-28690-8

Physics

OPTICAL RESONANCE AND TWO-LEVEL ATOMS, L. Allen and J. H. Eberly. Clear, comprehensive introduction to basic principles behind all quantum optical resonance phenomena. 53 illustrations. Preface. Index. 256pp. 5⅜ x 8½. 0-486-65533-4

QUANTUM THEORY, David Bohm. This advanced undergraduate-level text presents the quantum theory in terms of qualitative and imaginative concepts, followed by specific applications worked out in mathematical detail. Preface. Index. 655pp. 5⅜ x 8½. 0-486-65969-0

ATOMIC PHYSICS (8th EDITION), Max Born. Nobel laureate's lucid treatment of kinetic theory of gases, elementary particles, nuclear atom, wave-corpuscles, atomic structure and spectral lines, much more. Over 40 appendices, bibliography. 495pp. 5⅜ x 8½. 0-486-65984-4

A SOPHISTICATE'S PRIMER OF RELATIVITY, P. W. Bridgman. Geared toward readers already acquainted with special relativity, this book transcends the view of theory as a working tool to answer natural questions: What is a frame of reference? What is a "law of nature"? What is the role of the "observer"? Extensive treatment, written in terms accessible to those without a scientific background. 1983 ed. xlviii+172pp. 5⅜ x 8½. 0-486-42549-5

AN INTRODUCTION TO HAMILTONIAN OPTICS, H. A. Buchdahl. Detailed account of the Hamiltonian treatment of aberration theory in geometrical optics. Many classes of optical systems defined in terms of the symmetries they possess. Problems with detailed solutions. 1970 edition. xv + 360pp. 5⅜ x 8½. 0-486-67597-1

PRIMER OF QUANTUM MECHANICS, Marvin Chester. Introductory text examines the classical quantum bead on a track: its state and representations; operator eigenvalues; harmonic oscillator and bound bead in a symmetric force field; and bead in a spherical shell. Other topics include spin, matrices, and the structure of quantum mechanics; the simplest atom; indistinguishable particles; and stationary-state perturbation theory. 1992 ed. xiv+314pp. 6⅛ x 9¼. 0-486-42878-8

LECTURES ON QUANTUM MECHANICS, Paul A. M. Dirac. Four concise, brilliant lectures on mathematical methods in quantum mechanics from Nobel Prize-winning quantum pioneer build on idea of visualizing quantum theory through the use of classical mechanics. 96pp. 5⅜ x 8½. 0-486-41713-1

THIRTY YEARS THAT SHOOK PHYSICS: THE STORY OF QUANTUM THEORY, George Gamow. Lucid, accessible introduction to influential theory of energy and matter. Careful explanations of Dirac's anti-particles, Bohr's model of the atom, much more. 12 plates. Numerous drawings. 240pp. 5⅜ x 8½. 0-486-24895-X

ELECTRONIC STRUCTURE AND THE PROPERTIES OF SOLIDS: THE PHYSICS OF THE CHEMICAL BOND, Walter A. Harrison. Innovative text offers basic understanding of the electronic structure of covalent and ionic solids, simple metals, transition metals and their compounds. Problems. 1980 edition. 582pp. 6⅛ x 9¼. 0-486-66021-4

CATALOG OF DOVER BOOKS

A TREATISE ON ELECTRICITY AND MAGNETISM, James Clerk Maxwell. Important foundation work of modern physics. Brings to final form Maxwell's theory of electromagnetism and rigorously derives his general equations of field theory. 1,084pp. 5⅜ x 8½. Two-vol. set. Vol. I: 0-486-60636-8 Vol. II: 0-486-60637-6

QUANTUM MECHANICS: PRINCIPLES AND FORMALISM, Roy McWeeny. Graduate student-oriented volume develops subject as fundamental discipline, opening with review of origins of Schrödinger's equations and vector spaces. Focusing on main principles of quantum mechanics and their immediate consequences, it concludes with final generalizations covering alternative "languages" or representations. 1972 ed. 15 figures. xi+155pp. 5⅜ x 8½. 0-486-42829-X

INTRODUCTION TO QUANTUM MECHANICS With Applications to Chemistry, Linus Pauling & E. Bright Wilson, Jr. Classic undergraduate text by Nobel Prize winner applies quantum mechanics to chemical and physical problems. Numerous tables and figures enhance the text. Chapter bibliographies. Appendices. Index. 468pp. 5⅜ x 8½. 0-486-64871-0

METHODS OF THERMODYNAMICS, Howard Reiss. Outstanding text focuses on physical technique of thermodynamics, typical problem areas of understanding, and significance and use of thermodynamic potential. 1965 edition. 238pp. 5⅜ x 8½. 0-486-69445-3

THE ELECTROMAGNETIC FIELD, Albert Shadowitz. Comprehensive undergraduate text covers basics of electric and magnetic fields, builds up to electromagnetic theory. Also related topics, including relativity. Over 900 problems. 768pp. 5⅝ x 8¼. 0-486-65660-8

GREAT EXPERIMENTS IN PHYSICS: FIRSTHAND ACCOUNTS FROM GALILEO TO EINSTEIN, Morris H. Shamos (ed.). 25 crucial discoveries: Newton's laws of motion, Chadwick's study of the neutron, Hertz on electromagnetic waves, more. Original accounts clearly annotated. 370pp. 5⅜ x 8½. 0-486-25346-5

EINSTEIN'S LEGACY, Julian Schwinger. A Nobel Laureate relates fascinating story of Einstein and development of relativity theory in well-illustrated, nontechnical volume. Subjects include meaning of time, paradoxes of space travel, gravity and its effect on light, non-Euclidean geometry and curving of space-time, impact of radio astronomy and space-age discoveries, and more. 189 b/w illustrations. xiv+250pp. 8⅜ x 9¼. 0-486-41974-6

STATISTICAL PHYSICS, Gregory H. Wannier. Classic text combines thermodynamics, statistical mechanics and kinetic theory in one unified presentation of thermal physics. Problems with solutions. Bibliography. 532pp. 5⅜ x 8½. 0-486-65401-X

Paperbound unless otherwise indicated. Available at your book dealer, online at **www.doverpublications.com**, or by writing to Dept. GI, Dover Publications, Inc., 31 East 2nd Street, Mineola, NY 11501. For current price information or for free catalogues (please indicate field of interest), write to Dover Publications or log on to **www.doverpublications.com** and see every Dover book in print. Dover publishes more than 500 books each year on science, elementary and advanced mathematics, biology, music, art, literary history, social sciences, and other areas.